Lecture Notes in Artificial Intelli...

Subseries of Lecture Notes in Computer Science

LNAI Series Editors

Randy Goebel
University of Alberta, Edmonton, Canada
Yuzuru Tanaka
Hokkaido University, Sapporo, Japan
Wolfgang Wahlster
DFKI and Saarland University, Saarbrücken, Germany

LNAI Founding Series Editor

Joerg Siekmann
DFKI and Saarland University, Saarbrücken, Germany

Heather D. Pfeiffer   Dmitry I. Ignatov
Jonas Poelmans   Nagarjuna Gadiraju (Eds.)

Conceptual Structures for STEM Research and Education

20th International Conference
on Conceptual Structures, ICCS 2013
Mumbai, India, January 10-12, 2013
Proceedings

Springer

Series Editors

Randy Goebel, University of Alberta, Edmonton, Canada
Jörg Siekmann, University of Saarland, Saarbrücken, Germany
Wolfgang Wahlster, DFKI and University of Saarland, Saarbrücken, Germany

Volume Editors

Heather D. Pfeiffer
Akamai Physics, Inc., Las Cruces, NM 88005, USA
Email: hdp@cs.nmsu.edu

Dmitry I. Ignatov
National Research University Higher School of Economics
Moscow 109028, Russia
E-mail: dignatov@hse.ru

Jonas Poelmans
Katholieke Universiteit Leuven, Leuven 3000, Belgium
E-mail: jonas.poelmans@econ.kuleuven.be

Nagarjuna Gadiraju
Tata Institute of Fundamental Reseach
Mumbai 400088, India
E-mail: nagarjun@gnowledge.org

ISSN 0302-9743   e-ISSN 1611-3349
ISBN 978-3-642-35785-5   e-ISBN 978-3-642-35786-2
DOI 10.1007/978-3-642-35786-2

Springer Heidelberg Dordrecht London New York

Library of Congress Control Number: 2012954289

CR Subject Classification (1998):
I.2.3-4, I.2.6, I.2.8, G.2.2, H.2.4, H.2.8, H.3.4, I.5.4, I.5.1

LNCS Sublibrary: SL 7 – Artificial Intelligence

Typesetting: Camera-ready by author, data conversion by Scientific Publishing Services, Chennai, India

Printed on acid-free paper

Springer is part of Springer Science+Business Media (www.springer.com)

Preface

This volume contains the proceedings of the 20th International Conference on Conceptual Structures (ICCS 2013), the latest in a series of annual conferences that have been held in Europe, Asia, Australia, and North America since 1993. Details of these events are available at www.conceptualstructures.org, and www.iccs.info points to the current conference in this prestigious series. ICCS focuses on the useful representation and analysis of conceptual knowledge with research and business applications. It brings together some of the world's best minds in information technology, arts, humanities, and social science to explore novel ways that information and communication technologies can leverage tangible business or social benefits. This is because conceptual structures (CS) harmonize the creativity of humans with the productivity of computers. CS recognizes that organizations work with concepts; machines like structures.

ICCS advances the current theory and practice in connecting the user's conceptual approach to problem solving with the formal structures that computer applications need to bring their productivity to bear. Arising originally out of the work of John Sowa while at IBM and his work on conceptual graphs, over the years ICCS has broadened its scope to include a wider range of theories and practices, among them formal concept analysis (FCA), description logics (DL), the Semantic Web, the pragmatic Web, ontologies, multi-agent systems, concept mapping, relationships to uses in STEM education and more. Accordingly CS represent a family of approaches that builds on the successes of artificial intelligence (AI), conceptual modeling, information and Web technologies, and knowledge management.

The theme for this year's conference was "Knowledge Representation for STEM Research and Education." Science, technology, engineering, and mathematics (STEM) have, in recent decades, emerged as a lively new research areas. More and more data are being captured in these areas (particularly through the Web) and how to represent these data for useful research, searching, and education is a real challenge. These data now represent our business, economic, arts, social, and scientific endeavors to such an extent that we require smart applications that can discover the hitherto hidden knowledge and how to represent this mass of data. By bringing together the way computers work with the way humans think, conceptual structures align the productivity of computer processing with the ingenuity of individuals and organizations in addressing these highly mathematical data. The representation of these data can be used for both research areas and for collecting data for improved teaching techniques in these areas.

The ICCS papers that appear in this volume represent the rich variety of topics on CS. There were 43 submitted papers that were rigorously reviewed anonymously by at least three members of the Program Committee. An Editorial Board member oversaw each paper processed, and worked together with the organizers on making the final decisions. About 50% of submitted papers deemed relevant to the conference were accepted as both long and short papers. There were also three invited papers. As is evident in this volume, the number of accepted papers reflects the high quality of submissions, and the proceedings appear as volume LNAI 7735 of Springer's *Lecture Notes in Artificial Intelligence*, a subseries of the LNCS series. In addition to the ICCS 2013 main conference, there was an associated workshop—Workshop on Modeling States, Events and Processes (MSEPS). The papers from this workshop appear in their own proceedings.

We wish to express our thanks to all the authors of the submitted papers, the speakers, workshop organizers, and the members of the ICCS Editorial Board and Program Committee. We would like to thank Simon Andrews and Simon Polovina, who organized the anonymous reviewers of papers submitted by the ICCS Chairs. We wish to express our gratitude for the support we received from the Homi Bhabha Centre for Science Education, TIFR, by hosting the event and providing the facilities. With special thanks to the Dean of Science Education, Chitra Natarajan, and several personnel, Jayashree Ramadas, Madhavi Gaitonde, Rashmi Shrotri, Sumana Amin, Smitha Burli, Manoj Nair, Anil Kumar Shankhwar, and V.P. Raul. We also extend our thanks to the Local Organizing Chair, Meena Kharatmal, and to the staff of the Homi Bhabha Centre for Science Education, TIFR, for managing the production of the workshop proceedings. Lastly, we thank the very helpful people at Springer, to whom we owe our gratitude.

October 2012

Heather D. Pfeiffer
Dmitry I. Ignatov
Jonas Poelmans
Nagarjuna G.

Organization

Conference Chair

Nagarjuna G.   Homi Bhabha Centre for Science Education, TIFR, Mumbai, India

Local Organizing Chair

Meena Kharatmal   Homi Bhabha Centre for Science Education, TIFR, Mumbai, India

Program Chairs

Heather D. Pfeiffer   Akamai Physics, Inc., USA
Dmitry I. Ignatov   Higher School of Economics, Russia
Jonas Poelmans   Katholieke Universiteit Leuven, Belgium

Workshop and Tutorial Chair

Dmitry I. Ignatov   Higher School of Economics, Russia

Editorial Board

Simon Andrews, UK
Galia Angelova, Bulgaria
Madalina Croitoru, France
Frithjof Dau, Germany
Harry Delugach, USA
Bernhard Ganter, Germany
Ollivier Haemmerlé, France
Pascal Hitzler, Germany
Mary Keeler, USA
Sergei O. Kuznetsov, Russia

Peter Øhrstrøm, Denmark
Heather D. Pfeiffer, USA
Simon Polovina, UK
Uta Priss, Germany
Sebastian Rudolph, Germany
Henrik Scharfe, Denmark
John Sowa, USA
Rudolf Wille, Germany
Karl Erich Wolff, Germany

ICCS Program Committee

Babak Akhgar, UK
Jean-François Baget, France
Radim Belohlavek, Czech Republic
Peggy Cellier, France
Dan Corbett, USA
Aldo De Moor, The Netherlands
Juliette Dibie-Barthélemy, France
Pavlin Dobrev, Bulgaria
Florent Domenach, Cyprus
Gerard Ellis, Australia
Paul Elzinga, The Netherlands
Boris Galitsky, USA
Jan Hladik, Germany
John Howse, UK
Adil Kabbaj, Morocco
Mikhail F. Khoroshevsky, Russia
Markus Krötzsch, UK
Leonard Kwuida, Switzerland
Jérôme Lang, France
Ivan Launders, UK

Michel Leclère, France
Natalia Loukashevitch, Russia
Dickson Lukose, Malaysia
Philippe Martin, France
Carlo Meghini, Italy
Guy Mineau, Canada
Khalil Ben Mohamed, Malaysia
Bernard Moulin, Canada
Dmitry I. Mouromtsev, Russia
Sergei Obiedkov, Russia
Yoshiaki Okubo, Japan
Anne-Marie Rassinoux, Switzerland
Eric Salvat, France
Jeffrey Schiffel, USA
Dominik Ślęzak, Poland
Iain Stalker, UK
Francisco Valverde-Albacete, Spain
Martin Watmough, UK
Igor Zagorulko, Russia
Gq Zhang, USA

External Reviewers

Jim Burton, UK
Aidan Delaney, UK

Weng Onn Kow, Malaysia
Nikolay V. Shilov, Russia

Sponsoring Institutions

Homi Bhabha Centre for Science Education, TIFR

Table of Contents

Invited Talks

Accepted Papers

Conceptual Structures for STEM Data:

Linked, Open, Rich and Personal

Su White

Web and Internet Science,
ECS, University of Southampton, UK
saw@cs.soton.ac.uk

Abstract. Linked and open data is increasing being used by governments, business and administration. Awareness of the affordances and
potential utility of open data is being raised by the emergence of a host
of web-based and mobile applications.

Across the educational and research communities applications applying the principles linked data principles have emerged.

Systems developed and used by researchers and academics are most
likely to be predominantly in the hands of the early adopters and current developments found in higher education tend to be atomized, yet
there is potentially considerable advantage in associating and integrating
applications for organisational, educational and administrative.

This paper presents an argument for how we can move from early
adopters to early majority, and at the same time presents a roadmap
which will outline some of the significant challenges which remain to be
addressed.

Keywords: linked data, open data, semantic annotation, higher education, organizational change.

1 Introduction

A strong thread of the use patterns which have accompanied technological advances of computational machines has been their use for data processing. The
classic history of computers will inevitably acknowledge the use of Holerith Machines for the US census and the development of LEO to handle administration
for Lyons (a company famous in Britain for its chain of corner tea houses). It
will refer to the development of COBOL in response to apply computing power
to the problems of provisioning the US Navy and the subsequent widespread
growth of computer use for all aspects of business administration and record
keeping.

Universities, like any other large organisation, made use of computers for
administration and like the rest of the business world universities have integrated
the use of personal computers into their business processes over the past thirty
years. Further transformations in business and personal interactions with and use
of computers followed on from the introduction and subsequent refinements of

H.D. Pfeiffer et al. (Eds.): ICCS 2013, LNAI 7735, pp. 1–21, 2013.

the World Wide Web during the 1990s. It is common to refer to three generations of the Web.

- The vanilla web: early implementations, the web as a publishing device—a basic web of documents
- The social web: an enhanced web of documents, the read write web introducing blogs and wikis
- The semantic web: "an extension of the current web in which information is given well-defined meaning" [1],

Following this was much discussion of what was meant by the Semantic Web and how it could be realised. The discussion was between the purists who preferred the path of hard semantics to the more pragmatic approach of soft semantics. Beliefs and attitudes shifted and changed [2]. Five years on from the original publication in 'The Semantic Web revisited' Shadbolt et al [3] asserted

> The Semantic Web we aspire to makes substantial reuse of existing ontologies and data. It's a linked information space in which data is being enriched and added. It lets users engage in the sort of serendipitous reuse and discovery of related information that's been a hallmark of viral Web uptake. We already see an increasing need and a rising obligation for people and organizations to make their data available. This is driven by the imperatives of collaborative science, by commercial incentives such as making product details available, and by regulatory requirements.

Alongside the debate as to the instantiation of the semantic web, Berners-Lee was considering the nature of change which was inherent in the way that the web worked. Evolving the principle that the semantic web was concerned with better enabling computers and people to work in co-operation, he began to refer to the 'two magics'.

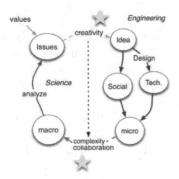

Fig. 1. Berners-Lee's science and engineering approach with magic modified to show complexity and collaboration

'Two magics' incorporates a generative interaction between social activity with the web and technological development. This model goes some way to explaining the way in which use and applications have taken off [4,5]. This was first described by Berners-Lee et al and then subsequently developed into the graphical form presented to the Web conference in 2007.

Alongside the changes related to the engineering of the web, the wider population were developing a conception of, relationship with, and reliance upon the web and its artefacts. Clay Shirky discusses the social web in his 2003 blog, and defines the social web as "software that supports group interaction' [6]. Evidence of the social web and social internet have a long history in discussion forums and Usenet groups. The social web, along with O'Reilly's observations of web 2.0 which can be formally dated from a 2005 blogpost and a 2007 paper [7,8] have become intermingled in the minds of the casual observer. This is of interest because if applications are to move from the early adopters to the early majority then they are more likely to succeed through the use of familiar metaphors (which are understandable and accessible).

In some ways we have been observing sets of memes being adopted by the general public. Just like the recommender systems deployed by Amazon and other commonplace shopping companies, everyday folk understand that applications which behave like applications you know and love are also likely to become applications which you know and love. Everyday folk may not understand the implications of big data, but they can begin to gain an idea of the mechanisms if they participate in citizen science, or see a news item featuring health benefits which have accrued from a massive genome data set collaboration. People are learning: learning from the technology—what it does and how it does it; learning from the people—what they do, how they use the technology. In academia we call it learning from good practice; gradually the concepts of open and linked data are seeping into the everyday consciousness. Alongside this comes some understanding of ontologies and semantic annotations. Concepts like domain models, reasoning and analysis may be more difficult to understand, but awareness is being raised. The educational domain is ready for change. It is being pressured, like so many other businesses, to streamline its processes. Academics are gaining hands-on experience of workplace tools in their research and are becoming ready to generalise these processes across their institutions. It will be interesting to see how these changes affect teaching and administration.

In the rest of the paper that follows, the Background section introduces an analysis of the means by which we can interact with a web of data rather than a web of documents, considering the scope of hard and soft semantics. These ideas form an important conceptual backbone to our understanding of the ways in which STEM data can be harnessed in university education. Through generic examples of implementation of open and linked data in 2.1 it will analyse current approaches and consider how everyday experience of open data shapes expectations and may therefore drive future developments. 2.2 presents a brief account of big data. The specific cases of linked and open data usage in Higher Education have been derived from the work of a number of communities, which are identified in 2.3. Specific educational approaches are examined in subsection 4 and the subsections which follow. The future directions of STEM education and the role played by conceptual structures in that future are then examined. The final section discusses the implications of the previous section in the light of challenges and opportunities in the educational domain and suggests some conclusions.

2   Background

The historical partnership between technology and administrative processes was traced in the introduction. The World Wide Web Consortium (W3C) and strong leadership from Tim Berners-Lee have been strongly influential in emerging standards for the Web. At the same time they have been taking forward the debate on ways in which infrastructure can be used and further developed. Much academic effort has been expended on the semantic web, following the 2001 article by Berners-Lee et al in Scientific American [1]. Widespread use of the Web has confronted users with the reality of shortcomings of the early systems. The early web was criticised as being a library where all the books had been thrown on the floor. As people used the web, understanding of what it could and could not achieve began to surface. The initial implementation was only a small part of the specification envisaged by Tim Berners-Lee and he has continued to work with W3C to realise the broader potential which he wanted to achieve.

Table 1. The Scope of hard and soft semantics

Hard semantics pure	Soft semantics pragmatic
Machine readable	Human readable
Rigorous modelling	Lightweight modelling

The story of the web is a story of engineering; it is a realisation of the difference between a model or proposal and the actual implementation. The web as we experience it has had structure imposed, after the fact. It is overlaid and there are many inconsistencies. As usage has developed in an ad hoc manner, although there are standards, they are many and varied. Early solutions to the combinatorial explosion which will surely follow any attempt to create a rich interlinked hypertext were mostly focused on abandonment of hyperlinking, and resorting to backend databases which serve pages engineered for delivery, but not for interaction. The social web went some way to creating a read-write web of the original conception. The social web has become a place for conversations and discourse. But it is the spam bots which demonstrate the power of machine processing of web pages over individual participation. As discussed in the Introduction, while the general public were becoming accustomed to the social web, the experts were discussing the nuances of the Semantic Web as summarised by Table 1; in particular whether hard semantic solutions were preferable to a more pragmatic approach of soft semantics.

2.1   Linked and Open Data

The UK government has become an enthusiastic supporter of Open Data. Early response to proposals were positive. Plans were initially cancelled with the incoming new Government in 2009, but have been fairly rapidly revived. In the UK, the Open Data Institute formally opened its doors in October 2012. Nigel Shadbolt commented on the ODI blog "Less than a year ago Tim Berners-Lee and I were writing a briefing note for Government outlining the opportunity for an Institute dedicated to realising the economic value of Open Data. Earlier in 2011 the Chancellor in a speech at the Google Zeitgeist stated "Our ambition is to become the world leader in open data. The economic impact of this open data revolution will be profound...". http://www.theodi.org/blog/ 1st October 2012.

The Open Data Institute was established with an objective of demonstrating the commercial value of open data. The UK government provides a set of case studies of Open Data on their web site http://data.gov.uk. Useful information on the economic impact of open data is available via the LinkedGov wiki which provides an index of a selection of peer reviewed papers. The UK government's 2011 Open Data White Paper [9] identified five agenda items for open data in the UK: i) building a transparent society; ii) enhanced access; iii) building trust; iv) making smarter use of data; v) The future transparent society. Examples of transparency and the benefits of it cover areas of transport, crime and spending. Making data available to citizens and businesses can enable government to more clearly account for their activities and spending, and provide information for feedback loops which can justify or promote changes in behaviours or responses. For government open data is key to understanding the nature of the businesses with which they are engaged. It also enables government to meet requirements of Freedom of Information legislation. The government proposition is that enhanced access leads to increased trust and thus to smarter use of data. There are a number of case studies provided to illustrate these arguments. The government has set itself a standard for information publishing and is promoting the five star scheme of data re-use originally proposed by Tim Berners-Lee. The government had already established a set of Information Principles.

Everyday experience of open data is typically mediated by mobile apps or web sites. Private companies and social collaborations, along with government, have produced and published a large amount of open data. Geographic data has been created in each of these three domains. In the UK the Ordnance Survey publicises its open data, but app developers use a range of different sources. However, people who use an app to display a map and overlay points of interest or to navigate between two locations will at the same time gain some kind of intuitive grasp of what may be possible, even if they do not yet understand the principles of publishing open data. Commonplace experience of apps like Tripit, LinkedIn, Open Street Map and Mendeley each in their different way help users build up an instinctive understanding of what is useful. System designers have learnt that making their API available is good for business and companies can compare the success of open and closed systems and draw their own conclusions

Table 2. Summarizing the semantic enhancements of Shotton et al.

Generic Enhancements	Adding value to the text
Providing access to actionable data:	Semantic annotations for key concepts
making the datasets available	Document summary and study summary
Data Fusion from Other Sources:	Tag tree and tag cloud
enriching the basic journal data	Supporting claims tooltip;
Making information more accessible	Various citation analysis tools
Provenance information	Alternative language abstract

as to the best way forward. In academia, researchers have been experimenting with open data. In 2009 Shotton et al report on an interesting experiment to see how it was possible to semantically enrich a traditional academic paper [10]. The account of this activity identifies a set of enhancements which can be seen in Table 2. The whole activity provided an immensely rich experience which if it could be produced in a replicable automated manner would add considerable value.

2.2   Big Data

The storage capacity of computer systems and the speed and power of data processing have enabled the collection of big data sets. Big data refers to massive datasets containing many billions of information items. Big data is too large to be analysed by conventional database tools. It comes from many different sources. These include:

i) data gathered by local and national governments as a result of providing services of systematic survey; ii) data gathered by businesses as a result of their interactions with clients and customers; iii) data from the natural world through scientific observations or experimentation; iv) collated and aggregated data sets which are gathered from diverse sources covering similar or identical subject areas; v) historic data sets;

Data sets may be examined to provide evidence and feed into businesses processes for bringing about change. Many large data sets are proprietary—owned and protected from wider use by copyright, privacy or business imperatives. Some large datasets are made available for distributed analysis—the SETI@home project was used to analyse large volumes of astronomical data. Big data can benefit from crowdsourced analysis just as big data sets may be assembled by crowdsourcing.

2.3   Educational Communities Around Linked and Open Data

Developer communities have a crucial role to play in the dissemination and sharing of ideas and helping establish good practice. In educational communities there are broadly four basic types of intersecting communities which are concerned with education specific linked and open data i) formal associations;

ii) institutional initiatives; iii) evangelist practitioners and researchers; iv) loose associations or communities of practice. Community has a very strong role to play in the development of standards in this area. All four types of communities, plus their respective associations of users of linked and open data, are involved in different ways [11]. The examples below are drawn from the UK, but similar development is taking place in many different countries across the world. Probably the most significant of these is the international OpenCourseWare Consortium.

Formal associations like XCRI (`http://www.xcri.co.uk`) can be seen as a combination of bottom up, specifications and demand arising from the community, met strategically with top down input from funding bodies to pursue a common objective. The XCRI initiative is funded by the JISC, the UK agency for technology infrastructure and development in Higher and Further Education. Formal associations produce tangible outcomes. For example, the XCRI initiative has developed and is now working to a standard model of course information.

Institutional initiatives are manifest in a number of ways. Some institutions, such as the UK Open University, pursue open and linked data because there is a strong business case in terms of managing administrative processes and gaining business intelligence. The University of Southampton has a very close link with the development of the semantic web. Tim Berners-Lee holds a chair at Southampton and in an initiative led jointly with Nigel Shadbolt has established the UK Open Data Institute. A number of other high profile institutions have this level of commitment.

Evangelist practitioners and researchers form loose associations irrespective of institutional ambitions. Researchers and application developers in universities often pursue their objectives through passion and academic interest. Often their collaboration is a mixture of face-to-face and online interactions supported by blogging and microblogging which support extended discussions and knowledge sharing.

The linked universities (`http://linkeduniversities.org/`) are a loosely coupled community of practice who work collaboratively on emerging standards. Being in academia, they can have a symbiotic relationship with the funding council through JISC funded standards related work (CETIS) and developers forums such as Dev8D. They have worked to establish a number of agreed vocabularies and to bring together a significant amount of expertise relating to linked data initiatives in the UK and across Europe.

All of these communities add to the common understanding by making visible their discussions and publicising their achievements. Further examples include the informative set of case studies made available via the UK XCRI-CAP web site, while the linked universities describe vocabularies and work in progress and present a collection of relevant publications.

Another significant educational community is that associated with Open Educational Resources. The OER community has two different manifestations. Some institutions have developed repositories which they are making open to share worldwide, while other repositories are shared efforts across institutions, sometimes with disciplinary groupings. Davis et al provide a comprehensive account of the roots of OERs and the experience of community building [12]. In many UK universities OER communities have strong ties with the open and scholarly publications community, and there is evidence in the literature that experience from one field sometimes informs the others. Open educational resources bring together those who take a resource based approach to learning (which survives from many of the early applications of hypertext) and those associated with formal learning design, working from an IMS perspective.

Open Educational Resources: (OERs) explicitly collected or assembled for sharing and reuse. Since 1990s standards have evolved which support and enable publication and re-use e.g., IEEE Learning Object Metadata (LOM) [14,15]. The standards enable resources to be found and provide systematic descriptions. Further standards such as IMS-CP support interoperability. It is possible to transfer sets of identified files and unpack them for use on another server. Further standards such as SCORM RTE and IMS-LD CopperCore [16] can be used to support sequencing and assembling resources for learners. OERs have a long pedigree. The MERLOT project was an international consortium which worked in 1997 to build a platform for open educational resources. In Europe, Rob Koper from the Open University in the Netherlands was influential in early work on learning objects specifying an educational modelling language [13]. Some major players in the OER community are also providers of OpenCourseWare. In 2005 the OpenCourseWare Consortium was established bringing together major international interests committed to open sharing of a range of educational resources presented as discrete courses. The OER and OCW community have been working in the area of modelling and linked and open data, but their activities have been driven from a bottom-up perspective of sharing and achieving interoperability rather than from a top down design approach.

Large-scale learning: "Learning analytics is essential for penetrating the fog that has settled over much of higher education" [17]. Just as business organisations use their big data on customers' behaviour to improve their bottom line, so learning analytics is being harnessed in HE. It is not surprising therefore that just as research into data mining has been influential in many approaches to big data, mathematical methods of examining large data sets in education have emerged as Learning Analytics. Universities and educators are much concerned at the micro level with checking whether learners understand whether learning has taken place, what feedback is needed and how progress may be measured. Data is collected per student and aggregated across cohorts. This data is analysed and reported internally and externally. Topics with which educators are particularly concerned include student achievement (measured through courseworks, assignments, tests and exams). Because institutions are concerned with awarding degrees, data on attainment contributes to progression and retention information which will be discussed and analysed internally and reported externally. Learning analytics brings the approaches of big data to many institutional agendas, particularly those of measuring learning, attainment, progression and retention. From a business process perspective this data may also be relevant to providing evidence for establishing financial cases for educational activities.

Larger institutions such as open universities may already have systematic data collection processes.

George Siemens is a leading thinker in the area of educational learning analytics and researchers at Athabasca and the UK Open University are prominent and influential in the field of online education termed Connectivism [18]. Siemens provides a personal participative link from Learning Analytics into MOOCs (massively open online courses) having collaborated with Stephen Downes to establish a MOOC in 2008 [19]. MOOCs provide an opportunity to gather learning analytic data, particularly that related to student behaviours. MOOCs provide a context for Open Educational Resources, while learners and participants provide a context for the data collected from a MOOC. Currently MOOCs are provided on a number of different platforms; for example Cousera, Udacity, MITx and EdX. In the US some high profile institutions are running MOOCs, perhaps for their advertising and reputation enhancing potential, or perhaps for their potential to collect learner data which can then feed back into the design of face to face learning activities.

2.4 Learning Approaches

There is a plethora of educational theory to which individual academics may refer, and many institutions will exhibit a broad range of approaches, some of which are explicitly designed within a given educational approach. Some relatively new institutions, especially large open universities may commit to a formal design process across the institution based on an acknowledged set of educational principles, but more often students experience diverse influences from a range of theoretical perspectives.

Figure 2 gives a much simplified representation of some of the key theoretical influences from education and technology which are prevalent in educational approaches in universities at the current time. For learners the network is playing an increasingly important role in learning. This is a reflection of the way in which technology infrastructure has become an intrinsic part of the fabric of everyday life. It also resonates with many educational theories such as constructivism and social constructivism. In Higher Education Laurillard has been highly influential and the conversational model of learning [20] has been credited with widespread impact, certainly amongst UK based educators engaged in technology based learning.

At the same time, work by Siemens has also been influential in proposing a model of connectivism [18] which emphasises the role of the network in shaping and determining the nature of learning and approaches which are relevant and effective. It proposes a model which is particularly relevant to the connected world. It has strong links to social learning theories and stresses the primacy of generative and transformative approaches to learning. Siemens subsequently has been involved with large scale learning activities at the Athabasca University in Canada. Following that work he also identifies specific links between connectivism and learning analytics [21], making connectivism a perspective which is particularly relevant to the scope of this paper.

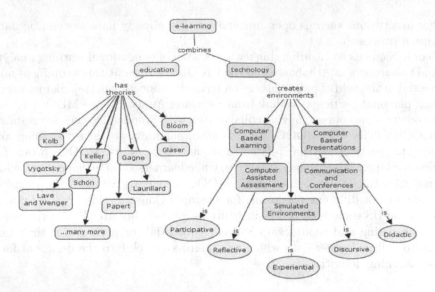

Fig. 2. Educational Approaches—a mass of theories

"Learning analytics currently sits at a crossroads between technical and social learning theory fields. On the one hand, the algorithms that form recommender systems, personalization models, and network analysis require deep technical expertise. The impact of these algorithms, however, is felt in the social system of learning. As a consequence, researchers in learning analytics have devoted significant attention to bridging these gaps and bringing these communities in contact with each other through conversations and conferences" [22].

There are many approaches to educational theory which do not take into account the online and connected world. One recent educator who has been influential in approaches to learning, but who does not deal specifically with technology and learning is Biggs [23]. His models of the student and effective ways of facilitating student learning have gained widespread currency, and there has been a growing interest in understanding what 'what the student does' as a means of modelling and enhancing education. Where Biggs seems to be particularly relevant to the emerging conceptual frameworks in STEM education is via recent interest in Digital Literacies. Much of the discussion within is aligned with Biggs' perspective. Students do things to learn, and can be expected to develop their learning skills whilst undertaking higher level study. Working with students to understand, develop and extend their digital literacies is an increasingly important agenda for Higher Education across the board. For students in STEM areas this will extend to mastering and understanding everyday work tools and to the specific sets of tools which predominate in their chosen specialisms.

Education has a long tradition of working with webs of documents. The role of text in education was a driver for early computer based systems, and in turn influenced models of understanding for approaches to technology based learning which differed from the more strongly industrially influenced approach of computer based training. Hypertext was implemented on mainframes through the Plato. On personal and distributed computers, Apple briefly led the way with their HyperCard system. Hypertext researchers looked at education, and much interest developed in personalisation, customisation and intelligent tutoring systems. Hypertext systems such as microcosm, although developed before the web came into use, used assembled collections of interlinked resources and it was that understanding of texts which informed the developments of early research and educational repositories such as ePrints and EdShare [24]. Different subject areas have tended to privilege different aspects of learning technologies; specifically those which are better suited to their individual disciplinary needs [25]. Independently, academics were programming web pages and producing specialised applications such as simulations, as well as using authentic data sets for structured tutorials.

At the same time commercial interests were promoting systems which has offered to manage the learning process, early systems often worked to extend the book metaphor pacing the learner through their material. It was natural that this work continued through the early years of web technologies, when systems which managed the learning experience Blackboard and WebCT for example were introduced. These systems often integrated with student management systems which the suppliers were also selling. Virtual Learning environments also encapsulate a number of different learning processes, for example tutorial plus questions plus simulations. Typically there are also analytics such as data logging and tracking. They have gained some popularity for their ability to sequence order and organise information and thus drive learning activities. The functionality of VLEs have evolved alongside technology in the wider world, and systems typically now incorporate aspects of the social read-write web, although their objective of being closed systems means that they cannot have the exact same affordances as the wikis, blogs and online discussions which are found on the wild web.

2.5   Semantic Technologies in Higher Education

The discourse and analysis of the potential for semantic technologies in Higher Education has strong links back to the research of the hypertext and adaptive hypertext communities and thus necessarily encompassed contributions from those concerned with AI agendas such as agents and Intelligent Tutoring Systems. The 2009 SemTech Report presents survey findings for semantic technologies in learning and teaching [26]. Things have moved on since, but some of the observations are still relevant. SemTech articulates the need to differentiate between soft and hard semantics in an educational context. Table 3 differentiates i) soft semantic technologies like topic maps and Web 2.0 applications, which provide lightweight knowledge modelling in formats understood by humans and ii) hard semantic technologies like RDF, which provide knowledge modelling in formats processable by computers.

There are many different ways in which semantic led approaches might contribute to educational activities. Table 4 differentiates between those which can

Table 3. Examples of hard and soft semantics in Higher Education

Hard Technologies Using machines to talk to each other	Soft technologies Helping people to organise knowledge
managing shared learning content	Tool to link learning and select careers
identify cross-curricular connections	Managing shared learning content
Support for personal learning	Support for personal learning
Search for people (people like me)	Developing reasoning skills and argument
Search for resources	Shared mind-maps/topic-maps

Table 4. Where semantic technologies can contribute to educational processes

Classroom administration	Visible data → dynamic analysis & feed-back
Assisting course creation	
Aggregate course and module information	Aggregate relevant resources & workflow
	Streamline accreditation & quality processes
Learning activities	Assessment, certification, counter-ing/detecting plagiarism
Critical thinking and argumentation support	Learning in the wild
Efficient personal & group knowledge construction	Informal learning
	Self-actuated learning
Authentic learning	Aggregation, personalisation, customisation
Group formation	

be considered to related to 'classroom' administration and those which might make up a direct component of learning activities.

2.6   Education in STEM Subjects, Some Scenarios

The challenge for educators is to provide the educational opportunities which enable the learner to familiarise themselves with, and then master the necessary knowledge, skills and understandings which can equip them to be competent in their chosen specialism. Can using smart technology help us address these ambitions? How is our model of student learning activities made more complex by these requirements?

At the same time as we address these aims, there is a widespread expectation that learning should be a transformative process, and that the students will be able and ready to make and sustain their contribution to their chosen workplace and career path. It is inevitable that such a trajectory necessitates a mastery which extends beyond academic subject specialism and into the applied discipline in a world where technology is an essential component. Disciplinary differences as investigated by Biglan [27,28] and subsequently Becher [29], play an important role in the nature of teaching in Higher Education. Disciplines determine needs

Table 5. Disciplinary needs of Hard Pure and Hard Applied Subjects [25]

	Curriculum/content	Assessment	Cognitive Purpose
Hard Pure e.g., Natural Sciences	Concepts and principles closely connected Content typically fixed and cumulative Quantitative Teaching and learning activities are focused	Specific and focused exam questions Object tests relying on quantitative nature of knowledge	Logical reasoning Testing of ideas in linear form of augmentation. Reliance on facts, principles and concepts
Hard Applied e.g., Engineering	Concerned with the mastery of the physical environment. Focus is on products and techniques. Knowledge is atomistic and cumulative. Emphasises factual understanding.	Preference for exam questions; especially problem solving.	Logical reasoning. Testing of ideas in linear form of argumentation. Reliance on facts, priniples and concepts

and establish context. The STEM subjects are considered to fall within the Hard Pure and Hard Applied disciplinary space. The context of disciplinary differences in e-learning has been investigated by White and Liccardi whose summary of the disciplinary needs relevant to STEM subjects is presented in Table 5. Alongside the disciplinary needs of learners, it is worth considering the affordances of different types of tools. Based on the observations of the SemTech report, and extrapolating from the experience of enriched publishing reported by Shotton et al. [10]. Table 6 builds on the activity gradient proposed by White and Liccardi, suggesting added value which might result if educational resources were semantically enriched.

The value of ontologies to planning the learning process has been recognised by researchers and educational specialists within some discipline areas [30]. This would come under the classroom administration category suggested in Table 4 which was previously discussed. In the SemTech survey [26] there was evidence of the purposeful use of semantic tools for authentic learning. Experience in the use of repositories and digital collections would suggest that they have strengths for the educator as well as the learner and would indirectly support informal learning, self actuated learning, personalisation and customisation. In the world of MOOCs specialist programs have been established which reflect an industry need; for example the solar power industry http://solpowerpeople.com/solar-courses/—which might also be appropriate to provide authentic resources for learners studying on a formal program in a relevant topic. At the time of writing, Coursera, a federation of OpenCourse-Ware, listed almost 200 courses with a duration of between 4 and 12 weeks. It

Table 6. Suggested benefits from semantically enriched learning in stem subjects

Teacher led/passive		
Resource	**Conventional use**	**Semantically enriched**
Notes on the web	Teacher: author, publisher. Student: consumer, viewer, use for reference.	Automatically linked to related resources. Dynamic annotation (semantic wiki). Use/integrate with OER.
Tests, questions	Teacher: author, publisher Student: participates/interacts.	Automatically linked. Dynamic and static generation of feedback. Dynamic linking to 'wild' resources. Use/integration of OER
Interactive tutorials. Incorporates learning activities and assumes structure.	Teacher: author, publisher.	Dynamically assembled, dynamically link.
Simulations. Incorporates learning activities and assumes structure.	Teacher: author, publisher. Student: participates/interacts Pathways dynamic/proxy for real world.	Distributed participants, use of authentic data.
'World Ware'	Teacher points to/requires use of authentic tools.	Real world datasets & tools.
Student created artefacts	Student freely utilises authentic tools.	Semantic publication/visibility
Online discussions. Blogs and Wikis.	Students engage is 'social' creation and 'social' learning.	Dynamic interlinking, semantic publication.
Student led/active		

claimed to have more than 1.7 million registered course participants. By far the majority of the subjects were in the stem subject area. High profile open courses have been offered by US Ivy League universities but are also being offered by informal networks of teachers, by individual academics and by small colleges.

The examples above considered the value of semantic enrichment and open and linked data from the perspective of purposeful course design, or as an adjunct to the educational administration. These two perspectives were the main line of analysis identified in the SemTech report. However, there has been considerable discussion across the community which argues for taking a personal learner perspective on educational resources, and to place the use of educational resources within a framework of a Personal Learning Environment. Within the framework of developing campus wide support for learning at the University of Southampton, the personal learning environment has been considered. Semantically rich environments provide ample opportunities for the interlinking and crafting of personal learning resources.

The world is changing and universities must respond to students' needs and expectations in agile and effective ways. Learners enter university with an inevitable diversity of technological familiarity and a mix of naïve and sophisticated approaches to using technology as a part of their learning. Students are using apps and becoming familiar with the potential of linked data. Just as they have learned how to Google for information and to look to Wikipedia as the first source of information, so they are also becoming familiar with technology behaviours which they might reasonably expect to appear in their study environment.

Using online services such as Facebook, Amazon, Delicious, Flickr, YouTube introduces them to a world where artefacts like integration and recommendation are an obvious part of the infrastructure. Familiarisation with these services shapes expectations and also prepares users to be adept at exploiting the affordances for their own reasons. Students develop skills and expectations. Familiarity with these specialised affordances of various common place yet separate applications, may result in students viewing the institutional provision of web sites and virtual learning environments (VLEs) as clunky and out of date. For their part, universities may feel themselves overloaded with the task of providing, maintaining and updating the necessary information needed to inform and educate their students and also to furnish and drive the workflows of their administrative processes.

Many universities are understandably proud of the historic heritage on which University system is based, and the historic roots on which their own institution is established. Yet these same roots and traditions are in some ways likely to be the source of some of the challenges which are faced by the University as an organisation.

Institutions are being changed by external factors. Siemens identifies an altered information cycle brought about by 'participatory technologies'. At the University of Southampton, four fundamental drivers for change were identified. i) support curriculum change and innovation; ii) address student expectations; iii) enable the university to remain credible in its support for learning and teaching—particular to be seen as fluent and innovative in the use of IT; iv) facilitate the adoption of a University-wide educational style. Students may want interconnectivity with external apps, actors in our systems may want to share data from internal info with external apps, but whatever else we certainly need to be able to share and reuse data from and between our internal apps, using a web2.0 approach using the web as a platform, exposing our data and devising services to enable apps to communicate is essential.

An early implementation of educational infrastructure to support teaching was developed in Southampton making use of linked data. We routinely use linked data for info within one part of the University ECS enabling all our info style pages to be generated dynamically. The mix of screen shots collected as a single figure illustrate this approach incorporating a personal page where data is associated to provide information based on real world relationships recorded as linked data. Academics are tutors; tutors have tutees; academics are lecturers; lecturers have teaching allocations, and so on. Assembled sets of relations generate informative and highly functional web pages. This approach creates official home pages and generates module pages for teaching activities. The resources page also accesses the institutional repository and using tags as filters populates the module page resources tab. A wiki is used to add and enter information; there is also a linked HTML web pages on the filestore. Dynamic content I added through retrieving delicious tags associated with course teachers, and a tag cloud index to the delicious data is generated.

We used RDF because it saves time; however, the hand crafted web sites will persist, and some colleagues use paper handouts. Automatically generated pages provide learners with a consistent backbone to which they can refer. Individual differemces will persist. Not everyone uses the EdShare repository, or edits the notes or student wiki.

Even with this proof of concept, the challenge remains, how to port it to the rest of the University? This system was introduced by those whose research is into linked data. Colleagues in Electronics, physic or chemistry might not regard the changes in the same light. Time and again we return to the issue that change is cultural, and individual responses and behaviours are mediated by skills and by available time, and willing priority. Open and linked data can be used by universities for business process management. The University of Southampton established an open data initiative in 2010. Full information can be found at the project's web site http://data.soton.ac.uk/.

Benefit can be gained from exposing and sharing the public and private capital of data and information within and across departments and institutions to enable workflows and promote and enable collaboration. Since its inception the project has supported, shown financial returns and won the support of senior administrators and managers who have particularly appreciated the way in which the data can be drawn upon at short notice to provide customised web sites (for example to support a student visit day). Among the achievements made by this initiative i) furnish components of a financial information system; ii) helped address external demands for information provision is a cost effective manner; iii) enable enhanced quality of data relating to room information; iv) increase the efficiency of the on campus catering provision; v) drive a mobile app detailing university and city wide bus services.

3   Discussion

The examples above demonstrate that education has more than one focus: learners, teachers, researchers, administrators; depending on the activity, some people are at the core, whilst some are at the periphery. Establishing a coherent approach to institution-wide change which incorporates technology and introduces new business practices is an ambitious challenge and it would not be surprising if a few challenges were encountered on the way.

Table 7. Southampton data sets available in autumn 2012

Apps using our data	Links to DisabledGo	Public Phonebook
Buildings and Places	Access Information	Published Accounts
Catering	Local Amenities	Services
Common Learning	Open Data Catalogue	Southampton Bus-routes
Spaces	Open Days July	Southampton Bus-stops
Extra Information	2011Organisation	Southampton Jargon
ECS EPrints Link set	Payments 2010-11 to	Dictionary
EPrints Repository	2011-01	Student Statistics
Easting/Northing	Photographs of	Students Union Events
EdShare	University of	Teaching Room Features
EdShare Video	Southampton Things	Transport Linkset
ECS EPrints Repository	Press Contacts	University of
Events Diary	Information	Southampton Profile
Facilities and Equipment	Programmes (2010-2011	Document
Food Hygiene Ratings	session)	Vending Machines
International Links	Programmes (2011-2012	WiFi
International Links	session)	iSolutions Workstation
DBPedia Data	Programmes (2012-2013	Clusters
JACS Codes	session)	

Surveying the use of Semantic Technologies in Education in 2009, Tiropannis et al observed of the the challenges faced by Higher Education that most could be addressed by querying across institutional repositories (databases, web pages, VLEs). Significant learning and teaching challenges can be addressed by accessing resources across departments, schools, institutions. The emergence of linked data fields across related repositories (seen in Table 7) will enable new applications relevant to identified HE challenges. They consider that the initial value of semantic technology will be in scale rather than reasoning and suggest that institutions will benefit from adopting a bottom-up approach starting from linked data which can be related to (layers of) ontologies later in the context of specific applications. The SemTech perspective quoted here focused on specific implications for learning and teaching, but from the material covered in this paper it would appear that a broader perspective would repay investigation.

This paper has looked at the roots of technology innovation which were created by the web and its technical developments incorporating linked and open data. It has considered a range of different technical innovations which can be found in the business and commercial domains and considered how they relate to educational domains through two mechanisms i) establishing patterns of use and user expectations through familiarity and perceived user benefit; ii) creating organisational gains either in terms of improved efficiency and effectiveness or through directly reduced costs. Furthermore, these changes have the potential to create indirect savings by streamlining processes and gathering valuable business intelligence which can help in strategic planning and direction.

Whilst much of the paper has discussed educational innovations, it may well be that for educational institutions the real gains which can be made are in the area of organisational efficiencies. Core business functions in educational institutions have much in common with commercial and business organisations, albeit there are some very particular constraints found in educational contexts because of the cyclical nature of the business, and the uneven tempo of the academic year. There has been some discussion of formal teaching related affordances which might be available to educators using linked and open data. However, semantic technologies do offer learners the means to independently craft and fine tune their own personal learning environments. The discussion and examples throughout this paper have referred to university education, but the case was made in the opening sections that this work is equally applicable to workplace learning at higher levels. There are particular constraints which apply to workplace learning which differentiate it from university learning.

The affordances of semantic technologies, open and linked data introduce potential for flexibility, dynamism and automation which may be particularly beneficial for those who are studying in a workplace context. Streamlining the ways in which we can assemble and inter-link content offers a considerable gain. This benefit will be as relevant to the work-based learner, topping up expertise or undertaking professional development. The models established in OpenCourse-Ware combined with the potential for personal learning environments appear to be particularly fruitful areas for future development. It also seems likely that academics, becoming familiar with technologies in their research activities, will find ways of introducing datasets and research practice into learning as authentic activities. The specific digital literacies of each learner within any given STEM area will be closely related to the authentic tools which are routinely used by practitioners (in the workplace or the research lab) associated with that specialism. Time savings may also accrue from a data based approach to gathering summary information about study programmes for accreditation. The curriculum is one area where the effort of building ontologies is beneficial. Institutions expend significant effort trying to gain broad-brush pictures across modules and programmes; work on knowledge modelling in this area could be fruitful.

The potential impact of widespread use of linked data in Higher Education is immense. Everyday understanding of the power derived by placing raw data in the public domain is growing. It promises to transform education, interconnecting administrative data, enriching and embellishing teaching resources while providing tools and resources for learners and researchers alike. Currently, semantic technologies are more widely and systematically used in research and administration than they are in teaching in higher education.

Having discussed the broad challenges and potential of greater use of data in an educational context, it might be constructive to suggest a way forward.

- Experience at Southampton has placed great value in purposefully constructing teams which incorporate a range of organisational perspectives. Existing literature on change processes identifies the need for champions and patrons.

Champions pursue agendas at a local level, while patrons support and visibly promote change at a strategic and trans-institutional level.
- Some of the simple demonstrators which helped disseminate the potential of using data in applications were developed speculatively on low budgets by student interns. Some apps were developed independently by students, mirroring the wider experience of making APIs available for those who will gain most to invest in.
- Many of our data clients were surprised by the simplicity of the changes they needed to make in order to publish their data and accrue additional benefit. A participant who saves money is a great advert and a willing advertiser of your hard work.
- Once the data has been published and example apps developed, new clients are more able to imagine what they want and what they might gain. This can then enable effective collaboration and co-creation of further apps which will in turn accelerate or refine future developments.
- Borrowing from business practices can be fruitful. Hackathons, BarCamps, UnConferences and Competitions can be ways of finding and pairing developers with clients and producing proof of concept apps in short timeframes. The energy created by these types of events will also sustain more measured developments.
- There is a wider understanding of what might be achieved by crowdsourcing: sharing the task of collecting data, or refining and correcting datasets.
- Publishing data is a wonderful way to distribute quality control tasks. Users can spot and correct published data. We found this worked particularly well in the case of our teaching room database, which was previously maintained on a PC and updated on an annual basis, often preserving mis-information from year to year.
- Shared initiatives lead to understanding and sharing organisational objectives
- Being a semantic squirrel may be rewarding. If there is an opportunity to collect data, the cost of storage will be small. If a means or motivation to analyse and use it emerges in the future, half of the job will already be done.
- Engaging in cool projects at your institution will make your techies happy, and give them things which they can go and brag about at developer events.
- Joint projects like the open data activities suggested here provide an opportunity to develop a local community of practice which will in turn enrich organisational knowledge

References

1. Berners-Lee, T., Hendler, J., Lassila, O.: The Semantic Web. Scientific American 284, 34–43 (2001)
2. Shadbolt, N.R., Gibbins, N., Glaser, H., Harris, S., Schraefel, m.c.: CS AKTive Space or how we stopped worrying and learned to love the Semantic Web. IEEE Intelligent Systems 19, 41–47 (2004)

3. Shadbolt, N., Berners-Lee, T., Hall, W.: The Semantic Web Revisited. IEEE Intelligent Systems 21, 96–101 (2006)
4. Berners-Lee, T., Weitzner, D.J., Hall, W., O'Hara, K., Shadbolt, N., Hendler, J.: A Framework for Web Science. Foundations and Trends in Web Science 1, 1–130 (2006)
5. Berners-Lee, T.: The process of designing things in a very large space: Keynote Presentation. In: WWW 2007 (2007)
6. Shirky, C.: A group is its own worst enemy - A speech at ETech. Clay Shirky's Writings About the Internet: Economics & Culture, Media & Community, Open Source (April 2003), http://www.shirky.com/writings/group_enemy.html
7. O'Reilly, T.: What Is Web 2.0 – Design Patterns and Business Models for the Next Generation of Software (2005), http://oreilly.com/web2/archive/what-is-web-20.html (last accessed June 2010)
8. O'Reilly, T.: What is Web 2.0: Design patterns and business models for the next generation of software. Communications & Strategies 1, 17–37 (2007)
9. Cabinet Office Open Data White Paper Unleashing the Potential Unleashing the Potential Open Data White Paper 52, London (2011), http://www.cabinetoffice.gov.uk/resource-library/open-data-white-paper-unleashing-potential
10. Shotton, D., Portwin, K., Klyne, G., Miles, A.: Adventures in semantic publishing: exemplar semantic enhancements of a research article. PLoS Computational Biology 5, e1000361 (2009)
11. Wilson, S.: Community-driven Specifications: xCrI, sword, and Leap2a. International Journal of IT Standards and Standardization Research 8, 74–86 (2010)
12. Davis, H.C., Carr, L.A., Hey, J.M.N., Howard, Y., Millard, D.E., Morris, D., White, S.: Bootstrapping a Culture of Sharing to Facilitate Open Educational Resources. IEEE Transactions on Learning Technologies 3, 96–109 (2010)
13. Koper, R.: Modelling units of study from a pedagogical perspective the pedagogical meta-model behind EML (2001)
14. IEEE Learning Standards Committee (LTSC) IEEE P1484.12 Learning Object Metadata Working Group; WG12, http://ltsc.ieee.org/wg12/
15. Campbell, L.M.: UK Learning Object Metadata Core Working Draft Version 0.3 1204 (2004)
16. Tattersall, C.: Comparing Educational Modelling Languages on a CaseStudy: An Approach using IMS Learning Design. In: Sixth International Conference on Advanced Learning Technologies, pp. 1154–1155 (2006)
17. Long, P., Siemens, G.: Penetrating the Fog: Analytics in Learning and Education. Educause Review 46, 31–40 (2011)
18. Siemens, G.: Connectivism: A Learning Theory for the Digital Age. International Journal of Instructional Technology and Distance Learning 2, 1–8 (2005)
19. Siemens, G., Downes, S.: Connectivism and connective knowledge: Course delivered at University of Manitoba (September-November 2008), http://ltc.umanitoba.ca/connectivism/
20. Laurillard, D.: Rethinking University Teaching: a Framework for the Effective Use of Educational Technology. Routledge, London (1993)
21. Siemens, G.: Learning Analytics: Envisioning a Research Discipline and a Domain of Practice. In: Second International Conference on Learning Analytics and Knowledge (LAK 2012), pp. 4–8 (2012)
22. Siemens, G., Gasavic, D.: Learning and Knowledge Analytics. Journal of Educational Technology & Society 15, 1–2 (2012)

23. Biggs, J.: Teaching for quality learning at university: what the student does. Open University Press in Association with The Society for Research into Higher Education, Buckingham (1999)
24. Hall, W., Davis, H.C., Hutchings, G.: Rethinking Hypermedia the Microcosm Approach. Kluwer, Boston (1996)
25. White, S., Liccardi, I.: Harnessing Insight into Disciplinary Differences to Refine e-learning Design. In: 36th Annual Frontiers in Education Conference, pp. 5–10 (2006), doi:10.1109/FIE.2006.322553
26. Tiropanis, T., Davis, H., Millard, D., Weal, M., White, S., Wills, G.: JISC - SemTech Project Report 28, Bristol (2009)
27. Biglan, A.: The characteristics of subject matter in different academic areas. Journal of Applied Psychology 57, 195–203 (1973)
28. Biglan, A.: Relationships between subject matter characteristics and the structure and output of university departments. Journal of Applied Psychology 57, 204–213 (1973)
29. Becher, T.: The Significance of Disciplinary Differences. Studies In Higher Education 19, 151 (1994)
30. Cassel, L.N., Davies, G., LeBlanc, R., Snyder, L., Topi, H.: Using a Computing Ontology as a Foundation for Curriculum Development. In: SW-EL 2008 Conjunction with ITS 2008 (2008)

Relating Language to Perception, Action, and Feelings*

Arun K. Majumdar and John F. Sowa

VivoMind Research, LLC, USA
arun@vivomind.com, sowa@bestweb.net

Abstract. The world is a continuum, but words are discrete. Sensory organs map the continuous world to continuous mental models of sights, sounds, and motions. Muscles and bones move in a continuous range of positions, postures, forces, and speeds. Internal feelings of hunger, thirst, pains, pleasures, fears, and desires have a continuous range of variation. But discrete words and patterns of words cannot faithfully represent the continuum of perceptions, actions, and feelings. Peirce's semiotics and Wittgenstein's language games provide a framework for relating language to the world and to perceptions and actions in the world. Peirce analyzed signs and transformations of signs in networks of discrete symbols and in patterns of continuous images. Wittgenstein showed how language is integrated with every aspect of human activity. To implement their insights, the discrete networks of symbols must be mapped to continuous mathematics. This article is a summary of the methods and applications for mapping natural languages to conceptual graphs and continuous transformations. Those methods have been used to analyze and classify plot twists in narratives and the structure of expository texts.

Keywords: conceptual graphs, natural language processing, semiotics, analogy, mental models, continuity, catastrophe theoretical semantics.

1   Language and Brain

The human brain is built on an ape-like plan with a greatly enlarged cerebral cortex. If the cortex were removed, the human brain stem and cerebellum would be hard to distinguish from those of a chimpanzee or gorilla. After reviewing the fossil evidence, Terrence Deacon [1] concluded that the mainstream of evolution from the apes to Australopithicus, Homo habilis, Homo erectus, and Homo sapiens was driven by "the co-evolution of language and the brain." Gradual changes in the vocal tract indicate an early shift toward more complex vocalization. The earliest language-like vocalizations, perhaps spoken by some Australopithecines, gave their speakers a competitive advantage over other primates. The need for a larger vocabulary and a more precise grammar drove the rapid increase in size and complexity of the brain. Figure 1 shows aspects of language that evolved in the past 6 million years on top of a foundation of 600 million years of evolution from primitive worms.

---

* An extended paper associated with this invited talk will appear in the Workshop Proceedings for the ''Workshop on Modeling States, Events and Processes (MSEPS)''.

H.D. Pfeiffer et al. (Eds.): ICCS 2013, LNAI 7735, pp. 22–28, 2013.
© Springer-Verlag Berlin Heidelberg 2013

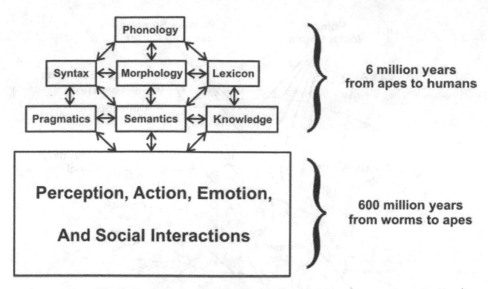

Fig. 1. Human language supported by an ape-like foundation

Yet Figure 1 has some questionable features. The labels on the language boxes correspond to traditional academic fields, but those boxes don't have a one-to-one mapping to modules in the brain. The box labeled *knowledge*, for example, includes information in different areas of the brain: language-independent images; concepts related to language, but independent of any specific language; and knowledge encoded in the patterns of a particular language. The box labeled *pragmatics* involves the use of language in all the activities of life. But knowledge of an activity is directly based on actions and perceptions, only indirectly on the words that express them. As Wittgenstein emphasized, words are always learned and used in the context of some activity: "The word *language-game* (Sprachspiel) is used here to emphasize the fact that the speaking of language is part of an activity, or of a form of life" [2: §23].

Nobody knows exactly how the brain works, but neuroscience has accumulated a great deal of evidence about the areas of the brain associated with various functions. For right-handed people, the left hemisphere of the cerebral cortex (LH) is critical for language. Figure 2 shows areas of LH involved in language processing [3, 4]. Broca's area, which generates the syntax and phonology of speech, is adjacent to the primary motor cortex for the mouth, face, tongue, and vocal tract. Wernicke's area, which relates language to semantics, is adjacent to the primary auditory cortex and close to the sensory areas for vision and touch. It is also directly beneath the parietal lobe, which maintains patterns that are variously called cognitive maps, frames, or schemata. Lamb [5, 6] argued that the primary nodes for concepts are located in the parietal lobe. Nouns, which map to images, are in the temporal lobe, close to both the auditory and visual areas. Verbs are in the frontal lobe, close to the motor areas that control actions.

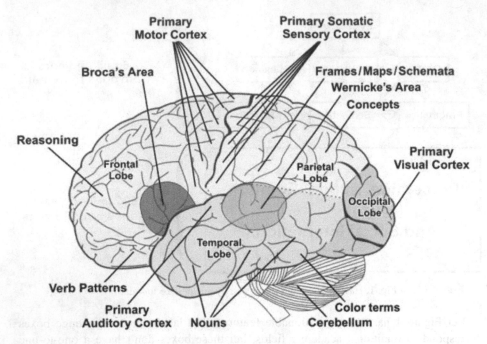

Fig. 2. Language areas of the left hemisphere

The cerebral cortex is essential for language and all complex reasoning. But the brain stem controls the basic functions necessary for maintaining life. It also integrates all inputs and outputs to and from the cerebral cortex. One part, called the *superior colliculus*, integrates vision with eye movements and head position. It also relates vision to the auditory inputs processed by the *inferior colliculus*. The *cerebellum* learns and controls fine tuned, but automatically executed skills. Using just the superior colliculus and the cerebellum, a frog can jump along the lily pads in a pond, track a fly in mid flight, and shoot out its tongue at just the right speed and direction to catch it. Humans use the same neural mechanisms for shooting a basketball, performing gymnastics, or playing the piano. For a person talking on a cell phone, the superior colliculus can bypass the cortex and enable the cerebellum to control walking without thinking.

As Figure 2 shows, language areas are distributed around the cortex. Instead of a dedicated language module, brain areas specialized for perception and action are also used to interpret and generate language. Broca's area, for example, overlaps two regions called Broadman's areas BA44 and BA45. BA44 is active in controlling the mouth for speech and eating. BA45 is active in producing both spoken and signed languages. It is also active in precise motor control for using tools.

The processes carried out by the brain stem and cerebellum are outside conscious awareness and independent of language. But they are essential for producing and maintaining the images in the cortex that are mapped to and from language. The superior colliculus is responsible for controlling eye movements, relating multiple fragmentary glimpses, and enabling the visual cortex to assemble them in a stable, panoramic image of the environment. But the frontal lobes can also provide the stimuli to generate an imaginary model of a planned, hypothetical, or desired environment. As the neuroscientist Antonio Damasio [4] said,

> The distinctive feature of brains such as the one we own is their uncanny ability to create maps... But when brains make maps, they are also creating images, the main currency of our minds. Ultimately consciousness allows us to experience maps as images, to manipulate those images, and to apply reasoning to them.

Although the details of information processing in the brain are major research problems, brain scans show the active areas, and anatomy shows the pathways that connect them. For language understanding and generation, Wernicke's area and Broca's area are connected by a thick bundle of fibers called the *arcuate fasciculus*. Those fibers transmit information in both directions to coordinate the semantic processing in Wernicke's area with the syntactic processing in Broca's area. But the verb patterns, which are located in BA47 close to Broca's area (BA45), are critical for relating sentence structures to actions, planning, and reasoning. The nouns that participate in those patterns are located in the temporal lobes. The arcuate fasciculus has branches that connect to those areas and to the parietal lobe with its concepts and cognitive maps.

The evidence from neuroscience confirms what linguists, psychologists, philosophers, and language users have known for millennia: nouns are linked to images, verbs are linked to actions, the links depend on background knowledge, and people with different backgrounds often misunderstand each other. For computer processing, it implies that language understanding requires methods that go beyond the discrete words and patterns of words that appear in texts. For any subject that has a continuous range of variation — anything except subjects like chess, Sudoku, or computer programs — the semantics requires continuous mathematics. The discrete patterns of language must be mapped to and from continuous fields and transformations.

2   Discrete and Continuous Processing

The early stages of language processing must analyze the discrete words and patterns of words that occur in speech and texts. For VivoMind software, the results of the analysis are translated to conceptual graphs (CGs) as the semantic representation [7, 8, 9]. But the graphs are still discrete. They must be translated to continuous fields for the next stage of analysis. Key to that translation is an insight by Charles Sanders Peirce, whose existential graphs (EGs) are the foundation for CGs [10].

Peirce was inspired by the graphs used to represent molecules in organic chemistry. He designed his EGs as a notation for representing "the atoms and molecules of logic." The continuous fields are forces like gravity or electromagnetism. In the verb patterns, each verb is a nucleus, and the nouns orbit the nucleus in a continuous force field. The verb *sleep*, for example, has one *actant* or participant; it has valence 1. The verb *hit* has 2, *give* has 3, and *buy* or *sell* has 4.

René Thom [11] was a mathematician who developed catastrophe theory and applied it to a variety of physical phenomena. As he broadened the range of applications, he discovered psychological and linguistic phenomena that displayed related patterns. He used the dependency grammar developed by Lucien Tesnière [12] to represent patterns of events. When he analyzed the dynamic evolution of those events and their consequences, he discovered that they fell into patterns that resembled the chaotic patterns he observed in physics. He showed that those patterns could be used to classify the typical kinds of plot structures found in literature. Thom's ideas were developed further by Petitot [13] and Wildgen [14]. Most linguists ignored those developments because they use mathematical computations that are unrelated to the usual linguistic theories. But Tesnière's dependency structures have a direct mapping to conceptual graphs [7]. That mapping enables the methods of catastrophe theoretical semantics (CTS) to be adapted to CGs.

3   Applications

VivoMind software has mostly been applied to nonfictional documents on subjects such as oil and gas exploration or rare earth magnetic materials [8, 9]. For those purposes, CTS proved to be valuable for classifying document types. It can distinguish documents that serve different purposes, even though they have similar vocabulary and ontology. Examples include chapters from a textbook, research reports, tutorials, and surveys that cover similar material. These methods can even distinguish routine or incremental research from novel papers and highly innovative speculation.

To illustrate the differences between a serious and a humorous text, we used the methods to compare the patterns of betrayal in two anecdotes: the BRUTUS story betrayal model [15] and a joke called "Meeting St. Peter." In both of them, the situation is painful for the victim. But the victim in the serious anecdote is a sympathetic character (a student); in the other, the victim is a stereotypical butt of humor (a salesman). Following are two stories of betrayal analyzed by the CTS methods.

"Betrayal" by BRUTUS

Dave Striver loved the university. He loved its ivy-covered clocktowers, its ancient and sturdy brick, and its sun-splashed verdant greens and eager youth. He also loved the fact that the university is free of the stark unforgiving trials of the business world — only this isn't a fact: academia has its own tests, and some are as merciless as any in the marketplace. A prime example is the dissertation defense: to earn the PhD, to become a doctor, one must pass an oral examination on one's dissertation. This was a test Professor Edward Hart enjoyed giving.

Dave wanted desperately to be a doctor. But he needed the signatures of three people on the first page of his dissertation, the priceless inscriptions which, together, would certify that he had passed his defense. One of the signatures had to come from Professor Hart, and Hart had often said-to others and to himself-that he was honored to help Dave secure his well-earned dream.

Well before the defense, Striver gave Hart a penultimate copy of his thesis. Hart read it and told Dave that it was absolutely first-rate, and that he would gladly sign it at the defense. They even shook hands in Hart's book-lined office. Dave noticed that Hart's eyes were bright and trustful, and his bearing paternal.

At the defense, Dave thought that he eloquently summarized chapter three of his dissertation. There were two questions, one from Professor Rodman and one from Dr. Teer; Dave answered both, apparently to everyone's satisfaction. There were no further objections.

Professor Rodman signed. He slid the tome to Teer; she too signed, and then slid it in front of Hart. Hart didn't move.

"Ed?" Rodman said.

Hart still sat motionless. Dave felt slightly dizzy.

"Edward, are you going to sign?"

Later, Hart sat alone in his office, in his big leather chair, saddened by Dave's failure. He tried to think of ways he could help Dave achieve his dream.

The next story is an updated and extended version of a joke that was circulated around the Internet.

Meeting St. Peter

A computer salesman died and went to meet St. Peter at the Pearly Gates.

St. Peter: Welcome to our reception hall. We've made some updates to our traditional procedures in order to speed up the process and make our guests feel more comfortable. Instead of the old book of sins, we now use an iPad.

Computer Salesman: That sounds great. I love the new technology.

Then St. Peter swiped the iPad and projected scenes from the salesman's life on the wall next to the Pearly Gates. The salesman began to squirm when he saw some of the long-forgotten events.

St. P: Relax. We got rid of the old trial because it takes too long. We developed new methods that predict the same results with six-sigma reliability. We just let people choose whether they would prefer to go to Heaven or Hell.

Then he took another swipe at the iPad and showed some scenes from Heaven. People in white robes were sitting on clouds, playing harps, and singing hymns.

C S: That looks boring.

Then St. Peter took another swipe at the iPad and showed scenes from Hell.

A toga party was going on. There was wild music, dancing, drinking, and carousing. Men and women in various stages of undress were engaged in every activity imaginable.

C S: That's fantastic. I choose Hell.

St. P: Done.

St. Peter took another swipe at his iPad, a trap door opened, and the salesman found himself sliding down a steel chute. He was rapidly accelerating down a well-worn path, polished by many previous travelers.

Finally, he flew through an open door into a huge cavern with fire and brimstone stinging his eyes. At once, a dozen little devils with pitchforks started prodding and pushing him toward a fiery pit.

C S: Hey, wait a minute! What happened to the party?

Then the chief devil walked over, stroking his beard.

Chief Devil: Ooooh. You must have seen our demo.

References

1. Deacon, T.W.: The Symbolic Species: The Co-evolution of Language and the Brain. W. W. Norton, New York (1997)
2. Wittgenstein, L.: Philosophical Investigations. Basil Blackwell, Oxford (1953)
3. MacNeilage, P.F.: The Origin of Speech. University Press, Oxford (2008)
4. Damasio, A.R.: Self Comes to Mind: Constructing the Conscious Brain. Pantheon Books, New York (2010)
5. Lamb, S.M.: Pathways of the Brain: The Neurocognitive Basis of Language. John Benjamins, Amsterdam (1999)
6. Lamb, S.M.: Neurolinguistics. Lecture notes for Linguistics, vol. 411. Rice University (2010), http://www.owlnet.rice.edu/~ling411
7. Sowa, J.F.: Conceptual graphs. In: van Harmelen, F., et al. (eds.) Handbook of Knowledge Representation, pp. 213–237. Elsevier, Amsterdam (2008), http://www.jfsowa.com/cg/cg_hbook.pdf
8. Majumdar, A.K., Sowa, J.F., Stewart, J.: Pursuing the Goal of Language Understanding. In: Eklund, P., Haemmerlé, O. (eds.) ICCS 2008. LNCS (LNAI), vol. 5113, pp. 21–42. Springer, Heidelberg (2008)
9. Majumdar, A.K., Sowa, J.F.: Two Paradigms Are Better Than One, and Multiple Paradigms Are Even Better. In: Rudolph, S., Dau, F., Kuznetsov, S.O. (eds.) ICCS 2009. LNCS (LNAI), vol. 5662, pp. 32–47. Springer, Heidelberg (2009), http://www.jfsowa.com/pubs/paradigm.pdf
10. Sowa, J.F.: Peirce's own tutorial on existential graphs. Semiotica 186(1-4), 345–394 (2010); Special issue on diagrammatic reasoning and Peircean logic representations
11. Thom, R.: Esquisse d'une Sémiophysique. InterEditions, Paris (1988)
12. Tesnière, L.: Éléments de Syntaxe structurale, 2nd edn. Librairie C. Klincksieck, Paris (1959)
13. Petitot, J.: Cognitive Morphodynamics: Dynamical Morphological Models of Constituency in Perception and Syntax. Peter Lang, Bern (2011)
14. Wildgen, W.: Process, Image, and Meaning: A Realistic Model of the Meaning of Sentences and Narrative Texts. John Benjamins Publishing Co., Amsterdam (1994)
15. Bringsjord, S., Ferrucci, D.: Artificial Intelligence and Literary Creativity. Lawrence Erlbaum, Mawah (2000)

PurposeNet: A Knowledge Base Organized around Purpose*

Rajeev Sangal, Soma Paul, and P. Kiran Mayee

Language Technologies Research Centre
International Institute of Information Technology
Hyderabad, India
{sangal,soma}@iiit.ac.in,
kiranmayee@research.iiit.ac.in

1 Invited Talk Summary

We show how *purpose* can be used as a central guiding principle for organizing knowledge about artifacts. It allows the actions in which the artifact participates to be related naturally to other objects. Similarly, the structure or parts of the artifact can also be related to the actions.

A conceptual base, architecture and implementation of a semantic knowledge base called *PurposeNet*, with an evaluation performed in comparison with other knowledge bases, shows that PurposeNet is a superior method in terms of coverage. Building an exhaustive knowledge base is a laborious and intense task, it needs human expertise and it needs good web data processing tools so that information from the web can be easily extracted in order to build the knowledgebase semi-automatically. In order to maintain the quality of the resource, it has been, till now, a case where the knowledge base was manually created. Nevertheless, creating such a huge resource completely in manual mode would be a time-consuming work. PurposeNet also makes it possible for automatic extraction of simple facts (or information) from text for populating a richly structured knowledge base.

Therefore artifact related information which is useful for our knowledge base is available in various resources such as WordNet, Wikipedia and other web corpora. Results are reported on conducting a few experiments on detecting and extracting purpose of artifacts from web corpus. An experiment in domain-specific question-answering from a given passage shows that PurposeNet used along with scripts (or knowledge of stereotypical situations), can lead to substantially higher accuracy in question answering. In the domain of car racing, individually they produce correct answers to 50% and 37.5% questions respectively, but together they produce 89% correct answers. These experimental results in domain-specific question-answering have produced promising results.

---

* An extended paper associated with this invited talk will appear in the Workshop Proceedings for the "Workshop on Modeling States, Events and Processes (MSEPS)."

H.D. Pfeiffer et al. (Eds.): ICCS 2013, LNAI 7735, pp. 29–30, 2013.
© Springer-Verlag Berlin Heidelberg 2013

References

1. Alani, H., Brewster, C.: Ontology ranking based on the analysis of concept structures. In: Proceedings of the 3rd International Conference on Knowledge Capture (K-CAP 2005), pp. 51–58. ACM Press, New York (2005)
2. Alani, H., Brewster, C.: Metrics for ranking ontologies. In: WWW 2006, Edinburgh, UK, May 22–26 (2006); Patton, H.D.: Physiology of Smell and Taste. Annual Review of Physiology 12, 469–484 (2006)
3. Aleman-Meza, B., Halaschek, C., Sheth, A., Arpinar, I.B., Sannapareddy, G.: SWETO: Large-Scale Semantic Web Test-bed. In: Proceedings of the 16th SEKE 2004: Workshop on Ontology in Action, Banff, Canada, June 21-24, pp. 490–493 (2004)
4. Bharati, A., Chaitanya, V., Sangal, R.: Natural Language Processing: A Paninian Perspective. Prentice-Hall of India, New Delhi (1995), http://ltrc.iiit.ac.in/downloads/nlpbook/nlp-panini.pdf
5. Bharati, A., Nawathe, S.A., Chaitanya, V., Sangal, R.: A New Inference Procedure for Conceptual Graphs. In: Proc. of 4th University of New Brunswick Artificial Intelligence Symposium (1991)
6. Cowell, E.B., Gough, A.E.: The Sarva-Darsana-Samgraha or Review of the Different Systems of Hindu Philosophy. Trubner's Oriental Series. Taylor & Francis (2001)
7. Lenat, D.B.: CYC: a large-scale investment in knowledge infrastructure. Communications of the ACM 38(11), 33–38 (1995)
8. Gangemi, A., Catenacci, C., Ciaramita, M., Lehmann, J.: Modelling Ontology Evaluation and Validation. In: Sure, Y., Domingue, J. (eds.) ESWC 2006. LNCS, vol. 4011, pp. 140–154. Springer, Heidelberg (2006)
9. Devi, G.: Padartha Vijnana made easy. Chaukhamba Sanskrit Pratishthan, Delhi (2007)
10. Miller, G., Beckwith, R., Fellbaum, C., Gross, D., Miller, K.J.: WordNet: An online lexical database. International Journal of Lexicography 3(4) (1990)
11. Iśvarakṛṣṇa, Sāṁkhyakārikā with Sankara Misra's commentary Sāṁkhyatattva kaumudi, Edited and translated into Hindi by Nigam Sharma. Parimal Prakashan, Varanasi (2007)
12. Mayee, P.K., Sangal, R., Paul, S.: Action Semantics in PurposeNet. In: Proceedings of 2011 World Congress on Information and Communication Technologies, IEEE WICT 2011, pp. 1299–1304 (2011)
13. Kulkarni, A.P.: Navya-Nyaya and Logic. MTech Thesis, IIT Kanpur (1994)
14. Liu, H., Singh, P.: ConceptNet: A Practical Commonsense Reasoning Toolkit. BT Technology Journal 22 (2004)
15. Nagaraj, A., Darshan, M.V., Prakashan, J.V.: Amarkantak (2003)
16. Praśastapāda, Padārthadharmasamgraha with Sridhara's commentary Nyāyakandali, Edited and Translated into Hindi by Sharma, S.D.J. Sampurnananda Sanskrita University, Vārānasī (1997)
17. Rao, R.S.: M. Ayurveda Padardha Vijnana (2003)
18. Sangal, R., Chaitanya, V.: An Intermediate Language for Machine Translation: An Approach based on Sanskrit Using Conceptual Graph Notation. Computer Science and Informatics Journal, Computer Society of India 17(1), 9–21 (1987)
19. Singh, N.: Comprehensive Schema of Entities: Vaiśeṣika Category System. Science Philosophy Interface 5(2), 1–54 (2001)
20. Sowa, J.F.: Conceptual Structures: Information Processing in Mind and Machine. Addison-Wesley, Reading (1984)
21. Sowa, J.F.: The Challenge of Knowledge Soup. In: Ramadas, J., Chunawala, S. (eds.) Research Trends in Science, Technology, and Mathematics Education, pp. 55–90. Homi Bhabha Centre, Mumbai (2005)

Classical Syllogisms in Logic Teaching

Peter Øhrstrøm[1], Ulrik Sandborg-Petersen[1], Steinar Thorvaldsen[2],
and Thomas Ploug[1]

[1] Department of Communication and Psychology, Aalborg University,
9000 Aalborg, Denmark
{poe,ulrikp,ploug}@hum.aau.dk
[2] Department of Education, University of Tromsø,
9037 Tromsø, Norway
steinar.thorvaldsen@uit.no

Abstract. This paper focuses on the challenges of introducing classical syllogisms in university courses in elementary logic and human reasoning. Using a program written in Prolog+CG, some empirical studies have been carried out involving three groups of students in Denmark; one group of philosophy students and two groups of students of informatics. The skills of the students in syllogistic reasoning before and after the logic courses have been studied and are discussed. The empirical observations made with the program make it possible to identify syllogisms which are found difficult by the students, and to identify others which the students find easier to handle. It is discussed why certain syllogisms are more difficult than others to assess correctly with respect to validity. The results are compared with findings from earlier studies in the literature. As in other studies, it is shown that the test persons have a tendency correctly to assess valid syllogisms as such more often than correctly assessing invalid syllogisms as such. It is also investigated to what extent the students have improved their skills in practical reasoning by attending the logic courses. Finally, some open questions regarding syllogistic reasoning are discussed.

1 Introduction

For centuries the Aristotelian syllogisms have been a crucial part of university courses introducing basic logic and human reasoning. In the medieval universities, syllogistics was regarded as an essential component of basic academic learning. At many modern universities this is still the view.

There is, obviously, a close relation between the ontological primitives (e.g., SubclassOf) and the categorical statements which are used in classical syllogistics. In fact, a number of syllogistic arguments can be inferred from the hierarchical structures used in formal ontology (see [4]). It appears to be evident that a proper understanding of conceptual structures in many cases will depend on the ability to handle basic syllogistic arguments correctly. Based on such observations syllogistic reasoning should still be considered to be an important prerequisite for the understanding of conceptual structures and indeed for

H.D. Pfeiffer et al. (Eds.): ICCS 2013, LNAI 7735, pp. 31–43, 2013.
© Springer-Verlag Berlin Heidelberg 2013

science in general. For mathematics, syllogistics can be said to form part of the foundation for mathematics as such. For engineering, a basic knowledge of syllogistics could enhance the potential for systematic reasoning about the artefacts being constructed. For this reason, it is our view that students working with science, technology, engineering, and mathematics – as well as students in the humanities – should be introduced to these basic forms of logic and reasoning.

From a modern point of view classical syllogistics may be seen as a fragment of first order predicate calculus. A classical syllogism corresponds to an implication of the following kind:

$$(p \wedge q) \supset r$$

where each of the propositions p, q, and r matches one of the following four forms

a(X, Y)	(read: "All X are Y")
i(X, Y)	(read: "Some X are Y")
e(X, Y)	(read: "No X are Y")
o(X, Y)	(read: "Some X are not Y")

These four functors were suggested by the medieval logicians referring to the vowels in the words "affirmo" (Latin for "I confirm") and "nego" (Latin for "I deny"), respectively. The classical syllogisms occur in four different figures:

$(x(M, P) \wedge y(S, M)) \supset z(S, P)$	(1st figure)
$(x(P, M) \wedge y(S, M)) \supset z(S, P)$	(2nd figure)
$(x(M, P) \wedge y(M, S)) \supset z(S, P)$	(3rd figure)
$(x(P, M) \wedge y(M, S)) \supset z(S, P)$	(4th figure)

where x, y, $z \in$ {a, i, e, o} and where M, S, P are variables corresponding to "the middle term", "the subject" and "the predicate". In this way 256 different syllogisms can be constructed. According to classical (Aristotelian) syllogistics, however, only 24 of them are valid. The medieval logicians named the valid syllogisms according to the vowels, {a, i, e, o}, involved. In this way the following artificial names were constructed (see [1]):

1st **figure:** barbara, celarent, darii, ferio, barbarix, feraxo
2nd **figure:** cesare, camestres, festino, baroco, camestrop, cesarox
3rd **figure:** darapti, disamis, datisi, felapton, bocardo, ferison
4th **figure:** bramantip, camenes, dimaris, fesapo, fresison, camenop

In these names the consonants signify the logical relations between the valid syllogisms, and they also indicate which rules of inference should be used in

order to obtain the syllogism in question from syllogisms which were considered
to be fundamental: barbara, celarent, darii, and ferio. – In fact, the system of
syllogisms may in this way be seen as the first axiomatic system ever (see [1] and
[3]). According to Aristotle a universal statement concerning an empty term cannot be true. For this reasoning an a-proposition must imply the corresponding
i-proposition. This view was rejected, when philosophers and mathematicians
began to pay more attention to the idea of an empty set. Without this Aristotelian view the number of valid syllogisms was reduced to 15 (leaving out the
9 syllogisms which have been underlined in the above list, i.e. the syllogisms
whose names contain either an x or a p).

The conceptual structures which form the foundation of the experiment conducted in this paper, may be illustrated by reference to the classical hierarchy
of categorization depicted in Figure 1.[1]

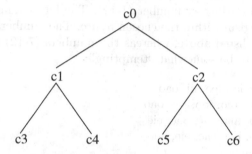

Fig. 1. A classical hierarchy of categorization

The relation between the concepts in this figure can all be described in terms of
the Aristotelian propositions which are used in syllogistics. For example:

- All c5 are c2
- Some c0 are c1
- No c1 is c2
- Some c0 are not c1

Even the properties of the IsA-relation can be said to correspond to the syllogistic
logic. One very simple example would be the transitivity of this relation, which
turns out to be equivalent with the syllogism called barbara, 1st figure. For
example:

All c3 are c1
All c1 are c0
Ergo: All c3 are c0

---

[1] It should be noted that in this paper, when we capitalize "Figure", we refer to an
illustration within the text. When, however, we do not capitalize "figure", we refer
to the four figures of the Aristotelian syllogisms.

For this reason, it seems obvious that the proper conception of the hierarchical structure of ontologies depends on a proper understanding of the logic embedded in classical syllogistics.

2 Data and Method

The data analysed in this study have been obtained using the system Syllog (described in detail in [17]), which has been implemented using an extended version of Prolog+CG (see [5–7]). Syllog presents the user with an arbitrary syllogistic argument and asks him or her to evalute the syllogism that appears on the screen as to its Aristotelian validity. The arbitrary arguments are generated by Syllog with S belonging to the set {"swedes","politicians","dentists"}, P belonging to the set {"halffools", "halfminded","redheaded"}, and M belonging to the set {"thieves", "crackpots", "fools"}. Syllog picks an arbitrary figure number (1-4) and an arbitrary number (1-12). The latter number corresponds to a particular syllogism within the chosen figure. The numbers, (1-6), point to the valid syllogisms listed above, whereas the numbers (7-12) stand for invalid syllogisms assumed to be somewhat "tempting":

1ˢᵗ **figure:** aia, oae, iai, ieo, iii, oao
2ⁿᵈ **figure:** oae, aoe, oio, ioo, ieo, oao
3ʳᵈ **figure:** aaa, iaa, aia, eae, oae, eie
4ᵗʰ **figure:** aaa, aie, iaa, eae, eie, aoe

Three groups of students in Denmark have been involved in the tests:

1. Two groups of second year University students in informatics (one in Copenhagen and one in Aalborg)
2. One group of first year University students in philosophy (in Aalborg).

All three groups were going to attend a basic course in logic – including Aristotelian syllogistics. The course was based on course material in Danish (corresponding to parts of [1] along with elements from [2]). The students were taught in basically the same way, to the same extent, and to a large extent using the same course material. The teacher of the Aalborg Informatics students also taught the course for the Philosophy students.

The students were asked to run Syllog individually or in groups of 2-4 both before and after their logic course (i.e. pre- and post-test). All these test results have been logged by Syllog.

The statistical analyses of the data were performed using standard methods from descriptive statistics and statistical testing. The chi-square test is applied to detect group differences using frequency (count) data, and also to look for significant differences between results from the pre-test and the post-test. We compared the pre-test results of the informatics-students in Copenhagen and in Aalborg, and found no significant differences between the two groups (p-value = 0.25, data not shown).

Table 1. The three 2x2 tables below summarize counts of how often students replied correctly and incorrectly when presented with valid and invalid syllogisms. The first table is for the pre-test of the whole student group, while the second and third table shows the same data separated in two subgroups. All three tables support strong statistical evidence against the presumption that student will handle valid and invalid syllogism equally well (p-values $< 10^{-5}$ by the two-sided chi-square test).

Syllogism	Correct reply? (all students, n=174)		Correct reply? (philosophy students, n=33)		Correct reply? (informatics students, n=141)	
	Yes	No	Yes	No	Yes	No
Valid:	814	318	307	80	507	238
Invalid:	697	538	283	179	414	357

Table 2. Two 2x2 tables summarizing counts of how often students replied correctly and incorrectly in the pre-test and in the post-test. The first table is for the philosophy students, whereas the second table shows the results for the Copenhagen group of informatics students.

Test	Correct reply? (philosophy students, n=33/n=17)		Correct reply? (informatics students, n=39/n=42)	
	Yes	No	Yes	No
Pre:	590	259	382	229
Post:	175	46	486	306

3 Results

The first online tests were carried out with all the students as a pre-test before their lessons in classical syllogistics. The students answered the exercises individually or in groups of 2-4. If less than 3 exercises were answered by a given individual or group, the record was excluded from the analysis to avoid data that were influenced by technical problems or unserious students. The final data from the pre-test consists of 2365 evaluated syllogisms, from n=174 groups (or individuals), with an average of 13.6 answered exercises. The results of the pre-tests are shown in the (a) parts of Figures 2-5 and in Table 1.

The group of informatics students in Copenhagen and the philosophy students in Aalborg also took part in a post-test under conditions similar to the pre-test. For the informatics students the post-test was performed four months after the pre-test, but before the exam, and for the philosophy students the interval between the tests was five weeks. The (b) parts of Figures 2-5 show the results obtained by the post tests, and Table 2 summarizes the results obtained by the pre- and post-tests. For the group of informatics students in Aalborg, the post-test were supposed to be done voluntarily at home, and only 6 students completed it, hence their results are left out.

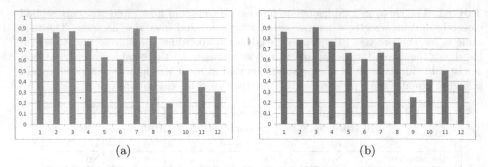

(a)   (b)

Fig. 2. Pre-test score (a) and post-test score (b) of syllogisms in figure 1. The first 6 syllogisms are valid according to Aristotelian syllogistics, whereas the 6 other syllogisms are invalid. It should be noted that the scores of syllogism no. 9 are very low (see the discussion in Section 4).

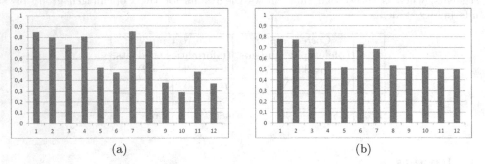

(a)   (b)

Fig. 3. Pre-test score (a) and post-test score (b) of syllogisms in figure 2. The first 6 syllogisms are valid according to Aristotelian syllogistics, whereas the 6 other syllogisms are invalid.

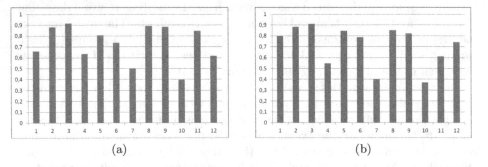

(a)   (b)

Fig. 4. Pre-test score (a) and post-test score (b) of syllogisms in figure 3. The first 6 syllogisms are valid according to Aristotelian syllogistics, whereas the 6 other syllogisms are invalid. Note that syllogism no. 3 has got very high scores (see the comments in Section 4).

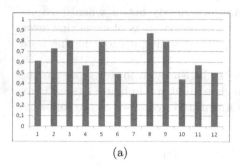
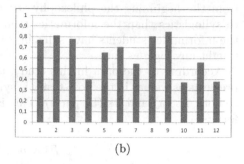

(a)   (b)

Fig. 5. Pre-test score (a) and post-test score (b) of syllogisms in figure 4. The first 6 syllogisms are valid according to Aristotelian syllogistics, whereas the 6 other syllogisms are invalid. Because of a minor programming error the results regarding syllogism 12 cannot be expected to be correct.

The student dropout is more than 50% in the post-test of the left part of Table 2, and hence the results are only indicative with p-value = 0.004 by the two-sided chi-square test.

The right part of Table 2 provides no significant evidence against the hypothesis that informatics students did not obtain better skills in syllogistic reasoning during the logic course (p-value = 0.66).

4   Discussion of the Results

The abilities of performing syllogistic reasoning have been studied earlier using other methods (see [9–16]). Our data can to some extent confirm the findings in these earlier studies. In addition, the present study also allows some new conclusions. The results in Table 1 show that students more often wrongly agree with invalid syllogisms than they wrongly disagree with a valid syllogism. The reason may be that the students find it more natural to agree with a difficult argument, than to disagree. In other words, it may be more natural for the human nature to be positive than to be negative. In this way there seems to be "a belief bias" in syllogistic reasoning (see [16]). The results listed in Table 2 may seem somewhat surprising. It seems that the philosophy students have in fact improved their skills in syllogistic reasoning during the course whereas there is no evidence that informatics students have made similar progress during their course. It is not clear how this difference should be explained. However, several comments and possible explanations may be considered. First of all, the philosophy students in general may see it as fascinating to reflect on the Aristotelian ideas and the notion of human reasoning as such, whereas the informatics students may not find the study and elaboration of Aristotelian syllogisms particularly interesting. In addition, it may be important that the philosophy students had their post-test immediately after the logic course, whereas the informatics students had to wait longer for their post-test. Furthermore, it should also be noted that right from the beginning the philosophy students are clearly better when it comes to syllogistics than the informatics students. However, it is obvious that even the informatics students have skills in syllogistic reasoning without having been taught any logic as such.

According to the data in this study, the syllogism with lowest score (in the pre-tests as well as in the post-tests) is syllogism number 9 in figure 1 (see Figure 2, both its (a) and its (b) parts). This is an invalid syllogism of the following form:

$$(i(M, P) \wedge a(S, M)) \supset i(S, P)$$

In Syllog the values of M, S and P are selected arbitrarily from certain fixed sets as mentioned above. The users may, for instance, be presented with the syllogism in the following way:

> Some crackpots are redheaded
> All Swedes are crackpots
> Ergo: Some Swedes are redheaded

A majority of users have mistakenly evaluated this syllogism as valid. This error may have occurred because the students have not fully understood the difference between "All Swedes are crackpots" and "All crackpots are Swedes". The syllogism

> Some crackpots are redheaded
> All crackpots are Swedes
> Ergo: Some Swedes are redheaded

is clearly valid. It is a disamis (in the 3rd figure). The difference between the two syllogisms can be made clear in terms of Euler circles. The point is that the information in the premises of the iai-syllogism in the 1st figure may correspond to the diagram in Figure 6(a). Here it is obvious that $i(S, P)$ cannot be concluded.

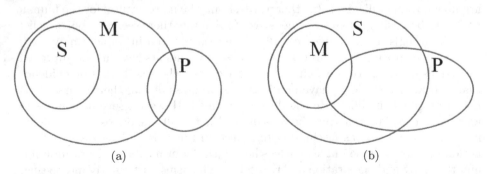

(a)   (b)

Fig. 6. Euler circles consistent with: (a) the information in the premises of the iai-syllogism in the 1st figure, and (b) the information in the premises of disamis in the 3rd figure

In disamis (in the 3rd figure), however, the information contained in the premises will have to be represented otherwise, as in Figure 6(b). Here, $i(S, P)$ obviously follows from the premises. If it is known that M and P have elements in common, and M is a subset of S, then it is evident that S and P must also have elements in common. It seems obvious that if the student immediately had been able to visualize the graphical relation between M and S when being given the premise (i.e. $a(S, M)$)), then they would not have taken the iai-syllogism in the 1st figure to be valid. This leads to a strong emphasis on the importance of the use of graphical representation in logic courses.

There are also interesting questions to ask concerning the data of the 24 syllogisms which are valid from an Aristotelian point of view. First of all one may focus on the syllogism which has got the highest score taking both pre-tests and post-tests into account, i.e. datisi in the 3rd figure:

$$(a(M, P) \wedge i(M, S)) \supset i(S, P)$$

Intuitively, it seems obvious that this structure is a valid syllogism: If all M's are P's, and if some M's are S's, the clearly some S's must be P's. However, disamis in the 3rd figure seems just as obvious:

$$(i(M, P) \wedge a(M, S)) \supset i(S, P)$$

In fact, datisi and disamis in the 3rd figure have got also the same score.

Another interesting question regarding the valid syllogisms has to do with the possible difference between the two subgroups of 15 and 9 syllogisms mentioned above. Given that the empty set is an integrated idea in modern thinking, it seems to be straight forward to expect that the 15 syllogisms would have a higher score the 9 Aristotelian syllogisms which have been questioned in modern logic. For this reason we have compared the scores for the 15 syllogisms which all modern logicians accept (i.e. figure 1 no. 1,2,3,4 & figure 2 no. 1,2,3,4 & figure 3 no. 2,3,5,6 & figure 4 no. 2,3,5) with the 9 syllogisms whose validity some modern logicians would question (i.e. figure 1 no. 5,6 & figure 2 no. 5,6 & figure 3 no. 1,4 & figure 4 no. 1,4,6). We compared the group of these 9 versus the 15 other valued syllogisms and the results appeared as shown in Table 3.

Table 3. Two 2x2 tables summarizing counts of how often students replied correctly to the subgroup of 15 and 9 valid syllogisms. The first table is for the pre-test, whereas the second table shows the results for the post-test.

	Correct reply? (Pre-test n=174)		Correct reply? (Post-test n=65)	
Syllogism	Yes	No	Yes	No
Valid in modern syllogistics	552	126	260	71
Questions in modern syllogistics	262	192	125	70

Table 4. Two 2x2 tables summarizing counts of how often students replied correctly to the asymmetric and symmetric syllogisms. The first table is for the valid syllogisms, whereas the second table shows the results for the invalid.

	Correct reply? (Valid n=174)		Correct reply? (Invalid n=174)	
Syllogism	Yes	No	Yes	No
Asymmetric (figures 1 and 4)	401	160	338	288
Symmetric (figures 2 and 3)	413	158	359	248

Table 3 shows data obtained in the pre-test of the whole student group, and also shows the data from the post-test. Both parts of the table support strong statistical evidence against the presumption that student will handle the two subgroups of valid syllogism equally well (p-value $< 10^{-5}$ and p-value $= 2 \times 10^{-4}$, respectively, by the two-sided chi-square test). However, we may also observe that the replies given to the subgroup of questioned syllogisms are still significantly different from random where we would expect equally many correct and not-correct answers (p-values 0.02 and 0.007 in the pre-test and post-test, respectively).

The results by Johnson-Laird and Bara [15] suggest that test persons will obtain a higher score for asymmetric syllogisms (figure 1 and 4) than for the symmetric syllogisms (figure 2 and 3). Our results based on the pre-test replies and listed in Table 4 do not confirm this suggestion. None of these tables support significant evidence against the presumption that students will handle the two types of syllogisms equally well (p-value = 0.75 and p-value = 0.068, respectively, by the two-sided chi-square test). – It should, however, be mentioned that our setup differs from the setup used in Johnson-Laird's and Bara's experiment in several respects, in particular regarding the temporal setup. So although their test seems to be analogous to ours, there may also be significant differences in the setup and methods used in the two experiments that can explain the deviation in the results. It is, however, an open question exactly how the discrepancy between their results and ours should be explained.

5   Future Research Agenda

As we have seen some syllogisms (e.g. number 9 in figure 1) have got very low scores in the present study. It is, however, an open question exactly which syllogisms are conceived as the most difficult to handle. Various hypotheses may be considered as potential answers. In order to investigate these hypotheses, more empirical studies will be needed. Looking at the data at hand there are indications of trends worthy of further investigation:

1. Syllogisms consisting of only existential functors, i.e. i- and o-functors (figure 1, syll. 11, figure 2, syll. 9 & 10) are among the syllogisms with the lowest rates of correct answers. Less than 40% of the students answer them correctly in the pre-test. The average rate of students answering these correctly in

the post-test is 0.49 compared to an average of 0.65 for all syllogisms. Of the syllogisms consisting of only universal functors, i.e. a- and e-functors (figure 1, syll. 1, 2, figure 2, syll. 1, 2, figure 3, syll. 7, 10, figure 4, syll. 2, 7, 10), the students do really well in the first four cases with 80% or more answering them correctly. These figures could indicate that syllogistic reasoning involving universal functors, i.e. a- and e-functors, are less hard than syllogistic reasoning involving existential functors, i.e. i- and o-functors. The current dataset does not allow any definite conclusions on this issue. In future research it would thus be of interest to test more systematically:

– For each of the figures how well syllogisms consisting of only a- and e-functors or only i- and o-functors are handled by the students.

2. Syllogisms consisting of only affirmative functors, i.e. a- and i-functors (figure 1, syll. 1, 3, 5, 7, 9, 11, figure 3, syll. 1, 2, 7, 8, 9, figure 4, syll. 1, 3, 7, 9) are generally handled well by the students. Only in three cases (figure 1, syll. 9, 11, figure 4, syll. 7) the rate of correct answers drops significantly below 50%. Unfortunately no syllogisms consisting of only negative functors, i.e. e- and o-functors, are included in the current study. In order to more systematically clarify the role of affirmative and negative functors in the reasoning of the students it would be of interest to investigate:

– For each of the figures how well are syllogisms consisting of only a- and i-functors or only e- and o-functors handled by students?

Note that research questions (1) and (2) may be combined such that the testing is directed at clarifying the role of quantification and affirmation/negation at the same time. This then requires testing of syllogisms consisting of only a-, e-, i- or o-functors.

3. One of the alleged benefits of formalization is a greater transparency of the logical structure of an argument. In the current study the syllogisms presented for the students are made up of natural kind terms such as "swedes", "politicians", "dentists", "halffools" etc., and the functors are expressed in natural language, e.g. the a-functor as "All ... are ...". The current study does not provide any data that may serve to determine the role of formalization for the ability to reason logically. Syllog may, however, easily be modified to accommodate such research interests. All that is required is for the post-test to be conducted with an on-screen presentation of the syllogisms in their basic form, i.e. with premises and conclusion in the form of e.g. a(S, P) etc. It seems as if the clarification of the role of formalization for the ability to analyse and determine the validity of arguments may have some important implications for the teaching of logic.

4. It would be interesting to measure the effect of not only formalizing the syllogisms being quizzed, but drawing diagrams in terms of Euler circles, Venn Diagrams, or Existential Graphs to support the student in deciding the validity of a certain syllogism. Measurements could be taken with or without the support of the diagrams, and comparisons could be made between groups answering with and without such support. This might lend support to the argument for the use of diagrammatic reasoning in teaching logic.

5. It should be noted that in the present study the students have been free to use as much time as they wanted on each evaluation. Since the evaluations of the syllogisms may depend on the temporal conditions given during the test, it would be interesting to investigate whether the results would be significantly different, if the students were asked to make a fixed number of evaluations within a given time limit. It may even be interesting to study how the results would be under mild stress.

6. Finally, it could also be interesting to carry out an experiment using our technique which as closely as possible resembles the experiment described in [15] about symmetric and asymmetric syllogisms, in order to investigate whether we can confirm the findings in [15] regarding symmetric and asymmetric syllogisms.

References

1. Parry, W.T., Hacker, E.A.: Aristotelian Logic. State University of New York Press (1991)
2. Sandborg-Petersen, U., Schärfe, H., Øhrstrøm, P.: Online Course in Knowledge Representation using Conceptual Graphs. Aalborg University (2005), http://cg.huminf.aau.dk
3. Aristotle: Prior Analytics. Translated by A.J. Jenkinson. The Internet Classics Archive (1994-2000), http://classics.mit.edu/Aristotle/prior.html
4. Panayiotou, C., Bennett, B.: Cognitive context and syllogisms from ontologies for handling discrepancies in learning resources. In: Bouquet, P., et al. (eds.) Workshop on Contexts and Ontologies, The 18th European Conference on Artificial Intelligence, Patras, Greece, pp. 21–25 (2008)
5. Kabbaj, A., Janta-Polczynski, M.: From PROLOG++ to PROLOG+CG: A CG Object-Oriented Logic Programming Language. In: Ganter, B., Mineau, G.W. (eds.) ICCS 2000. LNCS (LNAI), vol. 1867, pp. 540–554. Springer, Heidelberg (2000)
6. Kabbaj, A., Moulin, B., Gancet, J., Nadeau, D., Rouleau, O.: Uses, Improvements, and Extensions of Prolog+CG: Case Studies. In: Delugach, H., Stumme, G. (eds.) ICCS 2001. LNCS (LNAI), vol. 2120, pp. 346–359. Springer, Heidelberg (2001)
7. Petersen, U.: Prolog+CG: A Maintainer's Perspective. In: de Moor, A., Polovina, S., Delugach, H. (eds.) Proceedings of First Conceptual Structures Interoperability Workshop (CS-TIW 2006). Aalborg University Press (2006)
8. Leighton, J.P.: Teaching and assessing deductive reasoning skills. Journal of Experimental Education, Volume 74(2), 109–136 (2006)
9. Turner, P., Jamie, A., Thompson, V.A.: The role of training, alternative models, and logical necessity in determining confidence in syllogistic reasoning. Thinking & Reasoning 15(1), 69–100 (2009)
10. Monaghan, P., Stenning, K.: Effects of representational modality and thinking style on learning to solve reasoning problems. In: Gernsbacher, M.A., Derry, S.J. (eds.) Proceedings of the Annual Conference of the Cognitive Science Society, pp. 716–721 (1998)
11. Grossen, B.: The Fundamental Skills of Higher-Order Thinking. Journal of Learning Disabilities 24(6), 343–353 (1991)

12. Hoffman, B., McCrudden, M.T., Schraw, G., Hartley, K.: The Effects of Informational Complexity and Working Memory on Problem-Solving Efficiency. Asia Pacific Education Review 9(4), 464–474 (2008)
13. Bucciarelli, M., Johnson-Laird, P.N.: Strategies in syllogistic reasoning. Cognitive Science 23(3), 247–303 (1999)
14. Gilhooly, K.J., Logie, R.H., Wynn, V.: Syllogistic reasoning tasks, working memory, and skill. European Journal of Cognitive Psychology 11(4), 473–498 (1999)
15. Johnson-Laird, P.N., Bara, B.G.: Syllogistic inference. Cognition 16, 1–61 (1984)
16. Quayle, J.D., Ball, L.J.: Working memory, metacognitive uncertainty, and belief bias in syllogistic reasoning. Quarterly Journal of Experimental Psychology Section A – Human Experimental Psychology 53(4), 1202–1223 (2000)
17. Øhrstrøm, P., Sandborg-Petersen, U., Ploug, T.: Syllog – A Tool for Logic Teaching. In: Proceedings of Artificial Intelligence Workshops 2010 (AIW 2010), Mimos Berhad, pp. 42–55 (2010)

A Model to Compare and Manipulate Situations Represented as Semantically Labeled Graphs

Michał K. Szczerbak[1,2], Ahmed Bouabdallah[2],
François Toutain[1], and Jean-Marie Bonnin[2]

[1] Orange Labs, France Telecom R&D, Lannion, France
[2] Telecom Bretagne, Institut Mînes-Telecom, Cesson-Sévigné, France
{michal.szczerbak,francois.toutain}@orange.com,
{ahmed.bouabdallah,jm.bonnin}@telecom-bretagne.eu

Abstract. In our previous work we have introduced a novel social media that performs collaborative filtering on situations. This enhances user situation awareness with a collaborative effort to learn about importance of situations. In this paper we focus on defining a conceptual graph-based model used to represent situations in our system, so that it would (1) be consistent with existing formal definitions of situation, and (2) enable logical manipulations on situations, namely their detection and semantic generalization, which we employ in the system. In particular, we show how the latter can be accomplished thanks to situation lattices, which we adapt for the model.

Keywords: Situation awareness, situation theory, conceptual graphs, semantics, specialization / generalization, graph hierarchies, situation lattices.

1 Introduction

In the domain of interpersonal communication, we identify a potential in being able to communicate easily one's situation with one another. Users are already given web tools to exchange their availability statuses, location coordinates, moods, applications used, etc. And they use them willingly to share different pieces of information with whole groups of friends. However, we argue that enabling communicating one's complete and meaningful situations could result in more informed decisions on user interactions.

In [24] we introduce a context phonebook application to enable (1) sharing several context dimensions between contacts and (2) defining situations concerning those contacts that users wish to be notified of. We anticipate a stronger communication exchange need among close friends and family members. Therefore, we wish to assist in user situation awareness regarding their close ones. The KRAMER system employs collaborative filtering to suggest its users with notifications found to be important by others.

We model situations in our system with conceptual graphs [23]. Not only does this model make situation representations graphically pleasant and human-readable, but above all it enables reasoning on similarity of situations. Furthermore, we have tested

H.D. Pfeiffer et al. (Eds.): ICCS 2013, LNAI 7735, pp. 44–57, 2013.

logical manipulations on such semantically labeled graphs as KRAMER performs semantic generalization on situations it finds similar.

In this paper we elaborate a model of situations to use in our system. We justify it with a consistency with a theory of situation awareness [11] and the situation theory [3, 16]. We refer to several other works dealing with defining and modeling a situation.

Later, we place our model in a situation lattice, a hierarchical structure introduced in [28] for it maintains naturally the dependence relations between situations. We point out that Sowa's graph reasoning [23] can be seen as traversing such a lattice, which simplifies the process of situation generalization and specialization.

Finally, we focus on two main semantic operations implemented in the KRAMER system, namely situation detection and generalization of situation sets. We have already given the details, in particular of the latter in [25], but from the algorithm implementation point of view. In this paper we explain how do they employ the specialization / generalization reasoning inherited from conceptual graph reasoning.

The remainder of the paper is structured as follows. In Section 2, we gather the theoretical approaches towards defining what a situation is in a technical sense. In Section 3, we discuss different ways to model a situation present in the literature. We show that conceptual graphs are expressive in that manner and mix several good features of other models. Basing on this, we focus on the situation model used in the KRAMER system in Section 4. In Section 5, we explore the problem of comparing situations, as it is the basic problem in situation awareness. Afterwards, in Section 6, we discuss other logical operations that can be performed on situations in order to reason on them. We show how specialization and generalization operators can order situations in conceptual graph hierarchies. In Section 7, we explain how do we apply such hierarchies in our system to (1) detect situations and (2) perform semantic generalization on situations. We conclude and give future work directions in Section 8.

2   Theory of Situation (-Awareness)

A family of context-aware systems is very vast and rich. It gathers all systems that adapt their behaviors in function of changing context in general. However, context in its raw, low-level form is known to be often meaningless, trivial, uncertain and vulnerable to small changes [28]. Intelligent context-aware systems are more and more often interested in identifying situations as processed, more abstract context data, which provide a direct input to determine systems' adaptations and reactions.

As a result, situation-awareness is a property with a crucial impact on decision making and performance of both human and artificial system [11]. In fact, Endsley presents a model of human situation-awareness as ability to percept surrounding elements, comprehend their meaning and project their status into the near future. The author argues for that ability to require a much more advanced level of understanding than just being aware of numerous pieces of data.

The very same approach should be applied to artificially intelligent systems. In [28] situations are defined as semantic interpretations of context. They are more abstract than low-level context and in turn they are also more stable, more certain, and, most of all, more meaningful to context-aware systems.

The effort to better grasp the concept of a situation has been made in the situation theory [3]. Situation theory is an interdisciplinary theory of meaning, which combines perspectives of philosophical discussion, mathematical rigor, and implementation practicality [16]. Indeed, Devlin states that it is not possible to define a situation in terms of familiar mathematical concepts, whereas they can be modeled as such [8]. The real situations are, therefore, distinct elementary abstract objects.

Having argued that, Devlin draws a line between those real situations and abstract situations individuated by agents. The latter are imprecise representations, models that one can create to reason about the real situations that he picked out [8]. In fact, this individualization represents only a part of the reality, as limited was situation aware-ness in [11]. And this common understanding of a situation to be a relevant subset of the state of the universe [9] is used in situation-aware researches either explicitly, e.g. [17], or implicitly, e.g. [2].

3   Situation Models

Situations learned by systems are limited to a part of what is really going on in real situations [8]. Therefore, they can be structured and modeled. In return, systems would be able to process them, comprehend them, and reason upon them. This would finally make such systems truly aware of situations with respect to Endlay's model in [11]. In this section we survey different situation models.

There are many ways scientists model situations for their needs. The simplest re-presentation, as a straight forward attempt to capture real situations in a sense of situa-tion theory, is by elementary situation concepts. Each recognizable situation has its corresponding semantic concept, like "meeting", "running", etc. Frequently they are related with one another with "is-a" unidirectional relations, forming a taxonomy. For example, "business meeting" is a particular type of a "meeting" in [2]. The same au-thors give an impression in their following paper [1] that their situation taxonomy can be treated as a taxonomy of activities ("checking e-mails", "meeting", etc.) quite dis-tinct from concepts of an agent or its context.

Nevertheless, taxonomies require for all specialized concepts to be disjoint from one another and cover all cases of the super-concept, which can be too limiting in terms of measuring similarity between concepts [13]. Moreover, defining a compre-hensive taxonomy for a complex domain of situations is merely impossible.

Therefore, authors of [2] seek expressivity in a model based on OWL-DL[1] ontolo-gy language. Ontologies support more relations between concepts. This enables mod-eling situations as interrelated concepts of diverse context taxonomies, namely spatial, temporal, artifact, and personal. In fact, a situation is said to involve a composition of such different concepts connected by "AND" logical operator. Existential and quanti-ficational restrictions are also introduced in this model.

---

[1] http://www.w3.org/TR/owl-ref/

Having a set of context values is consistent with the definition of an abstract situation in situation theory. There, situation is a collection of infons that it supports [4]. Infons are the elementary informational items of a form $\ll R, a_1, \ldots, a_n, \{1,0\} \gg$, where R is n-ary relation of objects $\{a_i\}$. These pieces of information about different context dimensions are called characteristic features in [17]. The situation is, therefore, modeled as a set of such features in a given time interval.

Padovitz's Context Spaces [21] can be seen as a graphical representation of such composed situations. Each context type has its own dimension in the space and different values on scales are characteristic to different situations. As a result, situations are subspaces within the whole space. As such, they can be compared in terms of a geometric distance. Furthermore, one might see an equivalent of taxonomy relations as subsumption in space. The latter is harder for a vector representation of context in [7].

[27] presents another attempt to ontologically model situations. It is explicitly shown that a concept of a situation is on a different layer of the ontology than the concept of context. Any composite situation is a logical (conjunction, disjunction, negation) or temporal composition of atomic situations. The latter are further extended by three concepts: context type, boolean operator, and context value. This means that a situation can be seen as a combination of such triplets. Whereas those triplets are nothing else than infons restricted to binary relations.

Costa et al. notice further that context is only meaningfull with respect to an entity, whose concept is fundamentally different from the concept of context [5]. Therefore, context can be treated as a moment inhered in a substantial – an entity. As a result, a situation is a composition of such pairs of entities and their context. The authors introduce a graphical notation for situations involving different types of contexts and formal relations. This model is explained to be more applicable to context-awareness.

In consequence, we have investigated conceptual graphs. Such graph notations were developed at first to represent first-order logic and to create a mapping between queries in natural language and relational databases. In general, they are graphical representations of logical expressions, conjunctive first order logic formulas [20], and semantically-rich knowledge. If restricted to binary relations conceptual graphs become directed. Similarly to graphs in [5], they can be labeled with both entity and context value concepts.

Nodes and associated edges would form context type-value pairs, similar to [27], while multiple edges directed to one node would stand for "AND" operator, as in [2], matching multiple context moments with one entity [5]. Conceptual graphs can also represent if-then relations and negations, and might be extended with temporal relations. In the remainder of this paper we shall consider only conceptual graphs without cycles, namely conceptual trees. Their origin and definition are explained in the following section.

4   Situation Model in KRAMER

We start defining our model with a meta model inspired by a CONON upper ontology [26]. We distinguish, however, substantials (entities) from moments (context description) as proposed in [5]. As a result, we define concepts of a context entity and its context state. Computation entity and person are subclasses of an entity. Location and activity are domain specific moments.

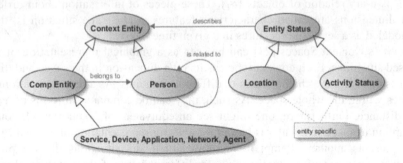

Fig. 1. Meta model of situation

We also introduce relations between moments and substantials, and between entities and a person to model the fact that one's situation is in fact one's context along with context of his or her close ones and his or her devices, services, etc. As a result, by instantiating concepts representing a situation and by inferring the respective meta-concepts relations, we receive a conceptual graph, a conceptual tree to be accurate. This conceptual graph has a "me" concept in its root. We say, that this complex situation exists with respect to one particular entity, a person who perceives the situation.

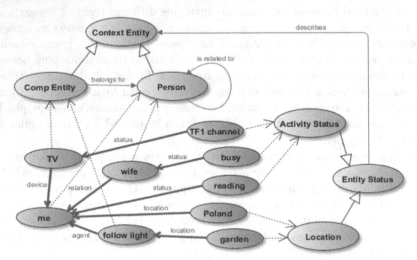

Fig. 2. Instanciating a conceptual graph from the meta model

Following the notation in [20] and [6] we define a situation conceptual graph (tree) *SCG* used in the KRAMER system along with its support. It should be noted that concept types are of four kinds (four concepts in the meta model) and relations are connecting either two entities or an entity with its context (red arrows in Fig. 1).

Definition 1. A support is a 5-tuple $S = (T_H, \{T_E\}, \{T_L\}, \{T_S\}, T_R)$, where:

- T_H is a finite, ordered set of human relations types $(T_H, \leq)$;
- $\{T_E\}$ is a set of finite, ordered sets of entity types $(T_E, \leq)$, e.g. services, devices, application, etc.;
- $\{T_L\}$ is a set of finite, ordered sets of location types $(T_L, \leq)$ specific the entity type;
- $\{T_S\}$ is a set of finite, ordered sets of status types $(T_S, \leq)$ specific the entity type;
- T_R is a finite set of binary relation types divided into two categories: those connecting entities to other entities $T_R^e = \{relation, device, service, agent, \dots\}$, and those connecting entities to statuses $T_R^s = \{status, location\}$.

Definition 2. A situation conceptual graph is a 3-tuple $SCG = [S, G, \lambda]$, where:

- $S = (T_H, \{T_E\}, \{T_L\}, \{T_S\}, T_R)$ is a support;
- $G = (V_C, V_R, E_G, l)$ is an ordered, directed graph having edges $E_G = (c_1, r, c_2)$:
  $\forall e \in E_G \; c_1^e \in T_H \cup \{T_E\}, \; r^e \in T_R^e \Leftrightarrow c_2^e \in T_H \cup \{T_E\}, r^e \in T_R^s \Leftrightarrow c_2^e \in \{T_L\} \cup \{T_S\}$
  and meeting a condition: $\forall c \in T_H \cup \{T_E\} \; \exists e \in E_G : c_1^e = c \wedge c_2^e \in \{T_L\} \cup \{T_S\}$;
- λ is a labeling of the nodes of G with elements from support S:
  $\forall c \in V_C \; \lambda(c) \in T_H \cup \{T_E\} \cup \{T_L\} \cup \{T_S\}; \; \forall r \in V_R \; \lambda(r) \in T_R$.

Every concept in nodes of such conceptual graphs is a semantic concept taken from a respective taxonomy. Taxonomies model different context dimensions: human relations, types of devices, locations, etc. For example, Figure 3 presents a situation, for which being located in Poland is more relevant than being in any city in particular. As a result, these semantically labeled graphs become more expressive than situation taxonomies as presented in [1]. This will also enable logical manipulation presented further in this paper.

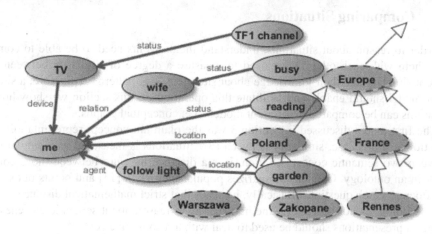

Fig. 3. Graphs are built of concepts from respective taxonomies

Moreover, our conceptual graph-based model is consistent with a definition of an abstract situation in situation theory [8]. Indeed, graphs represent only a part of the reality, of the real situation. In fact, every other entity taking marginal part in the situation can be represented as "any" concept, extended further by "any" concept for its context. "Any" is a root concept for every context taxonomy used in our model and is omitted in a situation representation.

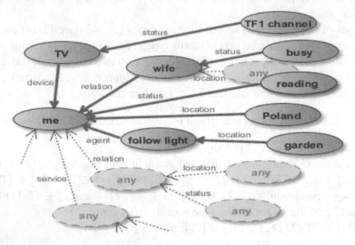

Fig. 4. Conceptual graphs model a part of a real situation

The motivation for us to select conceptual graphs as a model representing situations was its expressiveness, but also an easy comparison of conceptual graphs, which enables logical operations on situations, i.e. their generalization. We focus on those mechanisms in the following sections.

5   Comparing Situations

In order to reason about situations, understand them, agents need to be able to compare them with each other. They need to measure a degree of similarity between a current situation and their knowledge about situations, e.g. patterns. Therefore, a situation model should enable and facilitate this operation. In this section we show how situations can be compared in different models, i.e. conceptual graphs.

The first model discussed in Section 3 was the plain uni-concept representation of situations. In this case, similarity between two situations is measured as a similarity between two semantic concepts. Often, it is a distance measure between those concepts in an ontology. Gandon summarizes popular metrics in [13] and points out several open research questions. Basically, he states that strict mathematical distance on a static ontology is not necessarily the human way to reason about semantic closeness. Richer representations should be used to deal with a resolution error.

In that sense, having situations as a composition of several context dimensions could help. Scalar difference is a measure of distance between two points in Context Spaces theory [21]. Situation subspaces can also be compared, either by the distance measure or by the intersection operator, which finds if two subspaces have a common part. A context space makes detecting a situation extremely intuitive.

The introduction of context-operator-value triples [27] or substantial-moment pairs [5] combines the two preceding approaches. Comparing two situations requires measuring similarities of semantic concepts for each context dimension separately and calculating their weighted mean. The same principle applies to conceptual graphs that represent such multidimensional semantic spaces.

Even though optimal algorithms for matching graphs in general are reported to be exponential with respect to the number of nodes in either graph [14], we should remember that abstract situations do not represent the whole knowledge [8]. Instead, the number of nodes is limited to what is necessary for an agent to detect a situation. For instance in [17], "travelling" situation is defined only by using any transportation mean and by a fact of moving significantly.

Furthermore, Mugnier reports in [20] that many inter- conceptual graph operations become polynomial, should the involved graphs be restricted to trees. As shown in Section 4, situations are indeed considered to be represented by conceptual trees. In [29] one of the implementations of conceptual trees matching is reported to be polynomial. Furthermore, [6] gives an ontology similarity measure based on a projection between conceptual graphs, and [18, 19] present comparison of two conceptual graphs as a calculation of their overlapping parts with and without semantic subsumption.

6   Logical Manipulations on Situations

Most of the conceptual graph-based comparison algorithms mentioned in the previous section exploit the fact that concepts in nodes are structured in taxonomies per context dimension. As a result, "chasing an animal" is supposed to be matched with "chasing a mouse" [19], rather than "travelling by train", as concept of a "mouse" is a specialization of an "animal". On the other hand, in [22] the authors seek for the most interesting common generalization of two graphs in order to evaluate "thematic" similarity between two conceptual graphs.

In fact, according to [20], generalization and specialization are said to be the key computational notions in every reasoning concerning conceptual graphs. Sowa discusses 6 canonical formation rules as semantic graph-based operators for equivalence (copy, simplify), specialization (join, restrict) and generalization (detach, unrestrict) of conceptual graphs [23]. These operators can be interpreted by either logical subsumption or graph morphism. Just a negation operator needed to be added to handle full first-order logic.

Different researches make use of specialization rules, for instance [15] employs maximal join operator to perform high-level fusion on heterogeneous information represented by conceptual graphs. In our work [24], we are more interested in generalizing situations, and therefore generalizing associated with them conceptual graphs. We present the procedure and its motivation in Section 7.

Mugnier explains in [20] that for one graph to be a specialization of another, there needs to be a projection from the second graph to the first. Projection is a sequence of graph morphisms in a classical graph theory sense but implying equality of relation types and taxonomic specialization of concept types. As a result, a specialized graph is a super-graph of the original one (external join operation) with possibly semantically narrowed labels (restrict operation).

This makes the specialization relation a preorder because it is not anti-symmetric as redundant graphs are still possible. Should the injectivity constraint be introduced and internal join operator forbidden, the relation becomes a full order. Therefore, conceptual graphs can form a hierarchy, like in [10]. As a result, reasoning about relation between two graphs can be transformed into a problem of traversing such hierarchy. One graph is a generalization of another, if it is an ancestor of that other graph.

Considering that conceptual graphs represent situations, reasoning about similarity of two situations is reduced in a way to semantic distance measures as presented in [13]. Moreover, finding more abstract / detailed situations implies traversing the hierarchy upwards or downwards. Ye et al. introduce this idea in a concept of situation lattices [28]. Although they model situations as simple unitary concepts, similarly to [2], they notice that this organization reflects the internal structure of situations and is beneficial in identifying situations.

We argue that situations modeled as conceptual graphs can naturally form situation lattices. Situation awareness would benefit from seeing a situation space as such order structures. In the following section we show how our system implements the explained situation model.

7   Situation Operations in KRAMER

In [24] we present an overview of our system, KRAMER. The ambition there is to empower users with spreading information about importance of situations, which is established in a collaborative manner. We model situations as conceptual trees, a special case for conceptual graphs (see Section 4). We say that situation is related, meaningful to an entity, the "me" concept in a tree root (see Fig. 2).

In previous sections we showed how this model is consistent with situation theory and how researchers have applied sophisticated relation structuring to ease reasoning about specialization / generalization of conceptual graphs. In this section we discuss (1) detecting complex situations and (2) generalizing them by the KRAMER system based on those approaches.

7.1   Detecting Situations in KRAMER

Our system enables users' smart devices to sense the user context, share it among close phonebook contacts and fire programmed actions, i.e. notifications, should a set of conditions be fulfilled. This set of conditions concerns one's context and the context of his or her close ones. Simply put, an action is fired once a required situation matches the current one. Therefore, we need an efficient multiple situation detection mechanism.

For this matter we use a Rete production system [12]. This choice is well motivated in associating productions (decisions to perform an action) with complex set of conditions (situations). In order to introduce semantic reasoning, we enhanced our Rete implementation by replacing equality (=) condition with subsumption (≤) one in alpha network. Therefore, an enhanced Rete takes full advantage of situation lattices concept introduced in [28].

For example, wanting to be notified about a friend being in Poland requires a Rete condition C: (<x> ^location Poland). Admitting that a friend shares that he is currently in Warsaw, known to be a capital of Poland and a descendant of a "Poland" concept on a location taxonomy, one would expect the condition to be matched. Indeed, a situation "a friend is in Warsaw" implies a situation "a friend is in Poland". The actual situation would be, therefore, a specialization of the situation expressed by condition C.

One might notice that a condition is a context triple representing an atomic situation [27], always a context relative to an entity as in [5]. Should the situations become more complex, a set of conditions is introduced to a production system. From a Rete network point of view, conditions are connected by join nodes in beta network (see Fig. 5). From situation model point of view, atomic situations are logically connected with AND logical operator as in [1]. Complex situations form therefore a conceptual tree.

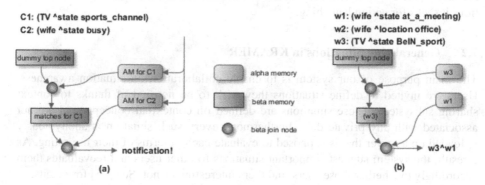

Fig. 5. Example Rete network, (a) structure, and (b) instantiation with working memory sets *w* for a situation "my wife is busy and my TV is on sports channel"

One might also notice that having redundant situations is not possible in the KRAMER system. Therefore, every possible situation for a given set of contacts in a phonebook forms a finite lattice. The supremum is the most abstract situation, "anything is going on". It stands for a trivial graph made of one node labeled "me". It might be extended to a full structure filled with "any" labels, see Fig. 4. The infimum would be therefore a set of most specialized full conceptual trees (see Fig. 6).

As a result, an abstract situation detecting problem can be transformed into determining whether a concrete current situation is a descendant of the first one. A situation the system needs to detect should be a generalization of the situation perceived. Therefore, there should exist a set of join (to merge multiple pieces of information coming from different sources of interest for a particular abstract situation) and unrestrict (to generalize current situations for particular context dimensions) operations from the complete situation to the abstract one.

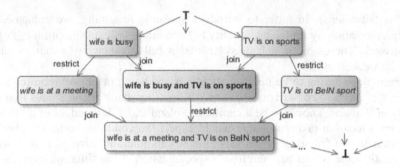

Fig. 6. Part of a situations lattice for the example in Fig. 5

Our Rete implementation performs a check for such set of operations. Firstly, it joins every atomic situation introduced separately to its network (in order to perform individual matching tests for each condition in its alpha network). Secondly, it introduces a semantic subsumption operator (restrict / unrestrict operation equivalent) rather than equality in Rete alpha network. As a result, it browses quickly a situation hierarchy [10, 28] to determine whether a current situation matches any of the situations it seeks for action launching.

7.2   Generalizing Situations in KRAMER

The main purpose of our system is to enable collaboration on situation awareness. Users are invited to define situations they wish to be notified of thanks to context sharing subsystem. Those situations are defined on contextual concepts and are not associated with any private data. Furthermore, every such situation is anonymously uploaded on a server that is supposed to evaluate each in terms of their use rating. As a result, the system suggests important situations to other users and reevaluates them accordingly to whether those users find them interesting or not. See [25] for details.

However, depending on the context granularity, it may be distorting to the situations rating if we treat two very similar situations as completely different ones. For example, a complex situation of having "a daughter leaving school and wife being busy, unable to take her home" might be essentially the same as having "a son leaving school and wife being busy, unable to take him home". It could only depend on what gender child one has. Therefore, our system is able to generalize similar situations and merge their ratings [25].

Our algorithm has, however, a couple of restrictions in order for two situations represented as conceptual graphs to be generalized. The first is a requirement for the graph structures to match. The other, for the corresponding concepts on those graphs to be semantically similar. We shall explain how do those restrictions relate to the situation hierarchy from [10].

As discussed in the previous subsection, every possible situation in a system and all corresponding abstractions exist in a common situation lattice. Therefore, finding a least common ancestor by mean of generalization for any two situations is always possible. For example, generalizing "wife is at work" and "wife is busy" results in trivial "wife (is anywhere and doing anything)". This does not make sense with respect to suggesting meaningful situations to system users. Therefore, our algorithm first groups situations by matching graph structures with respect to number of edges and their relation concept labels.

As a result, one might see this as an elimination of join operator from the situation hierarchy (like in Fig. 6) to create a family of smaller structures based on restrict operator only. The second step of the algorithm will be performed within the scope of each small hierarchy separately, see Fig. 7 for one example.

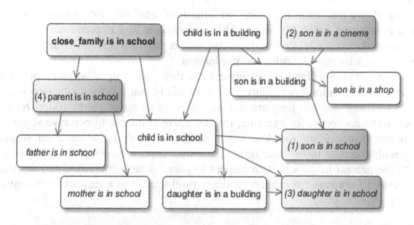

Fig. 7. Part of a hierarchy of situations with restrict-only relations

Having found two or more situations sharing the same graph structure, the algorithm proceeds to finding opportunistic generalizations of those that are similar. We define one requirement for a set of situations to be generalized into one: there cannot be a series of restrictions from the abstract situation to any leaf that does not pass through any of the situations in a set. One might say that the abstraction is the least common ancestor of situations in a hierarchy that covers all of those situations and none of the situations that does not subsume any of the situations in the set.

For example, let's consider the following set of situations: (1)"son is in school", (2)"son is in a cinema", (3)" daughter is in school", and (4)"parent is in school. A part of a restriction hierarchy for a situation matching scheme <a person> is in <a location> is presented in Fig. 7. As a result, a situation can be generalized from (1), (3) and (4) into "a member of a close family is in school". Meanwhile, the situation (2) remains not generalized, because there are many situation nodes missing if we were to generate an abstract situation "son is in a building", for example.

As a result, the KRAMER's algorithm performs two steps on a common situation lattice. First, it eliminates any join / detach operation connectors, which splices the structure into a family of hierarchies, each involving only one situation graph structure. Second, it scans all generalized situations (products of unrestrict operation) so that they are common for a subset of given situations while not having any restrict operation chain that would not lead to any given situation. This approach is found to be efficient from the computational complexity point of view in [25].

8   Conclusions

In the KRAMER system we perform two main semantic operations on situations: their detection and generalization. In this paper we present those operations as logical manipulations on situation hierarchies, lattices that constitute a space of all possible situations. We show that such hierarchies are a natural product of specialization / generalization relations between situations. To assure expressiveness of situation comparison we model situations as conceptual graphs. In addition, we discuss that model to be consistent with the situation theory.

As a result, we transform reasoning on situations similarity into a problem of traversing a conceptual graph hierarchy. This approach is less implementation centric, much more situation model driven. Nevertheless, it provides a set of straight-forward logical directives for our algorithms to implement.

For future works, we plan to investigate its further impact upon situation prediction and reasoning in situation uncertainty. It is very likely that a sensed situation is similar to the searched one but they are not the same in terms of subsumption relation. However, this may mean, for example, that an agent is unable to perceive some context dimensions necessary for detecting a situation, or that the searched situation would possibly appear in the near future. In either case, narrowing the situation hierarchy to the nearest neighbors with respect to join / detach operators, and discovery of their restrict / unrestrict operation products might result in a respective measure of probability.

References

1. Anagnostopoulos, C.B., Ntarladimas, Y., Hadjiefthymiades, S.: Reasoning about Situation Similarity. In: International IEEE Conference on Intelligent Systems, pp. 109–114 (2006)
2. Anagnostopoulos, C.B., Ntarladimas, Y., Hadjiefthymiades, S.: Situation Awareness: Dealing with Vague Context. In: ACS/IEEE International Conference on Pervasive Services, pp. 131–140 (2006)
3. Barwise, J., Perry, J.: Situations and Attitudes. Bradford Books, The MIT Press (1983) ISBN 0-262-02189-7
4. Cooper, R., Kamp, H.: Negation in Situation Semantics and Discourse Representation Theory. In: Situation Theory and Its Applications, vol. 2. Stanford University (1991)
5. Costa, P.D., Guizzardi, G., Almeida, J.P.A., Pires, L.F., van Sinderen, M.: Situations in Conceptual Modeling of Context. In: 10th IEEE International Enterprise Distributed Object Computing Conference Workshops, p. 6 (2006)
6. Croitoru, M., Hu, B., Dashmapatra, S., Lewis, P., Dupplaw, D., Xiao, L.: A Conceptual Graph Based Approach to Ontology Similarity Measure. In: Priss, U., Polovina, S., Hill, R. (eds.) ICCS 2007. LNCS (LNAI), vol. 4604, pp. 154–164. Springer, Heidelberg (2007)
7. Delaveau, L., Loulier, B., Matson, E.T., Dietz, E.: A vector-space retrieval system for contextual awareness. In: IEEE International Multi-Disciplinary Conference on Cognitive Metheods in Situation Awareness and Decision Support, pp. 162–165 (2012)
8. Devlin, K.J.: Situations as Mathematical Abstractions. In: Situation Theory and Its Applications, vol. 2. Stanford University (1991)
9. Dey, A.K.: Providing architectural support for building context-aware applications. PhD thesis, Georgia Institute of Technology (2000)

10. Ellis, G., Levinson, R.: Multi-Level Hierarchical Retrieval. Knowledge-Based Systems, Conceptual Graphs Special Issue 5, 233–244 (1992)

11. Endsley, M.R.: Toward a Theory of Situation Awareness in Dynamic Systems. Human factors 37, 32–64 (1995)

12. Forgy, C.L.: Rete: A Fast Algorithm for the Many Pattern/Many Object Pattern Match Problem. Artificial Intelligence 19, 17–37 (1982)

13. Gandon, F.: Graphes RDF et leur Manipulation pour la Gestion de Connaissances, Ch. 4: Graphes comme espaces métriques, HdR, Nice Sophia-Antipolis (2008)

14. Jiang, X., Bunke, H.: Graph Matching. SCI, vol. 73, pp. 149–173 (2008)

15. Laudy, C., Ganascia, J.G., Sedogbo, C.: High-level Fusion based on Conceptual Graphs. In: 10th International Conference on Information Fusion, pp. 1–8 (2007)

16. Mechkour, S.: Overview of Situation Theory and its application in modeling context, Seminar Paper, University of Fribourg (2007)

17. Meissen, U., Pfennigschmidt, S., Voisard, A., Wahnfried, T.: Context- and Situation-Awareness in Information Logistics. In: Lindner, W., Fischer, F., Türker, C., Tzitzikas, Y., Vakali, A.I. (eds.) EDBT 2004. LNCS, vol. 3268, pp. 335–344. Springer, Heidelberg (2004)

18. Montes-y-Gómez, M., Gelbukh, A., López-López, A.: Comparison of Conceptual Graphs. In: Cairó, O., Cantú, F.J. (eds.) MICAI 2000. LNCS, vol. 1793, pp. 548–556. Springer, Heidelberg (2000)

19. Montes-y-Gómez, M., Gelbukh, A., López-López, A., Baeza-Yates, R.: Flexible Comparison of Conceptual Graphs. In: Mayr, H.C., Lazanský, J., Quirchmayr, G., Vogel, P. (eds.) DEXA 2001. LNCS, vol. 2113, pp. 102–111. Springer, Heidelberg (2001)

20. Mugnier, M.L.: On Generalization / Specialization for Conceptual Graphs. Journal of Experimental & Theoretical Artificial Intelligence 7, 325–344 (1993)

21. Padovitz, A., Loke, S.W., Zaslavsky, A.: Towards a Theory of Context Spaces. In: 2nd IEEE Conference on Pervasive Computing and Communications Workshops, pp. 38–42 (2004)

22. Poole, J., Campbell, J.A.: A Novel Algorithm for Matching Conceptual and Related Graphs. In: Ellis, G., Rich, W., Levinson, R., Sowa, J.F. (eds.) ICCS 1995. LNCS, vol. 954, pp. 293–307. Springer, Heidelberg (1995)

23. Sowa, J.F.: Conceptual Graphs. Foundations of Artificial Intelligence, vol. 3, pp. 213–237 (2008)

24. Szczerbak, M.K., Toutain, F., Bouabdallah, A., Bonnin, J.M.: Collaborative Context Experience in a Phonebook. In: 26th IEEE International Conference on Advanced Information Networking and Applications Workshops, pp. 1275–1281 (2012)

25. Szczerbak, M.K., Bouabdallah, A., Toutain, F., Bonnin, J.M.: Generalizing Contextual Situations. In: 6th IEEE International Conference on Semantic Computing (to be published, 2012)

26. Wang, X.H., Gu, T., Zhang, D.Q., Pung, H.K.: Ontology Based Context Modeling and Reasoning using OWL. In: 2nd IEEE Annual Conference on Pervasive Computing and Communications Workshops, pp. 18–22 (2004)

27. Yau, S.S., Liu, J.: Hierarchical Situation Modeling and Reasoning for Pervasive Computing. In: 4th IEEE Workshop on Software Technologies for Future Embedded and Ubiquitous Systems, pp. 5–10 (2006)

28. Ye, J., Coyle, L., Dobson, S., Nixon, P.: Using Situation Lattices to Model and Reason about Context. In: 4th International Workshop on Modeling and Reasoning in Context, pp. 1–12 (2007)

29. Zhong, J., Zhu, H., Li, J., Yu, Y.: Conceptual Graph Matching for Semantic Search. In: Priss, U., Corbett, D.R., Angelova, G. (eds.) ICCS 2002. LNCS (LNAI), vol. 2393, pp. 92–106. Springer, Heidelberg (2002)

Analyzing Clusters and Constellations from Untwisting Shortened Links on Twitter Using Conceptual Graphs

Emma L. Tonkin[1], Heather D. Pfeiffer[2], and Gregory J.L. Tourte[3]

[1] UKOLN,
University of Bath, Bath, UK
e.tonkin@ukoln.ac.uk
[2] Akamai Physics, Inc.
Las Cruces, New Mexico, USA
hdp@cs.nmsu.edu
[3] School of Geographical Sciences,
The University of Bristol, Bristol, UK
g.j.l.tourte@bristol.ac.uk

Abstract. The analysis of big data, although potentially a very reward-ing task, can present difficulties due to the complexity inherent to such datasets. We suggest that conceptual graphs provide a mechanism for representing knowledge about a domain that can also be used as a useful scaffold for big data analysis. Conceptual graphs may be used as a means to collaboratively build up a robust model forming the skeleton of a data analysis project. This paper describes a case study in which conceptual graphs were used to underpin an exploration of a corpus of tweets re-lating to the Transportation Security Administration (TSA). Through this process we will demonstrate the emerging model built up of the data landscape involved and of the business structures that underlie the technical frameworks relied upon by microblogging software.

Keywords: Conceptual Graphs, Twitter, Microblogging, Models.

1 Introduction

The increasing prominence of Twitter as a social site in the last years has led to a great deal of interest in the way in which the site is used, as well as the technical enablers underlying the site and its applications. Of course, there is a significant existing body of literature describing various aspects of the site and the characteristics of its use, e.g., [11,12], so a researcher looking to analyse data taken from Twitter should naturally begin by reviewing that information. However, one aspect of Twitter is its apparent inconsistency across topics [30] and across cultures (see for example [39]). Another is the changing landscape of technologies and implementation decisions: as a developing platform seeking to marketize effectively, Twitter, like most services, evolves over time as a result

H.D. Pfeiffer et al. (Eds.): ICCS 2013, LNAI 7735, pp. 58–74, 2013.

of various motivating factors. As a consequence there is a need for exploratory data analysis [22].

Typically, the exploratory analysis of data (EAD) involves the use of information visualisation tools, cluster analysis, data mining approaches and so forth [22], which permits domain experts to begin to develop an understanding of the dataset at hand. This permits them to develop testable hypotheses. However, as Perer notes, one difficulty with this approach is that it is typically somewhat scattershot—discoveries made in this way are typically opportunistic. Yet an entirely systematic approach risks undermining the knowledge-driven, insight-led research pattern of domain experts. Generally, Perer suggests, systematic approaches do not always suffice when faced with real problems. Thus, Perer suggests, a series of design goals should be considered when developing data exploration interfaces: most relate to the ability to track actions already taken, to see available actions not already taken, to annotate actions, to retrace existing steps taken, and so on. Particularly interesting is design goal 6: the need to share progress with other users.

We begin this paper by exploring knowledge representations through which information learned about the entities, agencies, interactions and underlying infrastructure of the Twitter environment can be stored and shared within a team to support EAD, explaining why we chose to use conceptual graphs for the purpose of supporting a text mining application. We explain the development philosophy underlying the EAD approach taken and its limitations, and we provide a brief introduction to the literature surrounding Twitter and findings resulting from a preliminary exploration of certain aspects of Twitter infrastructure.

2   Method

Involving each member of a research group into an iterative process of data model development requires both appropriate communication channels and sufficiently useful proxies (e.g. imagery, model diagrams, etc) on which to work. There are many candidates for this process, of course, ranging from pen and paper or whiteboard to a shared collaborative space online such as a wiki, Google Doc, or a version or revision control system such as Subversion. However, it is important to separate the collaborative space that is used from the actual representation that is employed within that workspace, and to recognise that such aids to teamwork, whilst innately prerequisite, typically provide neither formalism nor guidance. It is for this purpose that a formal knowledge representation structure becomes of importance; according to Davis et al [6], KR may be described in terms of five roles: a KR may act as a proxy through which via thought-experiment the effects of an action may be deduced; a KR represents a series of ways of thinking about an entity; a KR represents a formalism expressed in terms of sanctioned and recommended inferences; a KR can be seen as 'a computational environment for thinking' and as 'a medium of human expression'. KRs may be classified into five categories: pictorial, symbolic, linguistic, virtual and algorithmic [19].

Given these five categories of KR the internal structure of the data must be able to hold not only factual data, but the conceptual dependencies between the elements so that their relationships are defined within the data structure. These structures can hold scripts [33] of information that is represented textually and can be formed into a story. This story line can then be structurally stored into a commonsense database of records. This database structure could hold three of the categories—symbolic, linguistic, and algorithmic—by using language theory. The two other categories virtual and pictorial would be lacking because the text basis of the scripts. However, the conceptual graphs structure, especially with time and space extensions [25], does not have this textual limitation, but does give the relational structuring between conceptual dependency and can therefore process all five categories.

Conceptual graphs (CGs) provide a formal visual approach to knowledge representation, closely linked to natural language [37] which have been found to be accessible by team participants from varying specialities in the past, including for example visual designers and managers [28], developers [16], engineers [4] and so forth. The graphical representation provided by the CG formalism is an aid to understanding that has in the past been shown to be effective in multidisciplinary team environments [4].

This graphical format of CGs can be represented in textual expression or as links to other types of conceptual information such as URL addresses to photos, videos, games, etc. The CGs as a set of partial models do not have to maintain truth as with other representation so they may contain conceptual relationship or dependencies that are in opposition to other graphs within the same model set. This is because partial models are snapshots in time [25]. When the final model is built all inconsistencies will be resolved.

2.1   Conceptual Graphs in Text Mining

Text mining, an area that remains relatively youthful, is a research area based on the detection/discovery of interesting patterns within textual corpora. Whilst the majority of text mining applications are essentially focused on relatively simple representations—key words/phrases, or even in some cases 'bag of words' representations, the use of conceptual graphs in text mining problems is well represented in the literature. Cao [2] describes the use of conceptual graphs alongside fuzzy logic as a means of extending Semantic Web technologies to approach human expression and reasoning more effectively; conceptual graphs are here used as a means of representing natural language sentences. Montes-y-Gomez et al. [10], for example, describe the use of conceptual graphs to represent a series of text, permitting the detection of rare patterns and local deviations (occurring at specific contexts and generalization levels) within the textual corpus. Spasic et al. [38] identify Daraselia et al's [5] use of conceptual graphs as a representation of a number of ontological frames, permitting them to be queried or for further text mning work to be completed against them. Shehata et al (2006) describe the use of conceptual graph representations to capture in detail sentence-level semantics, in order to improve the quality of text retrieval and indexing [34].

In general terms, then, text mining and conceptual graphs are demonstrably viable companions. However, by no means should this be taken to mean that the problem of mining a research corpus such as the Twitter corpus described here reduces to the use of an existing software package or service. There is significant variation between corpora; Twitter, for example, limits users to a small number of characters per utterance, typically resulting in a telegraphic, abbreviated style.

2.2 Agile Development

In this instance the proposal is to use conceptual graphs within the team to build up information about the various aspects of the dataset under investigation. We separated this work into two broad phases, the first one of which is exploratory in nature, and is intended to enable us to rapidly build up a basic model of the domain. For this purpose we use a variant of the agile software development methodology, conceptually linked to the Rapid Application Development models [15]. For the second, we link the conceptual graphs built up during the exploratory work in order to create a single composite knowledge representation, and explore its use as a basis structure on which to develop research questions about the dataset.

We begin by briefly reviewing literature relating to use of agile methodologies in exploration of scientific datasets, and move on to the development of fragmentary conceptual graphs though research findings.

2.3 Agile Methodologies for Scientific Datasets

Agile software development methodologies are designed to prioritise certain aspects of the software development process. The agile manifesto [9] expresses the methodology's practices as follows:

- *Individuals and interactions* over processes and tools
- *Working software* over comprehensive documentation
- *Customer collaboration* over contract negotiation
- *Responding to change* over following a plan

and states that 'while there is value in the items on the right, we value the items on the left more'.

The use of agile development methodologies for the purpose of development of scientific software is a concept that has been explored elsewhere; for example, Lane [13] describes a theory-driven methodology that encodes scientific knowledge and natural processes within an implemented piece of software. The importance of the computational model is clearly stated by Lane [13]: computational models, it is argued, are amongst other things able to clearly and rigorously lay out the components of the scientific theory under discussion, allow the derivation of testable predictions, and provide a useful mechanism to facilitate making sense of rich and dense datasets. Accepting the value of a computational model, it is therefore reasonable to consider the question of the quality, relevance and accuracy of its implementation. To successfully resolve these queries it is necessary to establish appropriate tests—that is, what makes a model 'good', or 'accurate', in the context of our research?

2.4 Building and Testing a Conceptual Model to Underlie Research

Due to the idiosyncratic composition of any research team it is reasonable to expect that the precise research interests/requirements of the individuals involved are likely to have an impact on the features highlighted, perhaps even on the inclusion of features. This is not uncommon; indeed, the process of mining a text is typically starkly reductive—reduction of entropy/compression of a text may be expected to have at its core a model of the aspects of that text most clearly of use. Features of the text that are not contained within that model may or may not survive the reduction. An obvious example of this is the previously mentioned bag-of-words model, that is, reduction of a text into its component words; the details of the syntax, the presentation, etc., cannot be expected to survive this process. If the research requirements of the team may be satisfied by the use of such a model, however, there is no pressing need to turn away from it. Thus the participants must be at the core of model development.

3 Mining a Twitter Corpus

The use of Twitter as a data source for various forms of data/text mining is well established. The data is usable for a variety of purposes, perhaps most easily classified according to the technologies used. Twitter is famously used as a resource for sentiment analysis and for opinion mining [20], with a variety of purposes in mind, including product/service/company profiling, marketing purposes, political analysis and opinion polling, for example, with activist aims in mind [20], but also for stock market prediction [1], disaster alerts [7], level of interest in news articles [26] and so forth. Explicitly topic-oriented mining is of use for various purposes, such as tracking public health trends [27,21], earthquake monitoring [31], news tracking [14,23] and so on. The very public nature of the service renders it of interest to spammers, and therefore another research topic in the text mining field is that of identifying and mining spam. Shekar et al. [35] demonstrate a mechanism for identifying spam from Twitter data through an initially manually input list of key terms, followed by the use of a Naive Bayesian algorithm and a J48 decision tree classifier. As well as straightforward use as a text corpus, Twitter's sharply time-based, turn-based and telegraphic nature introduces the need to consider issues that perhaps would not be as prominent where other types of information, such as perhaps academic papers or even blogs, are concerned.

Yet the content of users' tweets is certainly not the only aspect of Twitter that may be of interest, and the existing body of research certainly reflects this. Twitter's popular classification as a social network is well established despite senior Twitter executives' protestations that Twitter is 'a news network and not a social network' [18]. Certainly the findings of Kwak et al [12] bear out the assertion that trending Twitter topics are, in the majority of cases, either 'headline or persistent news' in nature, whilst their topology analysis suggests that Twitter is not a pure social network, as the distribution of followers and low reciprocity does not closely resemble the typical social network. Yet other studies suggest that this is the effect of noise; as Huberman et al. [11], social interactions exist within Twitter, as a sparse subset of the broadly declared set of friends and followers. It appears likely that Twitter, sharing aspects both of social networks and the emerging concept of a 'news network', must be modelled in such a way as to satisfy both definitions.

The technical infrastructure underlying Twitter's functionality is also of interest to researchers.

3.1   Infrastructure Underlying Twitter

A particularly important tool for Twitter users in the past has been the URL shortener [3]—a tool, often web based or built in to the application used by the individual to post their remarks to Twitter. These are conceptually simple: a URL is provided to the shortening tool, which assigns to it a unique key; when presented with that key, the tool will then present the browser with some form of redirect (often a 301) to return the user to the original long URL.

The primary benefit for Twitter users was simply that a shortened URL does not eat significantly into the limited space available for each tweet (Twitter's famous 140 characters or less), leaving the user with more space to present their own ideas or opinions. There are also secondary benefits, of course, such as relative opacity (i.e., it is not usually possible to guess at the destination of a shortened URL), making it possible for users to forward readers to unexpected URLs, providing potential for practical jokes and for malicious reuse as well as fulfilling the more general purpose of compressing information.

3.2   Rationale: Construction and Maintenance—Relative Costs?

Since URL shorteners are not technically complicated, they are relatively easy to set up, and indeed a site that tracks URL shorteners in use has identified over a thousand individual services [42]. However, like many other such initiatives the attrition rate of URL shorteners over time appears to be quite high—according to yi.tl, the majority of shortening services identified have since closed. As we will discuss in this poster, the majority of shortened URLs from a given US-centric discourse during the spring of 2012 make use of one of a few major service providers, either directly or via aliases run by those providers. One clear advantage of making use of a URL shortener is the opportunity to gain information about the number of clickthroughs—how many people accessed the link that was posted, when, and from which broad geographic region. This is particularly useful to those for whom the distribution of links in a given venue forms part of a marketing strategy—a group in which Higher Education institutions are increasingly likely to count themselves, as market forces penetrate ever more deeply.

This reasoning also leads the construction of URL shorteners in some domains—indeed, it is not uncommon for parent enterprises to sell social media analytics services or provide free or paid analytics services. Yet, as with sentiment analysis, much of this activity deals with short-term, transitory events. Such analysis is typically bound to a relatively brief timescale—a few hours to a few days. Little financial benefit may exist in long-term provision of a 'long tail' of older redirects.

3.3   Preservation of Shortened URLs

Shortened URLs, once identified, can (if the underlying service is still available), trivially be resolved into the original destination URL. This is a useful step for many forms of analysis (e.g., content/contextual analysis of tweets on Twitter). The half-life of social services is often short, but a URL shortener is more intimately bound into our ability to follow a conversation than, for example, a news aggregation service might be. The loss of the news aggregation service potentially compromises our ability to identify the trigger for transitory interest in a given subject or resource. The loss of the redirect service means that the key resources referenced during a conversation can no longer be referenced, compromising our ability to understand the social or political context and underlying framing of the discussion. URL redirection increasingly offers a further challenge, for although the number of discrete services in popular use appears to be reducing, the penetration of these services into the user experience continues to increase. Twitter itself did not initially impose the use of a domain redirection service. Later, the service began to 'wrap' popular (frequently retweeted/referenced) URLs into Twitter's own domain redirection service, t.co. In late 2011 Twitter made this mandatory for all URLs [8]; therefore, any URL published through the Twitter service will be published in the form of a t.co/key alias. Since users' choice of URL redirection service typically relates to their choice of application (for example, HootSuite users will find that they are minting ow.ly URLs, which are inbuilt), this means that a user making use of HootSuite will have a characteristic 'fingerprint': t.co → ow.ly (→ previous source of link).

There are many reasons to look into URL redirection other than preservation, such as the need to identify spam [41], or an interest in conversation/discourse analysis and information propagation [29].

3.4   In Chains: Unwrapping the URL

The implementation of various services and applications leads to the 'wrapping' of existing URLs into one or more URL redirects. The effect is similar to taking a postcard, and placing it into an envelope addressed to the initial receiver care of an intermediary. Then that envelope is passed on to a courier service who insist on placing the mail into their own brand of envelope and addressing it to 'Original recipient, care of initial intermediary, care of the courier service's posting office'. By this means, each agency is able to collect statistics about visitors to that URL.

For the user, this carries the penalty that URLs are both opaque and somewhat slower to resolve. It also implies that the user is providing considerable information about their interests and activities to each agency in the redirect chain. However, for the researcher at least, it provides us with additional information about the pathway that this information took on its way from the originator to the author of the tweet.

3.5   Backtracking the Trackers

A simple URL redirect tracker was developed for the purpose of tracking each step of URL redirection, using Perl's LWP libraries to extract information about each step of domain resolution. This 'traceroute' application is able to generate information about a shortened URL by backtracking through each step and documenting each redirect. A sample result is given (see Table 1).

Table 1. A sample HTTP response chain

Short URL	Response Code	Redirect	Chain ID
t.to/example	301	ow.ly/example	1
ow.ly/example	301	bbc.in/example	2
bbc.in/example	301	news.bbc.co.uk/example	3

An aggregate view of the redirect landscape is shown in Fig. 1. As is visible from this graph representation, there are many redirect services in use. To begin to build up the CG model for analysis of the trackers of the retweets the definitional graph for the knowledge representation of the data from Table 1 and extended data is found in Fig. 2. This REDIRECT graph indicates that a shortened URL can be redirected into a modified URL. From the type hierarchy associated with this definitional graph we see that during a join of factual data later in the representation processing URLs that are either 'Short URL' concepts or 'Modified URL' concepts can be joined. However, as our literature review has indicated, many of these are 'vanity' domains, so it is inaccurate to think of each redirect as a separate service. Rather, it is suggested that many are simply aliases of an existing commercial service. The question of identifying individual business entities within this group is one that can be solved quite simply on a technical level, through the use of domain analysis tools to identify the site operator. The use of WHOIS information presents difficulties since domains registered through separate registrars/various countries have quite different recordkeeping conventions and access regulations. Instead, service-level information such as IPs may be used as a rough indicator, with results such as those shown in Fig. 3.

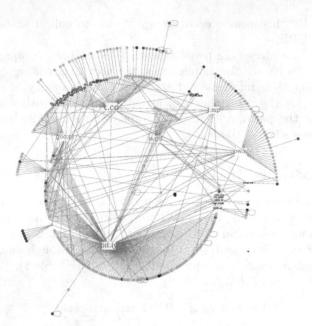

Fig. 1. The TSA web of redirects

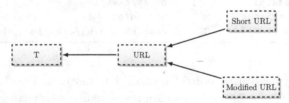

(a) Type hierarchy for REDIRECT schema

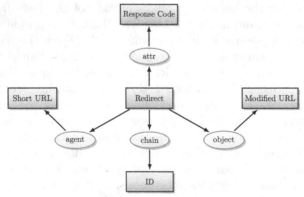

(b) CG definitional graph for REDIRECT schema

Fig. 2. CG schema for REDIRECT

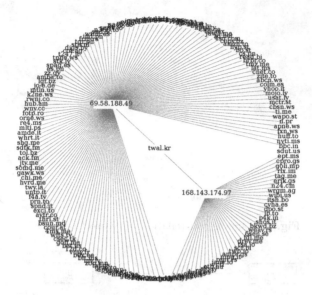

Fig. 3. The `bit.ly` infrastructure constellation

4   Results

4.1   Service Model

Network architecture relies on mapping relationships between conceptual entities such as businesses, service names and API endpoints and their representations on the network, i.e. IP addresses, ports, protocols and attributes. There is a great deal of information to record here (see Fig. 3); however, Fig. 4 presents a small part of the network with the essential information, prerequisite to further analysis—i.e., how the URLs are associated with their IPs. In Fig. 5, this basic conceptual information is given in a definitional CG. The underlying network (and, consequentially, business) relationships upon which the system depends can be instantiated into multiple factual CGs, in which Fig. 6 is an example.

4.2   URL Redirection Information Model

We presented in Fig. 2(b) a CG definitional graph representation of the information to be used to create a partial model for a URL redirection operation. This includes the URL originally provided (the short URL), the object that constitutes the direct object of that redirect (modified URL), and the agent responsible for the redirect (the redirect). Additional information typically retrieved during the URL resolution process is also indicated in this CG, such as the response code provided by the redirect service during the lookup process (instance metadata) and the position of this redirect object within the chain of redirects, which again differs by instance.

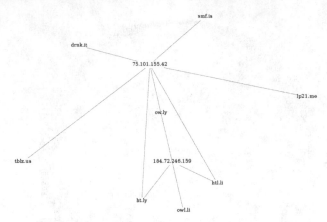

Fig. 4. A simple constellation of redirect services

Fig. 5. The service relationship definitional CG

Fig. 6. Instantiation of service relationship CG with single linkage from network

This CG graph can be instantiated with actual factual data producing, for example, the graph seen in Fig. 7. This information is processed from a bank of tweets, and can later be joined with the instantiated SERVICE CG already encountered (Fig. 6) using the Type Hierarchy from Fig. 2(a) to produce a graph with the service relationship from one CG tied to the instantiation of a redirection graph. This creates a partial model CG that has the original shortened URL linked to the service providing the actual disk space/web hosting service (see Fig. 8) by joining on the modified URL of the instantiated redirect CG.

4.3   Contextual Representation

This example, containing both instance data and modelled generalities, is contextual to the resolution process and outcome. By continuing on with the process discussed in the previous sub-section, the generation of the representation of Fig. 4 as a CG partial model is produced in Fig. 9. The ability to represent contextual information is a strength of conceptual graph theory and represents a core requirement for analysis of social data such as Twitter. This consideration becomes particularly important if the information is to be treated as elements of a broader discourse rather than orphaned utterances [17]. Extension to the model to deal with temporal and sequential aspects of Twitter discourse may be of benefit.

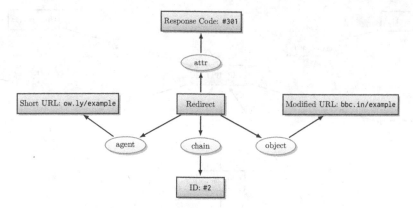

Fig. 7. Instantiation of REDIRECT definitional graph

Fig. 8. Join of fact SERVICE CG with fact REDIRECT CG

4.4 Individual and Chained Utterances

In reality there is a time-based aspect to the data in its originating context, that of conversations or interaction chains published on Twitter. There is also a mapping of the landscape of shared references upon which Twitter's message-passing depends. If the original redirect definitional graph also stored the relevant time information from the tweets [25], then a time line chart could be generated showing the impact on not only Twitter, but on the services providing storage. We wish to include this in future.

Social network graph representations are typically designed as directed graphs, showing self-declared relationships between individuals (i.e., 'friend', 'colleague', or—in the case of Twitter's '@'-reference, 'referent'. In a subset of cases, SNG representations are used that permit temporal reasoning (i.e., progression of the system's development through time). Consider for example Tang et al's proposed temporal distance metrics, designed to quantify the speed of information diffusion processes [40] in a manner that is sensitive to local and global network characteristics, Shekhar and Oliver's review of the challenges inherent in modelling time-aggregated graphs [36], or Santoro et al's judgment that '[m]ost instruments—formalisms, concepts, and metrics—for social networks analysis fail to capture their dynamics' [32].

Many research questions—particularly those linked to information propagation, reaction, etc. through Twitter—benefit from accurate and detailed modelling of temporal precedence. Research into the attractions of Twitter to its users are likely to focus on the reactions of its user community to different types of input. Investigation of the attractions of Twitter as a social news service (in comparison to a microblogging platform), will often focus on broad-grained

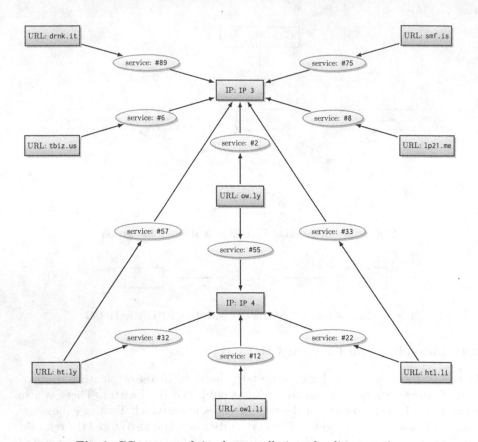

Fig. 9. CG storage of simple constellation of redirect services

metrics such as the overall proportion of tweets in any given locality that contain or refer to news items in some manner or another. However, a number of problems also exist that take a broader view of the information proliferation landscape, of which Twitter remains only a proportion, albeit at present an influential one. The relative significance of traditional media, 'new media' an social network services in information proliferation is an interesting subject and one which will undoubtedly continue to attract attention as the role of services in reflecting or even setting public opinion comes under scrutiny. Businesses continue to offer services intended to manage public opinion on social networking sites; mapping the territory is an important step in evaluating any such claim.

Much of the descriptive language from Santoro's paper is of direct relevance to our model; for example, Santoro et al [32] separate the concept of 'journey' from that of 'path'; that is, a type of path through a graph that includes waiting times at intermediate stages in travel through the graph. They also identify recent papers proposing temporal versions of the typical social-network metrics of proximity, betweenness, closeness and so forth.

4.5   Evaluation of the CG Representation as an EAD Research Tool

The exploratory analysis of a Twitter dataset described during this paper used CG representations as a backbone for representing information gathered about entities, agencies, interactions and infrastructure. This paragraph provides a brief review of this addition to the loose EAD methodology of visualisation, analysis and mining that we typically apply in the early stages of getting to grips with a large dataset. As we expected, conceptual graphs provided an accessible mechanism for knowledge representation within the team context. One team member was already familiar with the conceptual graph structuring. Another found that they were not intuitively readable, but was able to read them given appropriate guidance. It does seems necessary to have training before use.

5   Conclusion

Analysis of large datasets can be a tedious process. However new tools enhance the ability to process these datasets. These tools must be flexible while at the same time have a solid knowledge representation. We have discovered that graphical representation and graphics operators make building of the underlying models (and partial models) easier to visualize. Because conceptual graphs are both built on logic and graphical operators they can be used for this stable representation. They are also built such that time and space structure and process is built directly into the representation [25]. Microblogging creates many data records that are both similar and different at the same time. In particular re-tweets on Twitter can grow at a very fast rate and they are time dependent. Therefore the underlying representation needs to be easy to implement and fast to process [24]. The basic CG definitional and instantiated graphs for this case study has given us a good start on an over all processing graph set for discovering the data clustering and topology of the constellations from shortened links and service provider on Twitter. We can also use CGs as a teaching tool for learning how to define context with social network relationships.

References

1. Bollen, J., Mao, H., Zeng, X.: Twitter mood predicts the stock market. Journal of Computational Science 2(1), 1–8 (2011),
   http://www.sciencedirect.com/science/article/pii/S187775031100007X
2. Cao, T.H.: Fuzzy conceptual graphs for the semantic web. In: Proceedings of 2001 BISC International Workshop on Fuzzy Logic and the Internet, FLINT 2001, Berkeley, CA, USA (August 2001)
3. Carmody, T.: A Tangled Web of Shortened Links. A study of link shortening reveals hidden strands of the Web (2011), http://www.technologyreview.com/news/423170/a-tangled-web-of-shortened-links/ (retrieved May 16, 2011)
4. Carpenter, S.: Developing a measure to elicit and compare mental models of processes. Tech. rep., The University of Alabama in Huntsville (2007)

5. Daraselia, N., Yuryev, A., Egorov, S., Novichkova, S., Nikitin, A., Mazo, I.: Extracting human protein interactions from medline using a full-sentence parser. Bioinformatics 20(5), 604–611 (2004),
   http://bioinformatics.oxfordjournals.org/content/20/5/604.abstract
6. Davis, R., Shrobe, H.E., Szolovits, P.: What is a knowledge representation? AI Magazine 14(1), 17–33 (1993),
   http://www.aaai.org/ojs/index.php/aimagazine/article/view/1029
7. De Longueville, B., Smith, R.S., Luraschi, G.: OMG, from here, I can see the flames!: a use case of mining location based social networks to acquire spatio-temporal data on forest fires. In: Proceedings of the 2009 International Workshop on Location Based Social Networks, LBSN 2009, pp. 73–80. ACM, New York (2009), http://doi.acm.org/10.1145/1629890.1629907
8. dev.twitter.com: The t.co URL wrapper (2012),
   https://dev.twitter.com/docs/tco-url-wrapper (retrieved May 15, 2012)
9. Gelperin, D.: Exploring agile. In: Proceedings of the 2008 international workshop on Scrutinizing agile practices or shoot-out at the agile corral, APOS 2008, pp. 1–3. ACM, New York (2008), http://doi.acm.org/10.1145/1370143.1370144
10. Montes-y-Gómez, M., Gelbukh, A., López-López, A.: Detecting Deviations in Text Collections: An Approach Using Conceptual Graphs. In: Coello Coello, C.A., de Albornoz, Á., Sucar, L.E., Battistutti, O.C. (eds.) MICAI 2002. LNCS (LNAI), vol. 2313, pp. 176–184. Springer, Heidelberg (2002),
    http://dl.acm.org/citation.cfm?id=646402.691915
11. Huberman, B.A., Romero, D.M., Wu, F.: Social networks that matter: Twitter under the microscope. CoRR abs/0812.1045 (2008)
12. Kwak, H., Lee, C., Park, H., Moon, S.: What is twitter, a social network or a news media? In: Proceedings of the 19th International Conference on World Wide Web, WWW 2010, pp. 591–600. ACM, New York (2010),
    http://doi.acm.org/10.1145/1772690.1772751
13. Lane, P.C.R., Gobet, F.: A theory-driven testing methodology for developing scientific software. Journal of Experimental & Theoretical Artificial Intelligence 24(4), 421–456 (2012),
    http://www.tandfonline.com/doi/abs/10.1080/0952813X.2012.695443
14. Lerman, K., Ghosh, R.: Information contagion: an empirical study of the spread of news on digg and twitter social networks. CoRR abs/1003.2664 (2010)
15. Maurer, F., Martel, S.: Extreme programming. rapid development for web-based applications. IEEE Internet Computing 6(1), 86–90 (2002)
16. Mishne, G.: Source code retrieval using conceptual graphs. Master of logic thesis, Institute for Logic, Language and Computation, University of Amsterdam (2003)
17. Moulin, B.: Temporal contexts for discourse representation: An extension of the conceptual graph approach. Applied Intelligence 7, 227–255 (1997),
    http://dx.doi.org/10.1023/A:1008224616031
18. Oricchio, R.: Is Twitter A Social Network (2010),
    http://www.inc.com/tech-blog/is-twitter-a-social-network.html
    (retrieved September 16, 2012)
19. Owen, R., Horváth, I.: Towards product-related knowledge asset warehousing in enterprises. In: Proceedings of the Fourth International Symposium on Tools and Methods of Competitive Engineering, pp. 155–170. HUST Press (2002)

20. Pak, A., Paroubek, P.: Twitter as a corpus for sentiment analysis and opinion mining. In: Chair, N.C.C., Choukri, K., Maegaard, B., Mariani, J., Odijk, J., Piperidis, S., Rosner, M., Tapias, D. (eds.) Proceedings of the Seventh International Conference on Language Resources and Evaluation (LREC 2010), European Language Resources Association (ELRA), Valletta, Malta (May 2010)
21. Paul, M., Dredze, M.: You are what you tweet: Analyzing twitter for public health. In: International AAAI Conference on Weblogs and Social Media (2011), https://www.aaai.org/ocs/index.php/ICWSM/ICWSM11/paper/view/2880
22. Perer, A., Shneiderman, B.: Systematic yet flexible discovery: guiding domain experts through exploratory data analysis. In: Proceedings of the 13th International Conference on Intelligent User Interfaces, IUI 2008, pp. 109–118. ACM, New York (2008), http://doi.acm.org/10.1145/1378773.1378788
23. Petrović, S., Osborne, M., Lavrenko, V.: Streaming first story detection with application to twitter. In: Human Language Technologies: The 2010 Annual Conference of the North American Chapter of the Association for Computational Linguistics, HLT 2010, pp. 181–189. Association for Computational Linguistics, Stroudsburg (2010), http://dl.acm.org/citation.cfm?id=1857999.1858020
24. Pfeiffer, H.D.: The Effect of Data Structures Modifications on Algorithms for Reasoning Operations Using a Conceptual Graphs Knowledge Base. Dissertation, New Mexico State University (December 2007)
25. Pfeiffer, H.D., Hartley, R.T.: Temporal, spatial, and constraint handling in the conceptual programming environment, cp. J. Exp. Theor. Artif. Intell. 4(2), 167–182 (1992), http://dx.doi.org/10.1142/S0218001490000125
26. Phelan, O., McCarthy, K., Smyth, B.: Using twitter to recommend real-time topical news. In: Proceedings of the Third ACM Conference on Recommender Systems, RecSys 2009, pp. 385–388. ACM, New York (2009), http://doi.acm.org/10.1145/1639714.1639794
27. de Quincey, E., Kostkova, P.: Early Warning and Outbreak Detection Using Social Networking Websites: The Potential of Twitter. In: Kostkova, P. (ed.) eHealth 2009. LNICST, vol. 27, pp. 21–24. Springer, Heidelberg (2010)
28. Ribière, M., Matta, N., Cointe, C.: A proposition for managing project memory in concurrent engineering. In: International Conference on Computational Intelligence and Multimedia Applications, ICCIMA 1998 (February 1998)
29. Rodrigues, T., Benevenuto, F., Cha, M., Gummadi, K., Almeida, V.: On word-of-mouth based discovery of the web. In: Proceedings of the 2011 ACM SIGCOMM Conference on Internet Measurement Conference, IMC 2011, pp. 381–396. ACM, New York (2011), http://doi.acm.org/10.1145/2068816.2068852
30. Romero, D.M., Meeder, B., Kleinberg, J.: Differences in the mechanics of information diffusion across topics: idioms, political hashtags, and complex contagion on twitter. In: Proceedings of the 20th International Conference on World Wide Web, WWW 2011, pp. 695–704. ACM, New York (2011), http://doi.acm.org/10.1145/1963405.1963503
31. Sakaki, T., Okazaki, M., Matsuo, Y.: Earthquake shakes twitter users: real-time event detection by social sensors. In: Proceedings of the 19th International Conference on World Wide Web, WWW 2010, pp. 851–860. ACM, New York (2010), http://doi.acm.org/10.1145/1772690.1772777
32. Santoro, N., Quattrociocchi, W., Flocchini, P., Casteigts, A., Amblard, F.: Time-varying graphs and social network analysis: Temporal indicators and metrics. CoRR abs/1102.0629 (2011)
33. Schank, R.C., Abelson, R.P.: Scripts, Plans, Goals and Understanding: an Inquiry into Human Knowledge Structures. Lawrence Erlbaum, Hillsdale (1977)

34. Shehata, S., Karray, F., Kamel, M.: Enhancing text retrieval performance using conceptual ontological graph. In: Sixth IEEE International Conference on Data Mining Workshops, ICDM Workshops 2006, pp. 39–44 (December 2006)
35. Shekar, C., Wakade, S., Liszka, K., Chan, C.C.: Mining pharmaceutical spam from twitter. In: 2010 10th International Conference on Intelligent Systems Design and Applications (ISDA), pp. 813–817 (November-December 2010)
36. Shekhar, S., Oliver, D.: Computational modeling of spatio-temporal social networks: A time-aggregated graph approach. In: Proceedings of the 2010 Specialist Meeting on Spatio-Temporal Constraints on Social Networks (2010), http://www.ncgia.ucsb.edu/projects/spatio-temporal/docs/Shekhar-position.pdf
37. Sowa, J.F.: Conceptual structures: information processing in mind and machine. Addison-Wesley Longman Publishing Co., Inc., Boston (1984)
38. Spasic, I., Ananiadou, S., McNaught, J., Kumar, A.: Text mining and ontologies in biomedicine: Making sense of raw text. Briefings in Bioinformatics 6(3), 239–251 (2005), http://bib.oxfordjournals.org/content/6/3/239.abstract
39. Sullivan, J.: A tale of two microblogs in china. Media, Culture & Society 34(6), 773–783 (2012), http://mcs.sagepub.com/content/34/6/773.short
40. Tang, J., Musolesi, M., Mascolo, C., Latora, V.: Temporal distance metrics for social network analysis. In: Proceedings of the 2nd ACM Workshop on Online Social Networks, WOSN 2009, pp. 31–36. ACM, New York (2009), http://doi.acm.org/10.1145/1592665.1592674
41. Thomas, K., Grier, C., Song, D., Paxson, V.: Suspended accounts in retrospect: an analysis of twitter spam. In: Proceedings of the 2011 ACM SIGCOMM Conference on Internet Measurement Conference, IMC 2011, pp. 243–258. ACM, New York (2011), http://doi.acm.org/10.1145/2068816.2068840
42. Yi.tl: Url shorteners (2012), http://yi.tl/pages/urlshorteners.php (retrieved May 15, 2012)

Taking SPARQL 1.1 Extensions into Account in the SWIP System

Fabien Amarger, Ollivier Haemmerlé, Nathalie Hernandez, and Camille Pradel

IRIT, Université de Toulouse le Mirail,
Département de Mathématiques-Informatique, 5 allées Antonio Machado,
F-31058 Toulouse Cedex
fabien.amarger@gmail.com,
{ollivier.haemmerle,nathalie.hernandez,camille.pradel}@univ-tlse2.fr

Abstract. The SWIP system aims at hiding the complexity of expressing a query in a graph query language such as SPARQL. We propose a mechanism by which a query expressed in natural language is translated into a SPARQL query. Our system analyses the sentence in order to exhibit concepts, instances and relations. Then it generates a query in an internal format called the pivot language. Finally, it selects pre-written query patterns and instantiates them with regard to the keywords of the initial query. These queries are presented by means of explicative natural language sentences among which the user can select the query he/she is actually interested in. We are currently focusing on new kinds of queries which are handled by the new version of our system, which is now based on the 1.1 version of SPARQL.

1 Introduction

The amount of knowledge available on the semantic web increases everyday. Many OWL ontologies and RDF triplestores are put online, especially in the context of the linked open data initiative [1]. Accessing this knowledge is a real challenge since it is difficult for an end-user to handle the complexity of the "schemata" of these pieces of knowledge: in order to express a valid query on the knowledge of the semantic web, the user needs to know the SPARQL query language, the ontologies used to express the triples he/she wants to query on as well as the "shape" of the considered RDF graphs.

Extensive work has been carried out in order to help users express queries in graph formalisms (CGs, SPARQL...) during the recent period. The help provided for the user can rely on graphical interfaces such as [2] for RQL queries, [3] and [4] for SPARQL queries or [5] for conceptual graph queries. But such graphical interfaces need the end-user to be familiar with and, moreover, to understand the semantics of the expression of a query expressed in terms of graphs. The work presented in [6] aims at extending the SPARQL language and its querying mechanism in order to take into account keywords and jokers when the user does not exactly know the schema he/she wants to query on. Here again, such an approach requires that the user knows the SPARQL language.

H.D. Pfeiffer et al. (Eds.): ICCS 2013, LNAI 7735, pp. 75–89, 2013.

Other works aim at the automatic or semi-automatic translation of formal queries from user queries expressed in terms of keywords. The user expresses his/her information need in an intuitive way, without having to know the query language or the knowledge representation formalism used by the system. Some works have already been proposed to generate formal queries from keywords, resulting in different languages such as SeREQL [7], SPARQL [8,9] or conceptual graphs [10].

Our work belongs to this family of approaches. In [10], we proposed a way of building queries expressed in terms of conceptual graphs from user queries composed of keywords. In [11] we extended the system in order to take into account relations expressed by the user between the keywords he/she used in his/her query. In [12], we adapted our system to the Semantic Web languages instead of Conceptual Graphs. Such an adaptation was important for us in order to evaluate the interest of our approach on large knowledge bases. Since then, our system has taked into account queries expressed in natural language. Our work is based on two observations. First, end-users want simple query languages since they are used to this kind of querying on classic search engines on the Web. Second, in the main real applications, the submitted queries are variations of a few typical query families. We believe that each family can be prototyped and represented with a pattern. These observations led us to propose a mechanism allowing a user query expressed in terms of natural language to be translated into a SPARQL query built by adapting pre-defined query patterns chosen according to the natural language query.

The use of patterns allows us to avoid the step which consists in parsing the ontology in order to find potential relations which can be used to link the classes and instances identified in the natural language query, since the relevant relations appear in the pre-defined patterns. The process takes advantage of the relevant query families, which correspond to an actual information need. One of the main issues of our approach is therefore to select the pattern which best fits user needs. For the moment, patterns are built manually by domain experts.

The SWIP system presented in [12] was based on the 1.0 version of the SPARQL semantic web query language. At the beginning of 2012, SPARQL 1.1 [13] was released by the W3C. This new version features several improvements. We studied the new version and considered that the aggregates – which are more or less the same as in SQL – offer the possibility of handling new kinds of queries that were difficult to process in the previous version of SWIP. This article presents the extension of the SWIP system by taking into account the SPARQL 1.1 aggregates. For example, we are now capable of dealing with queries such as "How many artists are involved in a given film?", "The number of awards an artist won in a competition?" or "What is the average length of a film produced in 2012?".

In section 2, we present the SWIP system briefly. Section 3 introduces the aggregates in SPARQL 1.1. Section 4 details the extension of SWIP in order to implement the aggregates. Finally, section 5 describes the implementation of our system and presents a first experimentation.

2   The SWIP System

2.1   Overview

In order to allow end-users to express queries on knowledge bases expressed in the Semantic Web languages (RDF triples built on OWL ontologies), we propose a system by which a query expressed in natural language is translated into a SPARQL query. The SWIP[1] system was first presented in [11] in its preliminary Conceptual Graph based version, then in [12] in its SPARQL version. We do not present the SWIP system exhaustively in this article but briefly recall in this section the main features of the system and how it works, before focusing on the extensions of SWIP in section 4.

An overview of the process is presented in Figure 1. The global process of the system is as follows: (i) the user expresses his/her query in terms of a natural language query; (ii) the natural language query is then transmitted to a syntactic dependency analyzer which produces a dependency graph. This graph provides the main keywords of the query as well as the relations linking them ; they are represented by means of what we call a pivot query; (iii) the keywords and relations represented in the pivot query are matched to the elements of the ontology – concepts, instances and relations which seem to correspond to them ; (iv) the concepts, relations and instances obtained are mapped with the available query patterns; (v) explicative sentences corresponding to each possible SPARQL query are generated and proposed to the user, so that he/she can select the final query which best fits his/her information need; (vi) finally, the actual SPARQL query is generated and processed. In the following paragraphs, we present the process of the SWIP system in more details.

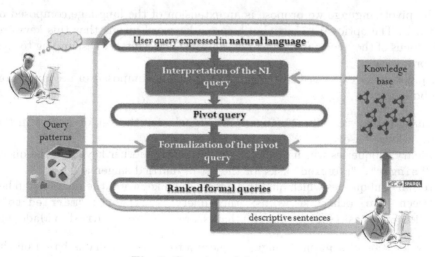

Fig. 1. Overview of the process

---

[1] Semantic Web Interface using Patterns.

2.2   From Natural Language Queries to Pivot Queries

Here we describe the first main step of the SWIP process. This step is illustrated in Figure 2. The user enters his/her query in natural language into the system. The first translation is performed to generate the query in a simplified and synthetic form that we call *pivot language*. This language is based on keywords connected with relationships which are more or less explicit. The detailed grammar of the pivot language is presented in [12]. We use this pivot language in order to facilitate the implementation of multilingualism by means of a common intermediate format.

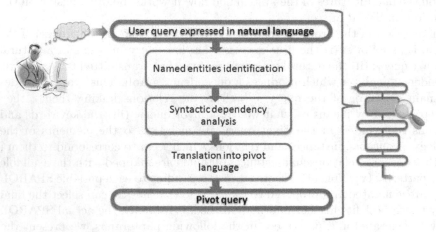

Fig. 2. Interpretation of the natural language query

The pivot language we propose is an extension of the language composed of keywords. The optional "?" symbol before a keyword means that this keyword is the focus of the query: we want to obtain specific results corresponding to this keyword.

A pivot query expressed in the pivot language is composed of a conjunction of subqueries:

- unary subqueries, like `?"singer"` which asks for the list of singers in the knowledge base;
- binary subqueries which qualify a keyword with another keyword: the query `?"singer": "married"` asks for the list of married singers;
- ternary subqueries which qualify, by means of a keyword, the relationship between two other keywords: the query `?"singer": "married to"= "Madonna"` asks for the singer(s) that is/are/was/were married to Madonna.

The translation of a natural language query into a pivot query is based on the use of a syntactic dependance analyzer which produces a graph where nodes correspond to the words of the sentence and edges to the grammatical dependencies between them. Before parsing the query with the analyzer, a first stage identifies in the sentence the named entities corresponding to knowledge base resources. These entities are then considered as a whole and will not be separated by the parser in the next stage. For example, in the sentence "what are the films of Jean Dujardin", "Jean Dujardin" will be considered as a named entity as it is the label of an instance of Actor in the knowledge base. This stage is particularly crucial when querying knowledge bases containing long labels, such as group names or film titles made up of several words or even sometimes of a full sentence.

Once the named entities are identified and the dependency graph is generated, a set of rules are applied to construct the pivot query. The different clauses of the sentence are considered. First, the head of the expression playing the role of the subject in the main clause is identified as the keyword corresponding to the object of the query. Then, if there is one, the expansion of the expression is used to complete a binary subquery. For example, for the query "what are the films of Jean Dujardin", the generated pivot query will be ?"film": "Jean Dujardin". If the clause is composed of a verb and a complement, a ternary subquery is constructed. For the query "what films were awarded prizes in Cannes?", the corresponding pivot query will be ?"movie": "awarded prizes in"= "Cannes". If the query contains relative clauses, the same rules are applied. The entity referenced by the relative pronoun is expressed in the subquery with the keyword used in the previous subquery. For example for the query "What are the awarded films in which Jean Dujardin played?", the corresponding pivot query will be ?"film": "awarded". "Jean Dujardin": "played in"= ?"film".   These rules might seem simple but we have observed that the structure of queries expressed by end-users is generally simple. Note that we use a specific syntactic analyzer, Maltparser [14], trained for each language we consider.

2.3   From Pivot to SPARQL

The second part of the process, which consists in the formalization of the pivot query, is illustrated in Figure 3. It is divided in four substeps which are described below. A more precise description of this step is given in [12].

Matching Keywords to Knowledge Base Entities. First, the ontology is used to determine which elements (concepts, instances or relations) are closest to the keywords appearing in the pivot query. This notion of closeness is based on the similarity measure between strings, as the one presented in [15]. However, in our system implementation, we do not use a classic method such as Levenshtein distance, but the Lucene score function, saving us the task of implementing a similarity measure and allowing us to benefit from the powerful Lucene indexation and "fuzzy matching" features, to handle different forms of lemmas and mistypings.

For example, with the pivot query ?"film": "Jean Dujardin'", SWIP searches for the labels corresponding to "film", then for the labels corresponding to "Jean Dujardin". For each element of the pivot query, SWIP returns a list of possible ontology entities weighted with respect to their relevance. The keyword "film" is associated with the concept "Film" with a weight of 1 since one of

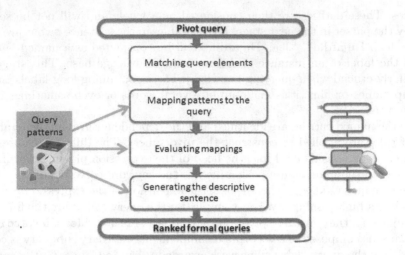

Fig. 3. Formalization of the pivot query

the labels of the concept "Film" is the string "Film". It is also matched with the concept "disaster film" because its label "disaster film' obtained a similarity measure of 0.625.

Mapping the Query Patterns to the Pivot Query. Once the ontology elements are identified, SWIP has to find one or more query patterns which fit these elements. The patterns are formally defined in [11,12]. Each pattern corresponds to a family of queries which can be asked on the knowledge base. Roughly speaking, each pattern is composed of: (i) a query graph which is a pre-written generic query expressed in SPARQL corresponding to an identified typical family of queries; (ii) a set of concepts and relations which belong to the query graph and correspond to the characteristic elements of the pattern; (iii) a model of an explicative sentence which will be used to generate the sentence in natural language allowing the user to understand the meaning of the SPARQL query. After the matching step, SWIP tries to associate the highest-weighted concepts with the different patterns in order to exhibit the patterns which seem to be used as a basis of relevant final queries.

Evaluating the Generated Mappings. The mapping step leads to a set of query mappings, each one corresponding to a possible query interpretation. These candidate interpretations must be ranked in order to present first to the user the queries which seem to be the most relevant. To this end, this step will allocate to each query mapping a *relevance mark R*, made up of several partial marks, each one taking into account a number of parameters that seem important to us:

– *Element mapping relevance mark R_{map}* represents how much we trust the different element mappings involved in the considered query mapping.

- *Query coverage relevance mark* R_{Qcov} takes into account the proportion of the initial user query that was used to build the mapping.
- *Pattern coverage relevance mark* R_{Pcov} takes into account the proportion of the pattern qualifying vertices that was used to build the mapping.

Generating the Query and the Explicative Sentence. The last step of the SWIP process consists in presenting the results to the user, and allowing him/her to query the knowledge base. For this, for each mapping between a set of keywords and a pattern we generate a sentence in natural language explaining the query represented by the mapping. We then present all the sentences to the user in decreasing relevance order. Thus, reading the explicative sentences, the user can easily understand the meaning of each query and choose the one matching his/her need. The system then formulates from the chosen mapping the final query expressed in SPARQL. Both operations, generating explicative sentences and formulating the query graph, are trivial, thanks to the explicative sentence attached to each pattern and to the graph architecture of each pattern.

For each mapping, the generation of an explicative sentence is carried out by taking the generic sentence attached to the mapped pattern and personalizing it, i.e. for each element mapping, the substring associated with the mapped pattern element is replaced by an appropriated string that comes from the matched label in the case of ontology elements or that is the string representation of the literal value in the case of literals. As regards pattern elements which are not involved in any element mapping, we keep the default associated substrings in the sentence.

The mapping generation is made more dynamic by adding the possibility of using regular expressions on parts of patterns, drawing ideas from [16]; some parts are omitted in the final mapping. This improves the readability of the explicative sentences of each mapping and makes patterns more generic (therefore less numerous).

The query graph of the selected query mapping is the pattern graph, apart from the odd detail; it is generated using the procedure presented in [10], except that relation vertices can be modified by specialization or generalization, in the same way as that for concept vertices.

3 SPARQL 1.1 Evolutions for Dealing with End-User Queries

3.1 SPARQL 1.1 Update

The W3C SPARQL 1.1 recommendation aims at improving the initial version of the language with incremental updates offering ascending compatibility with the first version [17]. The main issue is to ease the generation of queries by expanding the language syntax.

The most significant update for dealing with end-user queries is the possibility of using aggregate functions. In the initial version of SPARQL, it was possible to retrieve a solution set of triples according to a specific query pattern. With the use of aggregates, it is now possible to partition this set according to specified criteria and to compute a new solution using an aggregate on these partitions. When needed, aggregates were previously calculated by the applications. Now, however, they are calculated by the SPARQL engine.

The aggregate functions available in SPARQL1.1 are COUNT, SUM, MIN, MAX, AVG, GROUP_CONCAT and SAMPLE. As in SQL queries, aggregate functions can be used either as projection or as selection attributes.

COUNT

The COUNT aggregate enables the counting of the triples contained in the solution set as in the following query.

```
SELECT COUNT(?film)
WHERE
{
    ?film rdf:type cine:Film
}
```

The result of this query will be the integer corresponding to the number of instances of the class Film in the triplestore.

The aggregate can also be used in the condition defined to select the partitions to be considered as in the following query.

```
SELECT ?actor
WHERE
{
    ?film rdf:type cine:Film.
    ?actor rdf:type cine:Actor.
    ?film cine:hasForActor ?actor.
}
GROUP BY ?actor
HAVING COUNT(?film) > 5
```

This query will give all the actors who have played in more than 5 films. The "GROUP BY" clause specifies on which attribute the partitions are built and the "HAVING" clause specifies the condition that has to be satisfied by the partition.

SUM - AVG - MIN - MAX

The aggregates ("SUM" for sum, "AVG" for average, "MIN" for minimum and "MAX" for maximum) can be used in the same way as COUNT except that they have to be applied on numerical attributes.

GROUP_CONCAT - SAMPLE

These two aggregates are more specific to SPARQL 1.1. "GROUP_CONCAT" enables the generation of a string from all the triples contained in the solution set. SAMPLE will randomly select an element in the solution set. The query below will return the concatenation of the URI of each film contained in the dataset, separated with the character "|".

```
SELECT GROUP_CONCAT(?film ; separator="|")
WHERE
{
    ?film rdf:type cine:Film
}
```

3.2   Aggregates in End-User Queries

When generating SPARQL queries from natural language queries, considering aggregates can be interesting for two reasons. End-users may want to retrieve data that are not explicitly stated in the triplestore but can be calculated. Examples of queries can be: (i) "How many artists are involved in a given film?", (ii) "The number of awards an artist won in a competition?", (iii) "What is the average length of a film produced in 2012?". As this information may be calculated, they will most probably not be stated explicitly with an RDF triple linking the considered resource to a property and to the corresponding value. In its current state, this is what our system will look for. This kind of information will have to be retrieved by generating the corresponding SPARQL query using the aggregate function. For example, in the triplestore we consider for our experiments, the SPARQL query corresponding to query i) will have to count all the artists involved in a given film. However, for some queries the same question words will not always lead to the use of aggregate functions. For example, for the query "How many members are there in the jury of a given competition", the use of the aggregate "count" may not be relevant as a property can be explicitly stated. It is the case for this query as there is often no point in naming all the jury members, but only the number of members. As there are two ways of translating question words corresponding to aggregates in natural language into SPARQL, both possibilities will have to be considered when generating the SPARQL query. Aggregates will also be needed when the end-user wishes to select data according to comparison with other data. Examples of queries can be "What is the longest film ever made?", "Who are the actors that have obtained more than 2 awards", ...

In order to evaluate the added value of considering aggregates in the SWIP system, we asked 20 people to propose queries in natural language in the field of the cinema. The queries were manually translated into SPARQL1.1 to query a triplestore[2]. Among the 98 queries we collected, 9% of them implied the use of aggregates in the SPARQL query. Moreover, in all of these queries, aggregates are used as projection attributes. We thus decided to focus our work on this kind of query. By considering aggregates as projection attributes, our system is now able to deal with 99% of the queries asked by users. The remaining 1% corresponds

---

[2] The triplestore is described at http://ontologies.alwaysdata.net/cinema

to queries that contain terms that are not yet in the considered ontology (for example, SWIP cannot deal with the query "What is the filmography of Jean Dujardin?" as filmography is not in the ontology).

4 Evolution of SWIP 1.0 into SWIP 1.1

Several extensions to the SWIP system have been proposed in order to generate SPARQL queries with aggregate functions. The language proposed to represent pivot queries has been updated. Note that by considering pivot queries it is possible to dissociate treatments that are language-dependent from those that transform the pivot query into the SPARQL query. As the interpretation of the natural language query relies on the use of syntactic analyzers, we have improved the rules leading to the generation of the pivot query by taking into consideration typical words that may correspond to aggregates. The process generating the SPARQL query has also been improved.

Pivot Query Language. The language defined to represent pivot queries has been extended to take into account aggregate functions.

We propose to extend the pivot language by adding an optional subquery to the initial grammar described in this section.

This subquery corresponds to the case when an aggregate is used as a projection attribute. The subquery will be composed of one of the following keywords: COUNT, SUM, MIN, MAX, AVG. The semantics associated with the new subquery is that the aggregate function is applied on the projection attribute represented by the character ? in the query.

Definition 1. *q is a well formed query in the pivot language if it conforms to the following grammar, given in Backus-Naur form.*

```
query  ::=  subquerySet ( "." subquerySet )* ("." )? ( ("COUNT"
|  "MAX"  |  "MIN"  |  "AVG"  |  "SUM" ) ("." )? )?

subquerySet  ::=  keyword ( ":" keyword ( "="keyword ( ",
"keyword )*)? )?

keyword  ::=  ('a'..'z'  |  'A'..'Z'  |  '_') ( ( 'a'..'z'  |
'A'..'Z'  |  '0'..'9'  |  "_" ) )*
```

For example, the query "how many artists are involved in The Artist" corresponds to the pivot query : ?"artist":"involved in"="The Artist".COUNT

From the Natural Language Query to the Pivot Query. The pivot query is built automatically from the query in natural language with rules defined on the output of a syntactic analyzer.

To detect queries which could need the use of an aggregate once transformed in SPARQL, we have built a dictionary for each language we consider (French and English) stating, for each aggregate and condition, terms that may refer to them. The dictionary has been constructed by analyzing manually a corpus of approximately 100 queries in each language. A sample of the dictionary for English is presented in the following table:

Aggregate	Corresponding words
COUNT	"The number of", "How many", ...
SUM	"The sum", "add", "adding", ...
MIN	"The minimum", "minimum",
MAX	"The maximum", "maximum",
AVG	"The average", "average", ...

The words defined in the dictionary are searched in the syntactic dependencies graph given by the analyzer.

If one of them is found to be depending directly on the head of the expression playing the role of the subject in the sentence (often found after the question word), a subquery composed of the corresponding aggregate is added to the pivot query. Dictionary words can be found in the direct context of the subject in the following queries : "*How many* artists are involved in a given film?", "What is *the number of* awards an artist won in a competition?", "What is *the average* length of a film produced in 2012?". The remaining parts of the sentences are analyzed as in the previous version of SWIP.

In order to deal with the case when the data does not need to be calculated with an aggregate function because it is explicitly stated in the data set with a property and its corresponding value, a pivot query is also constructed without paying any specific attention to words from the dictionary. The pivot query is thus generated as in the previous version of SWIP. For example, for the query "What is the number of members in the Oscar jury?", the second pivot query will be `"member":"?number","member":"jury"="Oscar"`.

From the Pivot Query to the SPARQL Query. To generate the SPARQL query, the two pivot queries are mapped to the patterns. For the query containing the aggregate subquery, the mapping process is the same as in the previous version of the system as we have chosen to ignore the possible aggregate subqueries composing the pivot query during this phase. The main reason is that we consider than aggregate functions can be used on any of the qualifying concepts of the patterns and that the need for an aggregate will not be an indication for discriminating one pattern from another. Patterns represent information needs that can be expressed differently. The same pattern can map a query which may or may not need an aggregate in its interpretation.

Once the patterns are ranked according to the pivot query, the eventual aggregate subqueries are considered for completing the generated descriptive sentences. The goal is to show the user how the aggregate has been interpreted. The name of the aggregate is added in brackets before the term corresponding to the subject of the aggregate.

The SPARQL query is generated according to the sentence chosen by the user. The aggregate is added in the SELECT clause.

5 Implementation and Experimentation

5.1 Implementation

The improvements presented in the previous section have been implemented in the SWIP system. A graphical user interface has been added to the system, developed with the JQuery technology. We have implemented a simple interface with two text fields and very few buttons, as can be seen in Figure 4. The first text field is used so that the user can enter his/her query in natural language. The "translate" button generates the pivot language query related to the natural language query. The "search" button generates all the possible SPARQL queries.

When the user clicks on the search button, SWIP searches all the possible mappings to generate the best SPARQL queries, which are displayed on a dynamic table, as can be seen on Figure 5. All the results are displayed sorted by score (here, the "rel" column) as is the generated sentence associated with the SPARQL queries directly. To indicate the searched element on the query on the sentence, it is preceded by the "?" character. There are some selections available directly on the sentence to allow the user to modify the query. For example, if he/she wants to generalize a specific element. On each row there is a "+" button to display all the details of the query, the SPARQL query associated with it

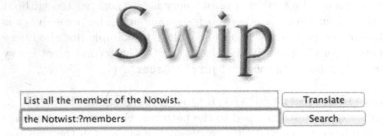

Fig. 4. SWIP interface

Fig. 5. SWIP interface - Generated SPARQL queries

and the mappings generated by SWIP. A double click on a row will execute the SPARQL query directly on the SPARQL endpoint to finally obtain the answer to the initial question. The answer is displayed on another table, next to the "Results" tab.

5.2   Experimentation

As explained in section 3, 9% of the queries collected on the field of the cinema implied the use of aggregates as projection attributes when translated manually into SPARQL.

In order to evaluate our approach, we compared the SPARQL queries generated by the new version of SWIP with the manually written query. For 100%, the "right" query appears in the first three propositions. This means that the generated pivot query is mapped relatively correctly with the right pattern. Moreover, the descriptive sentence obviously shows the right interpretation. This means that it is not difficult for the end-user to select the right query to generate.

We also considered the generated queries in which a dictionary word suggests that an aggregate function needs to be used but for which we have not used an aggregate in the manually built SPARQL query. This kind of query represents 4% of the total queries. For 50% of them, the first proposition made by SWIP is the correct one. When SWIP gives a wrong interpretation (i.e. an interpretation considering an aggregate), it is because the label of property that must be used in the query has not been identified correctly in the query. For example, in the query "how many members are there in the jury of the oscars" the property "numberOfMember" is mapped with too low a score to "how many members".

The evaluation is encouraging but we are currently looking for a larger set of queries to expand it.

6   Conclusion and Future Work

In this paper, we proposed a development of the SWIP system introduced in [10,11,12], for the system to take advantage of an enhancement of the SPARQL 1.1 query language: the aggregates. For this, we proposed an evolution of the pivot query language, and adapted the query interpretation process to make it take the aggregates into account.

The first evaluation results are very encouraging. We can now handle some queries that previously could not be processed.

The system has been provided with a graphical user interface which is very simple: the users are able to express their queries, and then choose the best query and modify it if necessary.

The preliminary step of building query patterns is done manually and thus requires a large amount of work. Moreover, this step must be repeated each time we want to address a new knowledge base. This is why the automatic or assisted pattern generation is the last important task we need to carry out to obtain a fully functional system. We have two potential leads for this purpose: building patterns covering a set of graph queries, or learning these patterns from a set of natural language queries. At first glance, the first method seems to be the easier to implement, since the input of this method consists of formal structures which are easy to handle. However, end user queries expressed in a graph formalism could be costly to obtain in practice, when natural language queries on a domain should be easy to find, looking on forums, FAQs, or simply asking users.

Moreover, we plan to extend our work in two other directions:

- we are currently looking for a larger set to evaluate our approach more precisely in order to identify its drawbacks; we are developing a partnership with the IRSTEA (a French institute on ecology and agriculture) in order to build a real application framework concerning French queries on organic farming.
- in the near future we intend to consider aggregate functions as selection attributes by means of nested subqueries which are also an evolution of SPARQL1.1.

References

1. Bizer, C., Heath, T., Berners-Lee, T.: Linked data - the story so far. Int. J. Semantic Web Inf. Syst. 5(3), 1–22 (2009)
2. Athanasis, N., Christophides, V., Kotzinos, D.: Generating On the Fly Queries for the Semantic Web: The ICS-FORTH Graphical RQL Interface (GRQL). In: McIlraith, S.A., Plexousakis, D., van Harmelen, F. (eds.) ISWC 2004. LNCS, vol. 3298, pp. 486–501. Springer, Heidelberg (2004)
3. Russell, A., Smart, P.R.: Nitelight: A graphical editor for sparql queries. In: Bizer, C., Joshi, A. (eds.) International Semantic Web Conference (Posters & Demos). CEUR Workshop Proceedings, vol. 401. CEUR-WS.org (2008)
4. Ferré, S., Hermann, A.: Semantic Search: Reconciling Expressive Querying and Exploratory Search. In: Aroyo, L., Welty, C., Alani, H., Taylor, J., Bernstein, A., Kagal, L., Noy, N., Blomqvist, E. (eds.) ISWC 2011, Part I. LNCS, vol. 7031, pp. 177–192. Springer, Heidelberg (2011)
5. CoGui. A conceptual graph editor. Web site (2009), http://www.lirmm.fr/cogui/
6. Elbassuoni, S., Ramanath, M., Schenkel, R., Weikum, G.: Searching rdf graphs with sparql and keywords. IEEE Data Eng. Bull. 33(1), 16–24 (2010)
7. Lei, Y., Uren, V.S., Motta, E.: SemSearch: A Search Engine for the Semantic Web. In: Staab, S., Svátek, V. (eds.) EKAW 2006. LNCS (LNAI), vol. 4248, pp. 238–245. Springer, Heidelberg (2006)
8. Zhou, Q., Wang, C., Xiong, M., Wang, H., Yu, Y.: SPARK: Adapting Keyword Query to Semantic Search. In: Aberer, K., Choi, K.-S., Noy, N., Allemang, D., Lee, K.-I., Nixon, L.J.B., Golbeck, J., Mika, P., Maynard, D., Mizoguchi, R., Schreiber, G., Cudré-Mauroux, P. (eds.) ASWC 2007 and ISWC 2007. LNCS, vol. 4825, pp. 694–707. Springer, Heidelberg (2007)
9. Tran, T., Wang, H., Rudolph, S., Cimiano, P.: Top-k exploration of query candidates for efficient keyword search on graph-shaped (rdf) data. In: ICDE, pp. 405–416. IEEE (2009)
10. Comparot, C., Haemmerlé, O., Hernandez, N.: An Easy Way of Expressing Conceptual Graph Queries from Keywords and Query Patterns. In: Croitoru, M., Ferré, S., Lukose, D. (eds.) ICCS 2010. LNCS, vol. 6208, pp. 84–96. Springer, Heidelberg (2010)

11. Pradel, C., Haemmerlé, O., Hernandez, N.: Expressing Conceptual Graph Queries from Patterns: How to Take into Account the Relations. In: Andrews, S., Polovina, S., Hill, R., Akhgar, B. (eds.) ICCS 2011. LNCS (LNAI), vol. 6828, pp. 229–242. Springer, Heidelberg (2011)

12. Pradel, C., Haemmerlé, O., Hernandez, N.: A Semantic Web Interface Using Patterns: The SWIP System. In: Croitoru, M., Rudolph, S., Wilson, N., Howse, J., Corby, O. (eds.) GKR 2011. LNCS, vol. 7205, pp. 172–187. Springer, Heidelberg (2012)

13. Harris, S., Seaborne, A.: Sparql 1.1 query language. w3c working draft (July 24, 2012), World Wide Web Consortium, http://www.w3.org/TR/sparql11-query

14. Nivre, J., Hall, J., Nilsson, J., Chanev, A., Eryigit, G., Kübler, S., Marinov, S., Marsi, E.: Maltparser: A language-independent system for data-driven dependency parsing. Natural Language Engineering 13(02), 95–135 (2007)

15. Levenshtein, V.I.: Binary codes capable of correcting deletions, insertions, and reversals. Soviet Physics Doklady 10, 707–710 (1966)

16. Alkhateeb, F., Baget, J.-F., Euzenat, J.: Extending sparql with regular expression patterns (for querying rdf). J. Web Sem. 7(2), 57–73 (2009)

17. Kjernsmo, K., Passant, A.: Sparql new features and rationale. World Wide Web Consortium, Working Draft WD-sparql-features-20090702 (2009)

System Architecture to Implement a Conceptual Graphs Storage in an RDF Quad Store

Khalil Ben Mohamed, Benjamin Chu Min Xian, and Dickson Lukose

MIMOS Berhad,Technology Park Malaysia, 57000 Kuala Lumpur, Malaysia
{khalil.ben,mx.chu,dickson.lukose}@mimos.my

Abstract. With the growth of interest in semantics around the world, we believe that conceptual graphs have an important role to play. However, from the best of our knowledge, there is a lack of conceptual graphs storage and retrieval engine capable of scaling up. In this paper, we propose to utilize the power of the RDF stores, and present a complete system and methods to implement an efficient conceptual graphs storage and retrieval engine in an RDF store. We translate conceptual graphs knowledge bases into RDF knowledge bases, create an external index of the conceptual graphs and use the index to efficiently retrieve a set of candidate conceptual graphs in response to an expanded user query. We also discuss several heuristics which aim to speed up the data retrieval process, and present preliminary experimental results using the different heuristics.

1 Introduction

Over the last decade, we have witnessed a tremendous growth of interest on the semantic web and semantic technologies in general. The Linked Open Data initiative[1] is an exceptional example of this seemingly unstoppable movement, containing as per as September 2011 more than 300 very huge datasets with over 31 billion triples of information. Consequently, the semantic web language RDF [Hay04] has become the main standard to model and engineer knowledge bases which are now emerging from all around the world. Besides, very large RDF data stores (e.g. triple or quad stores) have been designed, which have proven their efficiency in terms of scalability and retrieval speed (see for example [FCB12] and [GM10]).

The Conceptual Graphs (CGs), a well-established knowledge representation and reasoning formalism, were first introduced by J.F. Sowa in 1976 [Sow76]. They have been successfully applied at the end of the twentieth century to diverse Artificial Intelligence problems such as natural language processing and expert systems, and have continuously been enriched over the years [CM09]. Nevertheless the CGs community does not play a significant role in the semantic web movement as important as they should, as recently stated in a position paper by Rudolph et al. [RKH]. In 2005, Dieng-Kuntz and Corby had already outlined the advantages of the CGs formalism for the semantic web through several real-world applications [DKC05], and more recently Da Silva et al. made an attempt to answer the challenging question "How to store large knowledge bases to be able to scale up ontological conjunctive query answering?" [dBC11].

---

[1] http://linkeddata.org/

H.D. Pfeiffer et al. (Eds.): ICCS 2013, LNAI 7735, pp. 90–105, 2013.

To the best of our knowledge there exist no CGs data storages capable of scaling up, i.e. storing very large CGs knowledge bases (e.g. millions or trillions of data) while keeping good data retrieval performances, although several tools have been proposed over the years. The first attempt was made in 1992 by Ellis and Levinson [EL92] who initiated the PEIRCE project, an international conjoint effort with the aim of building an industrial strength conceptual graphs workbench, but the project suddenly stopped in its early stages. The second attempt was initiated by Haemmerlé in 1994 who created CoGITo, a conceptual graphs library. Since 1994 it has been continuously enhanced and renamed to CoGITaNT in 1998 [GS98][coga], and CoGui, a free graph-based visual tool using the CoGITaNT library has recently been developed [cogb]. Nevertheless using these tools as they are imply loading the entire datasets into memory for each searching-retrieving request.

Moreover, the lack of scalable storage and retrieval engines for conceptual graphs makes impossible the use of conceptual graphs in softwares designed for commercialization. Indeed, to enter the market a software must among others respond "in the blink of an eye" to user queries. We believe that this scalability issue plays an important role for the CGs not to be popular in view of the current concerns of dealing with very large amounts of data. In this paper we address this issue and propose to take advantage of the powerful and scalable RDF stores to implement a CGs storage and retrieval engine helping to filter the parts of the datasets in which answers to a user query may be found.

We propose a system and methods to implement an efficient CGs storage and retrieval engine in an RDF triple store. We first present the overall architecture of our system, with its components and the flows respectively for data indexing and data retrieval. In order to store and index CGs knowledge bases, we first translate them into an RDF notation. We rely on the translations from CGs to RDF and RDF to CGs proposed by Baget et al [BCG+10] that we extend to consider the *graph name* parameter. Then, we create three indexes (one for each object of a CGs vocabulary) by considering the concepts, relations and individuals in the graphs. To retrieve the stored data in response to a structured user query, we first expand the query according to a CGs vocabulary, and provide different retrieval heuristics to return a set of candidate CGs in which an answer to the query may be found. Finally, we show preliminary experiments using different benchmarks, regarding the system efficiency in terms of speed and accuracy while using different heuristics.

Paper layout. Section 2 recalls the basics of conceptual graphs, RDF and indexing techniques. In Section 3, we present the architecture of our conceptual graphs storage and retrieval engine, and the storage and retrieval methods that we have used. Section 4 presents preliminary experimental results of our system using different heuristics. The prospects of this work are outlined in Section 5.

2   Preliminaries

In this section, we present a brief overview of the CGs and the RDF formalisms through some fundamental notions which are used in the following, and a general overview of RDF stores and indexing techniques.

2.1 Conceptual Graphs

Conceptual Graphs (CGs) are a knowledge representation and reasoning formalism based on labelled graphs. It was introduced by J.F. Sowa in 1976 [Sow76], who wrote the first book on CGs in 1984 [Sow84]. Since 1984, it has been enriched and further developed by the CGs community (see [CM09] for a synthesis of the results obtained in the last decades). In this formalism, knowledge is represented as labelled graphs, and the reasoning mechanisms are graph operations. Moreover, the reasoning is logically founded, which guarantees the obtained results.

A conceptual graph (CG) is a bipartite graph G with two kinds of nodes: concept nodes and relation nodes. The former represents entities while the latter represents the relations between these entities. Each node has a type, e.g. "Car" or "Person" for a concept node and "belongTo" or "possess" for a relation node. In addition, a concept node is either *individual* or *generic*: it has respectively a marker which represents a known entity (e.g. "France", "John") or no marker which represents an unknown entity (noted $*$ in the figures). In the figures, the concept nodes are represented as rectangles and the relation nodes as ovals, and the edges linking a relation node r to its neighbor concept nodes are labelled from 1 to the arity of r (i.e. its number of neighbors). Figure 1 shows an example of two simple CGs. Intuitively, G_1 and G_2 can be respectively read as "the human A has as father the Man B who has as mother an unknown woman" and "the animal C belongs to the woman D and it holds something".

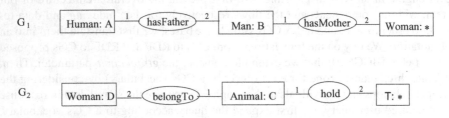

Fig. 1. Two basic conceptual graphs

The labels in the nodes (types and markers) are taken from a vocabulary $\mathcal{V}$ which can be more or less rich. A vocabulary $\mathcal{V}$ is composed of three parts: a specialization order on the concept types, a specialization order on the relation types, and a set of individuals. It defines the skeleton on which the conceptual graphs are built. In this paper we consider a fragment of the *Basic Conceptual Graphs Vocabulary* [CM09], more precisely a triple $(\mathcal{T}_C, \mathcal{T}_R, \mathcal{I})$ where (1) $\mathcal{T}_C$ and $\mathcal{T}_R$ are finite pairwise disjoints sets, (2) $\mathcal{T}_C$, the *concept types hierarchy*, is a set of concept types partially ordered by the relation $\leq$ (e.g. $Man \leq Human$ represents that Man is a specialization of $Human$), (3) $\mathcal{T}_R$, the *relation types (of arity 2) hierarchy*, is a set of relation types partially ordered by the relation $\leq$, (4) $\mathcal{I}$ is the set of *individual markers*. In the following such a triple is simply denoted *CGs vocabulary*. Figure 2 shows an example of a CGs vocabulary (the set of individuals is omitted).

A simple extension of the vocabulary is usually considered in order to restrict the domain (resp. the range) of a relation type to a concept type. We call *signature* of a

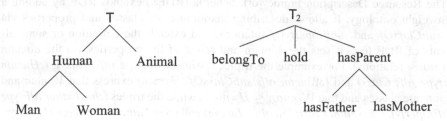

Fig. 2. A CGs vocabulary

relation type r the pair (c_1, c_2) where $c_1 \in \mathcal{T}_C$ and $c_2 \in \mathcal{T}_C$. For example *(Human,Man)* is a possible signature for the relation type *hasFather*.

Finally, we give some notions used in the following: we call *CGs knowledge base (CGs KB)* a pair $(\mathcal{V}, (F_1, \ldots, F_m))$ where $\mathcal{V} = (\mathcal{T}_C, \mathcal{T}_R, \mathcal{I})$ is a CGs vocabulary and $(F_1, \ldots, F_m)$ a list of CGs representing factual knowledge, denoted *CGs facts* (e.g. the pair composed of the vocabulary in Figure 2 and the CGs in Figure 1 forms a CGs KB). Let t be a concept or relation type and $\mathcal{V}$ a vocabulary, we call *descendants* of t w.r.t. $\mathcal{V}$ the types $t_1, \ldots, t_n$ s. t. $t_i \leq t$ for $i = 1 \ldots n$ (e.g. the descendants of $\top$ w.r.t. the CGs vocabulary in Figure 2 are *Human, Man, Woman* and *Animal*). We call *ancestors* of t w.r.t. $\mathcal{V}$ the set $t_1, \ldots, t_n$ s. t. $t \leq t_i$ for $i = 1 \ldots n$.

2.2 RDF/RDFS

The Resource Description Framework (RDF), introduced by the World Wide Web Consortium (W3C), is a metadata model designed as the standard for the semantic web [Hay04]. It enables the encoding, exchange and reuse of structured metadata (also called semantic annotations), as well as metadata interoperability. It is based upon the idea of making statements about resources (in particular Web resources) by a set of triples of the form *(subject, predicate, object)* (denoted *(s,p,o)* in the following). The subject denotes the resource, and the predicate denotes properties of the resource and expresses a relationship between the subject and the object. Figure 3 shows a set of RDF triples that "naturally" corresponds to the CGs in Figure 1, where labels starting with an underscore denote a *blank node*, i.e. an anonymous resource. Note that a set of RDF triples can also be visualized as a graph.

$$<: A > \; < rdf : type > \; <: Human > \quad <: A > \; <: hasFather > \; <: B >$$
$$<: B > \; < rdf : type > \; <: Man > \quad <: B > \; <: hasMother > \; \_b_1$$
$$\_b_1 \; < rdf : type > \; <: Woman >$$

$$<: D > \; < rdf : type > \; <: Woman > \quad <: C > \; <: belongTo > \; <: D >$$
$$<: C > \; < rdf : type > \; <: Animal > \quad <: C > \; <: hold > \; \_b_2$$
$$\_b_2 \; < rdf : type > \; <: \top >$$

Fig. 3. RDF triples corresponding to the example in Figure 1

The Resource Description Framework Schema (RDFS) extends RDF by adding a lightweight ontology. It allows describing taxonomies of classes and properties via the *subClassOf* and *subPropertyOf* relations, and extends the definition of some elements of RDF (e.g. it sets the domain and range of the properties via the *domain* and *range* relations). For example the triples *(:Woman rdf:type rdfs:Class)*, *(:Human rdf:type rdfs:Class)* and *(:Woman rdfs:subClassOf :Human)* express that *Woman* and *Human* are classes and that $Woman \leq Human$, while the triples *(:hasFather rdf:type rdf:Property)*, *(:hasFather rdfs:domain :Human)* and *(:hasFather rdfs:range :Man)* express that *hasFather* is a property with signature *(Human,Man)*.

In addition, a set of RDF triples (a graph) S can be clustered by assigning to each triple $(s, p, o) \in S$ a unique *graph name* (say g), i.e. we extend the triples to *quads* of the form *(s,p,o,g)*. The set of quads sharing the same graph name is called a *named graph*. Such named graphs are particularly useful for managing sets of RDF data within an RDF store (see Section 3.2).

2.3 RDF Store and Indexing Techniques

An RDF store (or triple store) is a system designed for the storage and retrieval of RDF triples. Like a relational database, triples are stored and retrieved via a query language (SQL for relational databases and SPARQL for RDF stores), but in contrast with a relational database, a triple store is optimized for triples and graph-based operations. A triple store usually contains an internal indexing of its content which is user settable, as well as RDF reasoning capabilities. We call *quad store* a triple store with the ability to deal with quads.

Over the last decade, very large triple/quad stores have been developed, and they have proven their efficiency in terms of scalability and retrieval speed (see for example [FCB12] and [GM10]). For example the quad store Allegro Graph (AG), which is equipped with internal indexing of the data and reasoning capabilities, is capable to scale to trillions of RDF triples while still providing outstanding data retrieval performances [Inc10]. The key of AG high performances partly resides in its internal index. For instance if the internal index has been built based on the *graph name*, when a quad $q = (s, p, o, gn)$ is searched, the system will automatically match gn to its internal index and only go through the portion of triples referring to gn.

More generally, in a data storage and retrieval management system the indexing of the data is one of the most critical processes in order to speed up the retrieval. According to the nature of the data (e.g. texts, graphs), many indexing methods have been proposed, and we only describe a few of them in the following. Inverted indexes are one of the most popular data structures used in document retrieval systems [ZMR98][BCC10]. An inverted index stores a mapping from content, such as words or numbers, to its locations in a set of repositories. It is used on a large scale for example in search engines. For graph-based indexing approaches, Yan et al. propose a triple store indexing method based on graph partitioning [YWZ+09]. They consider the triple store as a big graph, partition it into multiple subgraphs and index each subgraph by considering the atomic data (e.g. concept type) that appear in it. The main drawback of this method is that redundant information, more specifically the "cut points", is stored in several subgraphs to ensure the correctness of the method. Picalausa et al. propose to index the triple store according to the structure of its content [PLF+12]. This structural index is essentially a reduced version of the triple store content where nodes have been merged according to some notion of structural similarity. Zhao et al. propose an indexing method based on a specific logic graph structure, a star graph [ZQZ10]. Finally Sakr and Al-Naymat have recently exposed an overview of different techniques for indexing and querying graph databases [SAN11].

As a first version of our CGs storage and retrieval system we use the inverted indexing techniques to capture the appearance of terms (e.g. concept type) in the graphs (see Section 3.2). Other indexing methods would be tackled in the next developments of our system as mentioned in the prospects (see Section 5).

3   Proposed CGs Storage and Retrieval Engine

In this section, we present the overall architecture of our proposed CGs storage and retrieval engine, the methods to store and index a CGs KB, and the implemented data retrieval mechanisms.

3.1   System Architecture

The overall system architecture is shown in Figure 4. It consists of six main components: user interface, CGs to RDF translator, indexer, expander, selector and RDF to CGs translator. A brief description of these components follows:

1. *User Interface (UI)*: the user accesses the system via a UI allowing to store CGs KBs, to retrieve stored CGs via user queries and to display the results.
2. *CGs to RDF Translator*: it takes as input a CGs KB and translates it directly to an RDF KB (e.g. a CGs fact is translated into an RDF named graph) which is stored in an RDF quad store.
3. *Indexer*: it takes an RDF quad store as input and produces an index of the RDF named graphs content.
4. *Expander*: it takes as input a structured user query and expands it according to the vocabulary stored in the RDF quad store, i.e. it adds for each concept or relation type its descendants.
5. *Selector*: it takes as input the expanded user query, and selects a set of candidates RDF named graphs by exploring the created index.
6. *RDF to CGs Translator*: it takes as input a set of candidate RDF named graphs and translate them into CGs facts.

In the following subsections, each component is discussed in detail.

3.2   Storage and Indexing

This subsection describes the mechanisms used to store and index a CGs KB. It consists of translating a CGs KB into an RDF KB which is stored in an RDF quad store (CGs to RDF Translator) and indexing the RDF quad store (Indexer).

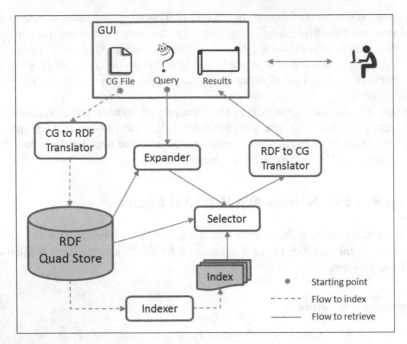

Fig. 4. Overall system architecture of the CGs storage and retrieval engine

Translation from Conceptual Graphs to RDF. The translation from a CGs KB to an RDF KB mostly relies on the work of Baget et al [BCG$^+$10]. In this work, the authors present sound and complete translations from CGs to RDF and from RDF to CGs. The vocabulary and the facts are translated into RDF triples and vice versa without loss of information (e.g. Figures 1 and 3).

We slightly adapt these translations by adding a new parameter, the *graph name*. Thus for example we can gather together all the triples related to the CGs concept types hierarchy into a named graph called "conceptTypeHierarchy". This clustering will speed up other processes (e.g. for the descendants of a concept type, we can quickly go through the portion of triples corresponding to the concept types hierarchy by using the quad store's internal index). In addition the graph name will be used for building an external index of the content of the named graphs for retrieval purpose. More precisely, we extend the translation of Baget et al. from triples to quads as following:

- CGs Vocabulary to RDF Ontology
    - Generate a unique graph identifier for the vocabulary's hierarchies, say *cth* for the concept types hierarchy and *rth* for the relation types hierarchy[2]
    - For all concept types c, add (c, rdf:type, rdfs:Class, *cth*)
    - For all concept types c_1 and c_2 s. t. $c_2 \leq c_1$, add (c_2, rdf:subClassOf, c_1, *cth*)
    - For all concept types c and its label l, add (c, rdfs:label, l, *cth*)

---

[2] Note that a unique graph identifier could be generated for the concept types, the relation types and the relations' signatures.

- For all relation types r, add $(r, \text{rdf:type}, \text{rdf:Property}, rth)$
- For all signatures (t_1, t_2) of a relation type r, add $(r, \text{rdfs:domain}, t_1, rth)$ and $(r, \text{rdfs:range}, t_2, rth)$
- For all relation types r_1 and r_2 s. t. $r_2 \leq r_1$, add $(r_2, \text{rdfs:subPropertyOf}, r_1, rth)$
- For all relation type r and its label l, add $(r, \text{rdfs:label}, l, rth)$
- CGs Fact g to RDF Named Graph
  - Generate a unique graph identifier for g (say $g\text{-}id$)
  - For all generic concept node c, assign to c a new blank node
  - For all individual concept node c, assign to c the URI corresponding to its individual marker
  - For all concept node of type c and its assigned term (blank node or URI) t, add $(t, \text{rdf:type}, c, g\text{-}id)$
  - For all subgraph $r(c_1, c_2)$ induced by a relation node r, add $(c_1, r, c_2, g\text{-}id)$

Figure 5 shows the quads obtained after applying the translation to the CGs in Figure 1.

$<: A > < rdf : type > <: Human > <: g1 >$ $<: A > <: hasFather > <: B > <: g1 >$
$<: B > < rdf : type > <: Man > <: g1 >$ $<: B > <: hasMother > \_b_1 <: g1 >$
$\_b_1 < rdf : type > <: Woman > <: g1 >$

$<: D > < rdf : type > <: Woman > <: g2 >$ $<: C > <: belongTo > <: D > <: g2 >$
$<: C > < rdf : type > <: Animal > <: g2 >$ $<: C > <: hold > \_b_2 <: g2 >$
$\_b_2 < rdf : type > <: \top > <: g2 >$

Fig. 5. RDF quads obtained by translating the BCGs in Figure 1

External Index Construction. The next step consists of building an external index to speed up the retrieval of the stored named graphs. We create three inverted indexes (one for the concept types, one for the relation types and the last for the individuals) to index the content of the named graphs. An inverted index is a two-column table, each row representing a pair *(term,* $\{d_1, \ldots, d_n\}$ *)*, where $\{d_1, \ldots, d_n\}$ is a set of data containing *term*. In the following we only detail the concept types inverted index (CTII). The first column of the CTII consists of concept type labels, and the second column consists of sets of named graph labels. Each row represents a pair (c, gns) where c is a concept type label and $gns = \{gn_1, \ldots, gn_k\}$ is a set of named graph labels in which c appears (note that we do not consider the named graphs associated to the ontological part). In the following an inverted index is viewed as a set of pairs (l, gns), where l is a label and gns a set of named graph labels.

Algorithm 1 performs the construction of the inverted indexes. It takes a set of named graphs $gns = \{gn_1, \ldots, gn_n\}$ as input and produces three inverted indexes namely concept type inverted index (CTII), relation type inverted index (RTII) and individual inverted index (III). The function isolatedNode takes a quad $q = \{s, p, o, gn_i\}$ as input and returns true if (s,p,o) represents an isolated node in the named graph (gn_i), false otherwise.

Table 1 shows the inverted indexes built from the named graphs g_1 and g_2 created from the CGs in Figure 1.

Algorithm 1. Inverted-Index-Builder(gn)

Input: a set of named graphs $gns = \{gn_1, \ldots, gn_n\}$
Result: Three inverted indexes
begin
    Let $CTII$, $RTII$ and III be three empty inverted indexes ;
    foreach *named graph* $gn_i \in gns$ **do**
        foreach *quad* $q = \{s, p, o, gn_i\}$ *s. t.* $p \neq$ rdf:type *OR isolatedNode*(q) **do**
            foreach *term* $t \in \{s, o\}$ **do**
                if t *is an individual* **then**
                    if $(t, gns_1) \in III$ **then**
                        Update III by replacing (t, gns_1) with $(t, gns_1 \cup \{gn_i\})$;
                    else
                        Add $(t, \{gn_i\})$ to III ;
                Let *type(t)* $\leftarrow o_1$ s. t. $q = \{s, rdf{:}type, o_1, g_i\} \in g_i$;
                if (type(t), $gns_2) \in CTII$ **then**
                    Update $CTII$ by replacing *(type(t)*, $gns_2)$ with $(t, gns_2 \cup \{gn_i\})$;
                else
                    Add *(type(t)*, $\{gn_i\})$ to $CTII$;
            if $p \neq$ rdf:type *AND* $(p, gns_3) \in RTII$ **then**
                Update $RTII$ by replacing (p, gns_3) with $(p, gns_3 \cup \{gn_i\})$;
            else
                if $p \neq$ rdf:type **then** Add (p, gn_i) to $RTII$;
    return $CTII$, $RTII$ and III ;
end

3.3 Retrieval Methods

In this section, we present the data retrieval methods which consist of expanding a user query w.r.t. a vocabulary and selecting a set of candidate named graphs (i.e. in which an answer may be found) by matching the expanded query to the inverted indexes. Several heuristics to speed up the expansion of the query and the selection of the candidate named graphs are also outlined.

Query Expansion. Once the knowledge base has been indexed, the end-user can start retrieving its content by sending queries. In this paper we consider positive conjunctive queries, which form a class of natural and frequently used queries and are considered as the basic database queries [CM77], and we see them as a set of positive subgoals $q = \{p_1, \ldots, p_n\}$. Each subgoal is of the form $c(x_1)$ (resp. $r(x_2, x_3)$) where c is a concept type (resp. r is a relation type) and x_1, x_2, x_3 are variables or individuals (in the following the variables start with a lowercase letter and the individuals start with a capital letter). For example the query $q = \{Human(x), Human(B), \top_2(x, B)\}$ asks for all the humans who have any relation with the human B.

Table 1. Inverted indexes built from the named graphs created from the CGs in Figure 1

CTII	Concept	Named_Graphs
	Human	$\{g_1\}$
	Man	$\{g_1\}$
	Woman	$\{g_1, g_2\}$
	Animal	$\{g_2\}$
	T	$\{g_2\}$

RTII	Relation	Named_Graphs	III	Individual	Named_Graphs
	hasFather	$\{g_1\}$		A	$\{g_1\}$
	hasMother	$\{g_1\}$		B	$\{g_1\}$
	belongTo	$\{g_2\}$		C	$\{g_2\}$
	hold	$\{g_2\}$		D	$\{g_2\}$

To take into account the partial order between the concept (resp. relation) types, there are two main methods:

1. Expanding $CTII = \{(c_1, gns_1), \ldots, (c_n, gns_k)\}$ and $RTII$ by propagating the named graphs in which a concept (resp. relation) type appears to its ancestors (e.g. if the concept type Man appears in the named graph gn and $Man \leq Human$ then we add the information "$Human$ appears in gn". The main drawback of this approach is that it is very time consuming since the slightest change in the type hierarchies will force to either re-index the whole knowledge base or process heavy procedures to rearrange the index.

2. Expanding the user query q, i.e. adding the descendants of the concepts and relations appearing in q, thus "melting" the partial orders into q. Algorithm 2 and its subalgorithm 3 give the details of the process. Note that Algorithm 2 takes as input a preprocessed query q, i.e. a list composed of three sublists containing respectively the concept types, relations types and individuals appearing in q. The function **descendants**(t, V) returns all the descendants of t w.r.t. V.

Before giving an example, let us consider the following notions:

Definition 1 (Expanded Type). *Let t be a type (concept or relation) and V a vocabulary. The* expanded type T *of t is the set containing t and the set of its descendants $t_1, \ldots, t_n$ w.r.t. V, i.e. $T = \{t, \{t_1, \ldots, t_n\}\}$ where $t_i \leq t$ for $i = 1 \ldots n$. We call t the* initial type *and $\{t_1, \ldots, t_n\}$ the* extension *of t.*

Definition 2 (Expanded Query). *Let $q = \{c_1, \ldots, c_m, r_1, \ldots, r_n, I_1, \ldots, I_p\}$ be a (preprocessed) query. The* expanded query EQ *of q is obtained from q by replacing each type t appearing in q by its expanded type T.*

Example 1. Let us consider the query $q = \{Human(x), Human(B), \mathsf{T}_2(x, B)\}$. It contains the concept type $Human$, the relation type T_2 and the individual B. We start expanding the concept type $Human$. According to the partial orders in Figure 2, the extension of the concept type $Human$ is composed of the concept types Man and $Woman$.

Algorithm 2. Query-Expander(Q,d)

Input: a query $q = \{c_1, \ldots, c_m, r_1, \ldots, r_n, I_1, \ldots, I_p\}$
Result: the expanded query EQ of Q
begin

    Let $C \leftarrow \emptyset$ and $R \leftarrow \emptyset$;

    foreach *concept type c appearing in q* **do**
        $C \leftarrow C \cup$ **Type-Expander**(c) ;

    foreach *relation type r appearing in q* **do**
        $R \leftarrow R \cup$ **Type-Expander**(r) ;

    Let $EQ \leftarrow \{\{C\}, \{R\}, \{I_1, \ldots, I_p\}\}$;

    return EQ ;

end

Algorithm 3. Type-Expander(t)

Input: a concept type or relation type t
Data: a vocabulary V
Result: t and all its descendants
begin

    $D \leftarrow$**descendants**(t,V) ;

    if D *is empty* **then return** t ;

    $all - descendants \leftarrow \emptyset$;

    foreach $t_i \in D$ **do**
        $all\text{-}descendants \leftarrow all\text{-}descendants \cup$ **Type-Expander**(t_i) ;

    return all-descendants ;

end

Thus we create the expanded concept $C = \{Human, \{Man, Woman\}\}$. We repeat the process for $\top_2$ and we obtain the expanded relation
$R = \{\top_2, \{belongTo, hold, hasParent, hasFather, hasMother\}\}$. Finally we obtain the expanded query $EQ = \{\{C\}, \{R\}, \{B\}\}$.

Candidate Graphs. The next step consists of "matching" the expanded query EQ to the inverted indexes in order to retrieve a set of candidate named graphs, i.e. in which an answer to the initial query may be found. It is roughly done through two main steps: (1) let t be a concept type or a relation type (resp. an individual) in EQ and $t_1, \ldots, t_k$ its extension. For each element $e \in \{t\} \cup \{t_1, \ldots, t_k\}$ (resp. $e \in \{t\}$), we retrieve from the index its assigned set of named graphs gns (empty if e does not appear in the index). Then we compute the union of all obtained sets of named graphs. We repeat this process for all concept types, relation types and individuals appearing in EQ and eventually we obtain n unions of sets of named graphs (say $U_1, \ldots, U_n$); (2) if there exists an empty $U_i \in \{U_1, \ldots, U_n\}$ we return the empty set (there are no candidate named graphs in which an answer to the initial query may be found, intuitively because at least an element of the query can not be matched to any named graph). Otherwise we return the set of candidate named graphs obtained from the intersection $\bigcap_{i=1}^{n} U_i$.

Example 2. Let us consider the expanded query $EQ = \{\{Human, \{Man, Woman\}\},$ $\{\top_2, \{belongTo, hold, hasParent, hasFather, hasMother\}\}, \{B\}\}$. The matching of each element of EQ to the indexes is given in Table 2. From the expanded concept type we obtain $U_1 = \{g_1\} \cup \{g_1\} \cup \{g_1, g_2\} = \{g_1, g_2\}$, from the expanded relation type we obtain $U_2 = \{g_1, g_2\}$ and from the individual B we obtain $U_3 = \{g_1\}$. Finally the set of candidate named graphs is $gns = U_1 \cap U_2 \cap U_3 = \{g_1\}$.

Table 2. Elements and the results of their matching to the inverted indexes

			$\top_2$	$\emptyset$		
Human	$\{g_1\}$		belongTo	$\{g_2\}$		
Man	$\{g_1\}$		hold	$\{g_2\}$		
Woman	$\{g_1, g_2\}$		hasParent	$\emptyset$	B	$\{g_1\}$
			hasMother	$\{g_1\}$		
			hasFather	$\{g_1\}$		

RDF to CGs Translator. This component takes as input a set of named graphs and translate them into CGs. This translation also relies on the work done by Baget et al. [BCG+10], excepted the manipulation of quads instead of triples. Then the candidate CGs are returned to the user.

3.4 Heuristics

Finally we discuss several preprocess functions and heuristics which aim to meet end-user requirements (e.g. response time) and speed up the retrieval processes. They concern the query expansion and the candidate named graphs retrieval.

Type pruning. Let us consider the query containing the concept types $c_1 = Human$ and $c_2 = Woman$. According to the concept type hierarchy in Figure 2, their extensions are respectively $ext_1 = \{Man, Woman\}$ and $ext_2 = \emptyset$. It is straightforward to see that the set of candidate named graphs retrieved by considering the concept types *Human,Man,Woman* (say U_1) will include the candidate named graphs (say U_2) retrieved by considering the concept type *Woman*. As $U_2 \subseteq U_1$, the result of the intersection function $U_1 \cap U_2$ is U_2. U_1 is useless and thus c_1 can be avoided without loss of information. More generally, let q be a query, for each concept (resp. relation) type t appearing in q, if a more specific concept (resp. relation) appears in Q then t can be avoided in the retrieval processes.

Ordering of EQ elements. Notice that for each element of EQ if the computed union (for expanded types) is empty then we can directly return the empty set, i.e. no candidate named graphs fulfill the requirements of EQ. Note also that the intersection function between two sets U_1 and U_2 is faster if one set is very small. Therefore in order to prune as fast as possible an EQ (i.e. returning the empty set) or dealing as soon as possible with a small set of candidate named graphs, the proposed heuristic consists of ordering the treatment of the elements of EQ. The most "critical" elements would be first analyzed. Individuals and the most specific concept types and relation types seem the most critical elements, since it would be more difficult to find many named graphs containing them than for example for a very general concept type with a big extension.

Type expansion depth. An option to speed up the retrieval process would be to restrict the query expansion depth to a predefined constant. However in spite of the fact that a more shallow expansion will increase the process speed, it will also imply a possible loss of information since avoided specialized types could have led to an answer to the query.

4   Experiments

In this section, we show a preliminary experimental work where we benchmark the two heuristics *Type Pruning* and *Ordering EQ* described in Section 3.4.

4.1   Initial Setup

We consider a CGs KB which comprises a total of 81777 concepts, 45 relations, 23282 individuals and 1000 CGs facts of different sizes (from few different relation nodes to more than 30). It has been automatically generated from a text document using an intern text to CGs tool. The translation of the knowledge base into AllegroGraph quad store took 18 seconds and the obtained quad store was set with an internal indexing on graph names. The creation of the three inverted indexes took less than 5 minutes, and the indexes were stored in three simple relational tables. In addition, we have randomly generated a total of 3000 queries of different sizes. Table 3 shows the details of the three generated query sets Q_1, Q_2 and Q_3, each set containing 1000 random queries.

Table 3. Details of the generated query sets

Query Set	Queries Content	Queries Size
Q_1	2 concept types, 1 relation type, 0 or 1 individual	3
Q_2	3 or less concept types, 2 relation types, 1 individual	5
Q_3	4 or less concept types, 3 relation types, 1 or 2 individuals	7

We measured the difficulty in terms of CPU time to expand the query, search the index and retrieve the candidate named graphs. For each value of the varying parameter (i.e. the query size), we considered 1000 queries and computed the mean search cost of the results (i.e. the CPU time) on these queries. The program is written in Java. The experiments were performed on a Dell Latitude E6410, equipped with a 2.6 GHz Dual-Core CPU and 4G of RAM, under Windows.

4.2   Experimental Results

Figure 6 depicts the results obtained using four different heuristics (*no heuristic, types pruning, ordering* and *types pruning+ordering*). As expected the increasing of the complexity of the query (its size) increases the response time as well, since more processes have to be done to expand, select and retrieve the candidate named graphs.

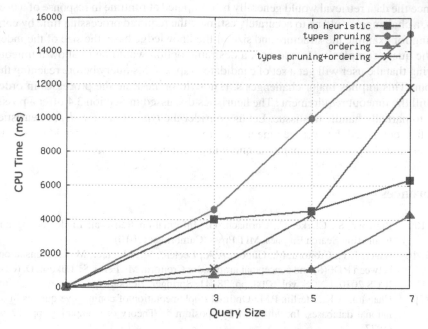

Fig. 6. Influence of the heuristics to the execution time

Moreover we observe that the *types pruning* heuristic produces the worst results. This is caused by the large concept types hierarchy, since the cost of checking if a concept type is a subtype of another depends of the concept types hierarchy size. Indeed, the difference is more obvious when there are more concept types in the query. For example when *query size = 7*, it takes 15 seconds for the retrieval process. In contrast, the *ordering* heuristic has the best response time for all the query sets. Relatively for all queries, the retrieval process will take less than 2 seconds.

Finally, note that these are only preliminary results; further experiments are needed in refining the results to provide a more comprehensive benchmarking evaluation.

5   Conclusion

In this paper, we propose a CGs storage and retrieval system architecture. On the one hand we translate a CGs KB into an RDF KB which is stored in an RDF quad store and create an external index of the CGs facts content, and on the other hand we retrieve a set of candidates CGs facts in response to a query by expanding the query and filtering the CGs facts based on several heuristics. Finally we provide a preliminary evaluation of the system.

A direct perspective would be to extend the experiments to different datasets in order to validate the preliminary results, to propose and validate new heuristics and finally to compare our system with existing storage and retrieval systems.

Since the data retrieval would generally be computed at runtime in response of a user query, a challenge would be to accurately estimate the required processing time by considering for example the structure and size of the knowledge base, the size of the index and the query complexity. In addition, a desirable option would be to allow a timeout ensuring that the user will get a set of candidate graphs to his query before reaching the timeout. This implies major challenges s. t. how to orchestrate the processes in order to fulfill the timeout requirement? The heuristics discussed in Section 3.4 offer a possibility to partially handle this issue. Let us consider the *type expansion depth* heuristic: if we first computed the average time needed to expand one level, we would be able to estimate how deep we can expand the query in a certain time-frame.

References

[BCC10]   Büttcher, S., Clarke, C., Cormack, G.: Information Retrieval: Implementing and Evaluating Search Engines. MIT Press, Cambridge (2010)

[BCG$^+$10]   Baget, J.-F., Croitoru, M., Gutierrez, A., Leclère, M., Mugnier, M.-L.: Translations between RDF(S) and conceptual graphs. In: Croitoru, M., Ferré, S., Lukose, D. (eds.) ICCS 2010. LNCS, vol. 6208, pp. 28–41. Springer, Heidelberg (2010)

[CM77]   Chandra, A.K., Merlin, P.M.: Optimal implementation of conjunctive queries in relational databases. In: 9th ACM Symposium on Theory of Computing, pp. 77–90 (1977)

[CM09]   Chein, M., Mugnier, M.-L.: Graph-based Knowledge Representation and Reasoning—Computational Foundations of Conceptual Graphs. Advanced Information and Knowledge Processing. Springer (2009)

[coga]   The cogitant library

[cogb]   Cogui: a graph-based visual tool for conceptual graphs

[dBC11]   da Silva, B., Baget, J.-F., Croitoru, M.: Ontological conjunctive query answering over semi-structured kbs. In: Proceedings of the 2011 IEEE 27th International Conference on Data Engineering Workshops, pp. 118–123. IEEE Computer Society (2011)

[DKC05]   Dieng-Kuntz, R., Corby, O.: Conceptual Graphs for Semantic Web Applications. In: Dau, F., Mugnier, M.-L., Stumme, G. (eds.) ICCS 2005. LNCS (LNAI), vol. 3596, pp. 19–50. Springer, Heidelberg (2005)

[EL92]   Ellis, G., Levinson, R.: The Birth of Peirce: A Conceptual Graphs Workbench. In: Pfeiffer, H.D., Nagle, T.E. (eds.) Conceptual Structures: Theory and Implementation. LNCS, vol. 754, pp. 219–228. Springer, Heidelberg (1993)

[FCB12]   Faye, D., Cure, O., Blin, G.: A survey of rdf storage approaches (2012)

[GM10]   Gaignard, A., Montagnat, J.: Survey on semantic data stores and reasoning engines. Research Report (2010)

[GS98]   Genest, D., Salvat, É.: A Platform Allowing Typed Nested Graphs: How CoGITo Became CoGITaNT. In: Mugnier, M.-L., Chein, M. (eds.) ICCS 1998. LNCS (LNAI), vol. 1453, pp. 154–164. Springer, Heidelberg (1998)

[Hay04]   Hayes, P.: Rdf semantics (2004)

[Inc10]   Franz Inc. Allegrograph rdfstore version 3.3 lubm benchmark results (2010)

[PLF$^+$12]   Picalausa, F., Luo, Y., Fletcher, G.H.L., Hidders, J., Vansummeren, S.: A Structural Approach to Indexing Triples. In: Simperl, E., Cimiano, P., Polleres, A., Corcho, O., Presutti, V. (eds.) ESWC 2012. LNCS, vol. 7295, pp. 406–421. Springer, Heidelberg (2012)

[RKH]   Rudolph, S., Krötzsch, M., Hitzler, P.: Quo vadis, cs?— on the (non)-impact of conceptual structures on the semantic web (position paper)

[SAN11]   Sakr, S., Al-Naymat, G.: An overview of graph indexing and querying techniques. In: Sakr, S., Pardede, E. (eds.) Graph Data Management, pp. 71–88. IGI Global (2011)

[Sow76]   Sowa, J.F.: Conceptual graphs for a data base interface. IBM J. Res. Dev. 20(4), 336–357 (1976)

[Sow84]   Sowa, J.F.: Conceptual Structures: Information Processing in Mind and Machine. Addison-Wesley (1984)

[YWZ+09] Yan, Y., Wang, C., Zhou, A., Qian, W., Ma, L., Pan, Y.: Efficient indices using graph partitioning in rdf triple stores. In: Proceedings of the 2009 IEEE International Conference on Data Engineering, pp. 1263–1266. IEEE Computer Society (2009)

[ZMR98]   Zobel, J., Moffat, A., Ramamohanarao, K.: Inverted files versus signature files for text indexing. ACM Trans. Database Syst. 23(4), 453–490 (1998)

[ZQZ10]   Zhao, B., Qian, W., Zhou, A.: Towards bipartite graph data management. In: Proceedings of the Second International Workshop on Cloud Data Management, pp. 55–62. ACM (2010)

Medical Archetypes and Information Extraction Templates in Automatic Processing of Clinical Narratives

Ivelina Nikolova[1], Galia Angelova[1],
Dimitar Tcharaktchiev[2], and Svetla Boytcheva[3]

[1] Institute of Information and Communication Technology,
Bulgarian Academy of Sciences, Sofia, Bulgaria
{iva,galia}@lml.bas.bg
[2] University Specialised Hospital for Active Treatment of Endocrinology,
Medical University Sofia, Bulgaria
dimitardt@gmail.com
[3] American University in Bulgaria, Blagoevgrad, Bulgaria
svetla.boytcheva@gmail.com

Abstract. This paper discusses the notion of medical archetype and the manner how the archetype elements are documented in hospital patient records. This is done by interpreting the archetypes as information extraction templates in automatic text analysis of clinical narratives. The extensive extraction experiments performed over thousands of anonymous discharge letters show the actual instantiation of the required and expected items in the narrative clinical documentation; in fact much tacit medical knowledge is implicitly presented in the real clinical texts. This fact suggests that the archetype approach to defaults and inheritance might need certain development.

Keywords: Clinical knowledge, Medical archetypes, NLP of clinical narratives, Information extraction, Template filling.

1 Introduction

Archetypes are chunks of declarative medical knowledge that are designed to capture maximally expressive and internationally reusable clinical information units. They encode knowledge about clinical observations, evaluations, actions and instructions in a coherent and holistic manner with the intension to present language-independent specifications. Archetypes are based on conceptual structures of medical knowledge and provide standardised clinical content. Medical ontologies conceptualise domain objects, actions and relationships among them; the archetypes, representing the blueprints of defined medical domains, are focused on capturing clinical information about the patient. Archetypes are not linked a priory to any medical terminology but they can refer to multiple external medical classifications (e.g. SNOMED) from where controlled vocabularies

H.D. Pfeiffer et al. (Eds.): ICCS 2013, LNAI 7735, pp. 106–120, 2013.

are incorporated as labels of archetype elements. The *open*EHR project[1] aims at the acquisition of a representative set of freely available archetypes thus enabling information sharing between clinical systems. Hundreds of archetypes in ADL (Archetype Definition Language) are publicly available via the *open*EHR Clinical Knowledge Manager. *open*EHR expresses health information systems and interoperability mechanisms in UML (Unified Modelling Language).

Automatic processing of free clinical texts, however, might reveal whether medical experts keep the requirements to document clinical units in a manner which ensures their unambiguous export to other clinical systems. Analysing the free text of 6,204 anonymous discharge letters of diabetic patients, we present empirical observations whether the slots of the diabetic-relevant archetypes, published by *open*EHR, are filled in by the necessary information of classification codes or free text. In a sense we discuss how the theoretical models of clinical knowledge are applied in practical settings when the medical case is documented.

The article is structured as follows. Section 2 overviews the notion of archetypes. Section 3 discusses the archetypes as Information Extraction templates applied in automatic text processing. Section 4 presents the experiments performed on a large corpus of discharge letters. Section 5 contains some discussion and the conclusion.

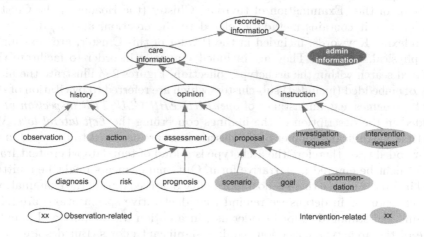

Fig. 1. The Clinical Investigator Record Ontology [1]

2   Archetypes as Conceptual Structures

Archetypes are designed during the last decade to make health information systems properly and safety interoperable [1]. They are based on the notion of "recording" in medicine. The health record content is likely "to be a small, selective choice of notes about real events, situations etc. intended for interpretation by other professionals rather than some more general notion of comprehensive fact representation". Analysing the important types of information in

---

[1] http://openehr.org, the openEHR Foundation.

the health care process, the authors propose the Clinical Investigator Record Ontology where the observations (evidences) and opinions (inferences) are different categories as shown on Figure 1. This taxonomy provides the categories in the Entry classes of the *open*EHR reference model. For our purposes we shall be interested in the archetypes capturing the observations (findings of examinations, measurement, questioning, or testing of the patient or related substance like blood, tissue etc.), because automatic information extraction from clinical narratives is most successful for declarative statements.

An archetype is a "computable expression of a domain content model in the form of structured constraint statements, based on a reference (information) model" [2]. Archetypes define conceptual items and relationships among them as well as constraints on the values of their instances: e.g. allowed types, ordering, cardinality, (referent) values etc. We are interested mostly in the conceptual background of the clinical archetype model which is defined together with the *open*EHR software development requirements.

The Clinical Knowledge Manager supports two major kinds of archetypes: the Electronic Health Record (EHR) Archetypes where patient-centered data is kept and the Demographic Model Archetypes. The EHR Archetypes are *Clusters*, *Compositions*, *Elements*, *Entries*, *Sections* and *Structures*. Figures 2–3 show the Items of the "Examination of thyroid" Cluster (the Header of the Cluster is skipped as it contains metadata related to the creation, author, date etc.) The indexing Keywords, included in the Header of this Cluster, are "examination, physical, thyroid". They are included manually in order to facilitate the advanced search within the archetype collection. Figures 2–3 illustrate the hierarchy of embedded (included) sub-clusters which are referred to by citation of the archetype names e.g. an instance of *openEHR-EHR-CLUSTER.inspection.v1* is included in the description of the findings concerning the *Left lateral lobe*. We note that a significant number of the descriptions are assumed to be typed in as free or coded text, therefore the archetype is a kind of template where text fragments might be entered in narrative form. Optional elements might be omitted in the instances in case there is no abnormality observed during the examination. Without entering in details we remind that declarative specifications are hard to define and standardize for broader use; in addition the support and maintenance of the archetype collection requires significant efforts. But despite these shortcomings it is clear that the Archetype model responds to the needs of establishing standards in the EHR content (and the clinical documentation practice in general) in order to ensure semantic interoperability between the healthcare systems.

In 2008 the archetype approach to structuring patient-related records was accepted as ISO standard 13606-2:2008. It specifies the information architecture required for interoperable communications between systems and services dealing with EHR data [3]. In this way ISO 13606-2:2008 defines how to organise hierarchically the EHR content, how to define the individual data items and their aggregations, what types of values or measurement units are appropriate and so on. Archetypes are viewed as a serialised representation, an exchange format

Structure: Cluster Occurrences: 1..1 (mandatory) Cardinality: 1..* (*mandatory, repeating, unordered*)		
Normal statements Cluster Occurrences: 0..1 (optional) Cardinality: 1..* (*mandatory, repeating, unordered*)	A group of statements about the normality of the examination.	
Normal statement Text, Occurrences: 0..* (optional, repeating)	A specific statement of normality.	Free or coded text
Clinical description Text, Occurrences: 0..1 (optional)	Textural description of the part examined.	Free or coded text
Findings Cluster Occurrences: 0..1 (optional) Cardinality: 1..* (*mandatory, repeating, unordered*)	Clinical findings.	
Visible abnormality Boolean Occurrences: 0..1 (optional)	There is a visible thyroid abnormality.	
Mobility on swallowing liquid Text, Occurrences: 0..1 (optional)	Description of thyroid mobility on swallowing liquid.	Free or coded text
Left lateral lobe Cluster Occurrences: 0..1 (optional) Cardinality: 1..* (*mandatory, repeating, unordered*)	Findings of left lobe of thyroid.	
Description Text, Occurrences: 0..1 (optional)	Text description of clinical findings.	Free or coded text
Left lateral lobe SLOT (Cluster) Occurrences: 0..1 (optional)	Detailed findings of left lobe of thyroid.	Include: openEHR-EHR-CLUSTER.inspection.v1 and specialisations *Or* openEHR-EHR-CLUSTER.exam-generic.v1 *Or* openEHR-EHR-CLUSTER.palpation.v1

Fig. 2. Items of the CLUSTER *"Examination of thyroid"*

Right lateral lobe Cluster Occurrences: 0..1 (optional) Cardinality: 1..* (*mandatory, repeating, unordered*)	Findings of right lobe of thyroid.	
T Description Text, Occurrences: 0..1 (optional)	Text description of clinical findings.	Free or coded text
Right lateral lobe SLOT (Cluster) Occurrences: 0..1 (optional)	Detailed findings of right lobe of thyroid.	**Include:** openEHR-EHR-CLUSTER.inspection.v1 and specializations *Or* openEHR-EHR-CLUSTER.exam-generic.v1 *Or* openEHR-EHR-CLUSTER.palpation.v1
Isthmus Cluster Occurrences: 0..1 (optional) Cardinality: 1..* (*mandatory, repeating, unordered*)	Findings of isthmus of thyroid.	
T Description Text, Occurrences: 0..1 (optional)	Text description of clinical findings.	Free or coded text
Isthmus SLOT (Cluster) Occurrences: 0..1 (optional)	Findings of isthmus of thyroid.	**Include:** openEHR-EHR-CLUSTER.inspection.v1 and specialisations *Or* openEHR-EHR-CLUSTER.exam-generic.v1 *Or* openEHR-EHR-CLUSTER.palpation.v1
Detail SLOT (Cluster) Occurrences: 0..* (optional, repeating)	More focused examination findings	**Include:** openEHR-EHR-CLUSTER.exam-generic.v1 and specialisations *Or* openEHR-EHR-CLUSTER.auscultation.v1 *Or* openEHR-EHR-CLUSTER.inspection.v1 *Or* openEHR-EHR-CLUSTER.palpation.v1 *Or* openEHR-EHR-CLUSTER.percussion.v1 *Or* openEHR-EHR-CLUSTER.physical_properties.v1
Image Multimedia Occurrences: 0..* (optional, repeating)	Drawing or image of the area examined.	image/gif, image/png, image/jpeg

Fig. 3. Items of the CLUSTER *"Examination of thyroid"* (Continued)

for communicating individual archetypes between archetype libraries. Current efforts of the *open*EHR-related community are dedicated to the definition of further archetypes at the optimal level of granularity and specificity in order to ensure their wide adoption. In this way more medical experts could be involved in the creation of archetype repositories. Best practices are sought to achieve multi-professional clinical consensus. Having in mind all the recent developments, we think that Natural Language Processing (NLP) of clinical narratives can help much in the tests whether archetypes are properly defined. The automatic text analysis might reveal the actual status of clinical event documentation and suggest potential drawbacks in the archetype definition. This paper presents such tests for some essential archetypes, related to diabetic patients.

Authoring and review of archetypes is viewed as a knowledge acquisition task with highest priority. An Archetype Editorial Group has been established as an expert clinical team to lead the authoring of archetypes within the *open*EHR community. The national eHealth programs in several countries (Australia, Denmark, Singapore, Sweden, and UK) include archetype-related initiatives in order to involve medical professionals, agencies and educational institutions into development activities. International agreements should be sought by international authorities (like the World Health Organisation and relevant standardisation bodies). Actually the unification of clinical narrative content is a long process which is still in its infantry. Nevertheless it is important that this process has started and an ISO standard has been adopted.

At the end of this section we present the data fields included in two other archetypes:

(*i*) *Blood pressure* (openEHR-EHR-OBSERVATION.blood_pressure.v1) and
(*ii*) *Body weight* (openEHR-EHR-OBSERVATION.body_weight.v1).

Extracting automatically these items from the discharge letters of diabetic patients we can check their availability and actual use in the clinical documentation.

3   Information Extraction Templates

Information Extraction (IE) is a popular technique for Natural Language Processing (NLP) which aims at partial text understanding in order to provide fast and efficient analysis of texts in specialised domains. The IE systems identify specific events or topics, searching for relevant information only and disregarding the remaining text fragments. IE typically extracts named entities and words referring to objects or events in order to recognise their roles in event descriptions. The identification is supported by the so called *templates* feature-value structures that capture the entities recognised by the text analysers. Most generally, the IE success is measured by the accuracy of filling in the template slots by proper words encountered in the text.

Table 1. Entities included in the *Blood Pressure* (BP) archetype

Entity name	Content	Value
Systolic	Peak systemic arterial BP	Units: mm[Hg]
Diastolic	Minimum systemic arterial BP	Units: mm[Hg]
Mean arterial pressure MAP	Average arterial pressure	Units: mm[Hg]
Pulse pressure	Difference between the systolic and diastolic pressure	Units: mm[Hg]
Comment	Comment about the measurement	Free or coded text
Position	Description	Standing; Sitting; Reclining; Lying; Lying with tilt to left
Confounding factors	Free or coded text: factors that may impact the measurement	For instance: level of anxiety; pain or fever
Exertion	Details about physical activity undertaken at the time of measurement	Includes openEHR-EHR-CLUSTER.level_of_exertion.v1 and specialisations
Sleep status	Supports interpretation of 24-hours BP measurement	Alert & Awake; Sleeping
Tilt	Surface craniocaudal tilt	Angle, plane, degrees
Cuff size	The size of the cuff used for the measurement	Adult thigh; Large adult; Adult; Small adult; Paediatric/Child; Infant; Neonatal
Location /cluster		
Location of measurement	Body site where BP is recorded	Right arm; Left arm; Left thigh; Right wrist; Left wrist; Right ankle; Left ankle; Finger; Toe; Intra-arterial
Specific location	Specific details about the site where the BP is recorded	Free or coded text
Method	Method of measurement	Auscultation; Palpation; Machine; Invasive
Mean arterial pressure formula	Formula used to calculate MAP	Free or coded text
Diastolic endpoint	Which Korotkoff sound is used	Phase IV; Phase V
Device	Details about the device used to measure BP	Includes openEHR-HER-CLUSTER.device.v1 and specialisations
Event	Description	Any relevant event
24 hour average	Estimate of the average BP	Math function Mean

Early IE papers consider the template design as an essential step in the IE system development. Templates are flat or object-oriented [4] and their design should satisfy a number of requirements:

- *descriptive adequacy* - the template should represent all the information necessary for the task at hand, having in mind that adding features often requires to add further features;

Table 2. Entities included in the *Body Weight* archetype, which is indexed by the keywords *weight, gain, loss, increase, decrease, mass, estimate, actual*

Entity name	Content	Value
Weight, quantity	Weight mass	Units: kg, lb
Comment	Comment about the measurement of weight	Free or coded text
State of dress	Description	Lightly clothed/Underwear; Naked; Fully clothed including shoes; Nappy/diaper
Confounding factors	Free or coded text: factors that may impact the measurement	For instance: timing of menstrual cycle, timing of recent bowel motion, noting of amputation
Device	Details about the weighing device	Includes openEHR-EHR-CLUSTER.device.v1 and specialisations
Event	Description	Any relevant event

- *clarity* - the ability to represent all the information in the template unambiguously;
- *determinacy* - there should be only one way of representing a given item or a complex of items;
- *perspicuity* - the degree to which the design is conceptually clear to the human analyst who will input or edit information in the template or work with the results;
- *monotonicity* - the template should reflect the data content monotonically or incrementally (adding a new value should not cause update, restructuring or removal of the values in other template slots);
- *application considerations* - the particular task might impose constraints e.g. evaluation metrics and further limitations; reusability the template objects should be potentially reusable in other domains and applications.

It is easy to see the similarities between the definition of *template* (a chunk of declarative knowledge automatically extracted from text) and *archetype* (an ultimate, universal chunk of clinical knowledge, to be declared manually and used as standard aggregation of atomic elements). Without loss of generality we can consider the attributes, listed at Figures 2–3 and Tables 1–2, as prototypical elements of flat templates to be used in IE from clinical texts. It is obvious that simple conceptual graphs [5] can capture the semantics of the feature-value pairs in Figures 2–3 and Tables 1–2. In the next section we shall present the results of IE experiments using the archetypes listed above.

It should be added that the notion of template evolves in the NLP field; recent papers suggest learning template structure automatically from raw text without using predefined template schemes [6].

4   Extracting Archetype Items from Clinical Texts

Here we report the results of experiments with 6,204 anonymised patient records (PRs) of diabetic patient and assessment whether the archetype elements are explicitly documented or not. Our attention is focused on the three archetypes that have been previously discussed: *examination of thyroid*, measurement of *blood pressure* (BP) and measurement of patient *body weight*. The experiments are performed using an IE environment that has been recently developed by the authors [7], [8].

4.1   Examination of Thyroid

More than 97% of the PRs in our corpus contain explicit descriptions of thyroid examination. Many PRs contain more than one discussion of thyroid because they include basic description in the Status section and more detailed tests (like echography) in the Clinical tests and/or Consultations sections. Due to this reason some 11,606 instances of the archetype are found in 6,058 PRs (see Table 3).

Table 3. Availability of *thyroid descriptions* in 6,204 discharge letters

Total PRs	6,204
PRs with no explicit data for thyroid	146
PRs containing description of thyroid	6,058
Total extracted records for thyroid	11,606

Table 4. Availability of *thyroid descriptions* in 6,204 discharge letters

	Visible abnormality	1,556
Items/Findings	Mobility of swallowing liquid	1,892
	Left lateral lobe	1,836
	Right lateral lobe	2,304
	Isthmus	1,846
Items/Normal statements	Normal statement	5,144

Our IE components identified text fragments describing certain abnormalities, the left/right thyroid lobe, the mobility of the swallowing liquids and the isthmus (see Table 4). More than 82% of the PRs (5,144 out of 6,204) contain a statement about normality which can be positive or negative. Comparing the available descriptions to the map view of the archetype in Figure 4 we see that almost all data items are regularly filled in.

4.2   Measurement of Blood Pressure

About 78% of the PRs in our corpus contain explicit BP values. Table 5 illustrates the findings. In the 2,111 PRs without explicit values, there could be

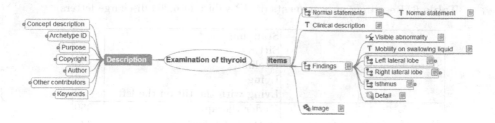

Fig. 4. Map View of the *"Examination of thyroid"* archetype

phrases referring to normal and default values like: *"Blood pressure in the norm"*, *"No data/signals for Arterial Hypertonic illness"* and so on. Some PRs contain more than one occurrence of BP values and this explains the fact that 4,841 items were extracted from 4,093 PRs.

Table 5. Availability of BP descriptions in 6,204 discharge letters

Total PRs	6,204
PRs with no explicit data about BP	2,111
PRs containing data about BP	4,093
Total extracted records about BP	4,841

Further details about available descriptions are given in Table 6. Only 47 PRs discuss the position when the BP measurement is performed (less than 0,01% of all PRs). About 12,6% of the PRs discuss confounding factors. Both systolic and diastolic values are given in the 4,841 particular measurements cited in the corpus. Some 8% of the PRs discuss the mean arterial BP. Pulse pressure occurs in 57% of the analysed discharge letters. The abbreviation (RR) in Protocol/Method denotes BP measurements taken with the technique of the sphygmomanometer invented by Scipione Riva-Rocci. It occurs in 26,6% of all PRs.

Comparing the extracted values to the map view in Figure 5, we see the elements that are rarely instantiated: most items in *State* section (position, exertion, sleeping status, tilt) and in *Protocol* section (cuff size, location of measurement, method, mean arterial pressure formula, diastolic endpoint).

4.3   Measurement of Body Weight

The absolute value of body weight is a factor when diagnosing with diabetes but even more important is the deviation from the patients ordinary body weight. For the professional it is necessary to know whether the patient has experienced any significant change in the weight during the recent months or year(s). Along with the thyroid gland, limbs and skin description, body weight change is one of

Table 6. Recording measurements of BP values in 6,204 discharge letters

State/ Position	Standing	25
	Sitting	3
	Reclining	0
	Lying	19
	Lying with the tilt on the left	0
State/Confounding factors	Under therapy	350
	Without Orthostatic Symptoms	428
	With Orthostatic Symptoms	6
Data/ Systolic - Diastolic		4,841 - 4,841
Data/ Mean Arterial Pressure	Usually/Average	501
	Max	456
	Min	150
Data/ Pulse Pressure		3,566
Protocol/ Method	RR	1,834

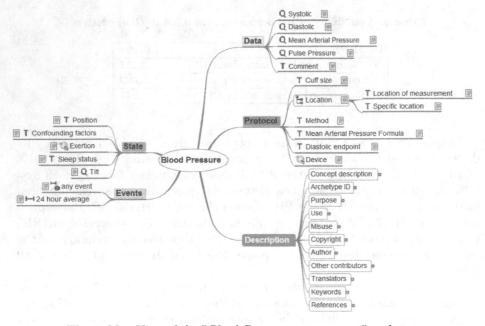

Fig. 5. Map View of the *"Blood Pressure measurement"* archetype

the most often met PR descriptions. Table 7 summarizes the number of events extracted from the patient records.

It is obvious from Table 7 that descriptions of increase or decrease of body weight are almost twice the mentions of exact weight in our document collection. Often when the weight is discussed in a PR, it is mentioned more than once describing the changes during the development of the disease and this also explains why the percentage of files containing exact weight mentions (62%) is quite close

Table 7. Available values of "*weight*" and "*weight change*" in 6,204 discharge letters

Total PRs	**6,204**
PRs containing data about exact weight	3,820
Total extracted occurrences of exact weight	3,884
PRs containing data about weight change	3,097
Total extracted occurrences of weight change	6,806
PRs containing data about increase of weight	2,613
Total extracted occurrences of increase of weight	5,533
PRs containing data about decrease of weight	1,083
Total extracted occurrences of decrease of weight	1,273

to the percentage of PRs containing weight change (52%). Mentions of increased body weight are almost 3 times more often than mentions of decreased weight.

Most weight-related expressions include references to quantities:

(*i*) body weight change which can be found in the *Anamnesis* or *Patient status* section and is expressed as an interval value, exact value or by an expression, all of them showing the *direction of the change*:

"*increased her body weight with about 10-12 kg in the last 6 months*"
"*reduction of body weight 15 kg for 2 years*"
"*overweight*"

(*ii*) *exact weight values* which can be found in the *Laboratory tests* section:

"*weight - 89 kg*"
"*170/86kg*"

(*iii*) *relative expressions* referring to previous conditions like:

"*succeeded to go back to his regular weight*"

which are hard to interpret in absolute values and to fill in into archetype slots.

Our corpus contains no weight-related expressions that can provide input for the archetype slots *state of dress, confounding factors, device*, and *event* (see slots at Figure 6). Obviously these are not a subject of interest in endocrinology.

4.4   Extraction Accuracy and Discussion

Our IE components work in the following manner:

- The English terms, available in Figures 2–3 and Tables 1–2, are translated to Bulgarian;

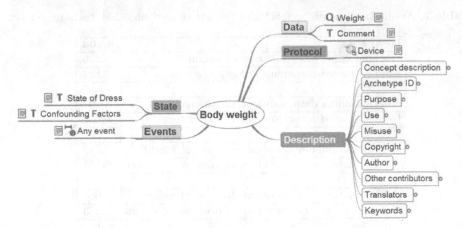

Fig. 6. Map View of the *"Body weight"* archetype

- Their synonyms (terms or paraphrases) are found in the dictionaries that we have developed in our previous research;
- Then the target terms for the selected archetypes are searched in the texts of the corpus PRs.

In this way we identify availability and type of the recognised descriptions. There might be other items, expressed by different words that remain unidentified; however, the observations centered on the terms mentioned in Figures 2–3 and Tables 1–2, deliver a relevant generalised view about text content.

Here are some examples how we capture thyroid gland descriptions in our data starting from the archetype description. The recognition modules are rule-based and are built on archetype keywords and slot descriptions. We know from previous experiments that the description of the Status of an anatomical organ is normally present in a single sentence or in consequent sentences in various. The rules are constructed to capture expressions starting from one mention of the anatomical part of interest (*thyroid gland* in this case) and try to find subsequent descriptions of the archetype slots. Below are given examples that include description of the *left lobe* and *thyroid gland properties*, which are listed one after another and separated from the anatomical part by hyphen:

щитовидна жлеза – увеличена ... левия лоб, с еластична консистенция
(*thyroid gland* - enlarged... left lobe with elastic consistency)

ехо на *щитов. жлеза* – уголемени размери, хипоехогенна
(echography of the *thyroid gland* - enlarged size, hypoechogenic)

щитовидна жлеза – увеличена, плътна консистенция, чувствителна при *палпация* (*thyroid gland* - enlarged, solid consistency, sensitive when *palpated*)

The performance accuracy of Information Extraction is measured by the precision (percentage of correctly extracted entities as a subset of all extracted entities), recall (percentage correctly extracted entities as a subset of all entities available in the corpus) and their harmonic mean (F-measure)

$$F = 2 * Precision * Recall/(Precision + Recall).$$

Table 8 shows the extraction accuracy in the present experiment. Due to variety of paraphrases and keywords in the *blood pressure* description, the precision is relatively low. In contrary, only few words and their abbreviations describe the *thyroid* and *body weight* in our training and test corpora, therefore the extraction accuracy is very high.

Table 8. Accuracy of extraction of archetype slots from clinical narratives

	Precision	Recall	F-measure
Thyroid	96.25%	93.42%	94.81%
Blood Pressure	71.37%	90.63%	79.86%
Body Weight	95.65%	94.02%	94.83%

Our IE module easily identifies expressions which contain terms used in the archetype definition. However, narratives such as comments are difficult to capture. They are free text fields and their arbitrary content does not allow suggesting any keyterms to search for. For exhaustiveness we rely on the linguistic peculiarities of our data which usually contain one body part description within a single sentence.

Summing up all finings of the experiment we see that medical doctors hardly explicate in clinical narratives:

- hospital-dependent implicit knowledge when reporting about patient cases, for instance type of devices (e.g. cuffs for blood pressure measurements);

- values that are irrelevant for the particular disease (e.g. exact weight of diabetic patients and conditions when it was measured, or location where the blood pressure is measured). Instead, they document relevant features like weight change for given period which should be included in the archetype as comment or event..

There is also tacit knowledge which holds in the respective domain and it is regularly omitted in the particular texts. These observations show the difference between theoretical information models in medicine and their practical application. The standards of writing clinical documentation do not affect quickly the established tradition in writing domain-specific texts.

5 Conclusion

In this paper we present evidences about availability of unified elements in clinical descriptions. It is clear that the conceptual structures, designed to capture patient-related clinical information in order to ensure its systematic representation, need a long period of development, standardisation and wide adoption in order to provide interoperable resources of clinical knowledge. Perhaps thinking in terms of archetypes and conceptual structures needs to be incorporated in the medical training as well. The Topic Maps, as illustrated in Figures 3, 5 and 6, are a suitable visualisation tool that might help to advertise the archetype methodology.

We propose that the archetype design process should integrate language technologies for information extraction which enable immediate verification whether the theoretical conceptual model is aligned to the clinical practice of reporting events and observations. For instance, if the checks show that the medical experts regularly omit device descriptions, then this element might be included in the specific archetype instance by default for the particular clinical units. In this way some information in the instantiated archetype might be imported from a separate hospital unit description without burdening the clinicians with too much documentation. Another possibility is to offer specific predefined menus for item selection that are contextualised for the hospital unit. Simplifying the documentation process will facilitate the wide archetypes adoption.

Acknowledgments. The research work presented in this paper is supported by grant DO 02-292 "Effective search of conceptual information with applications in medical informatics", funded by the Bulgarian National Science Fund in 2009–2012.

References

1. Beale, T., Heard, S.: An Ontology-based Model of Clinical Information. In: Kuhn, K., et al. (eds.) Proceedings MedInfo 2007, pp. 760–764. IOS Publishing (2007)
2. Beale, T., Heard, S. (eds.): Archetype Definitions and Principles. openEHR Report (March 2007)
3. ISO 13606-2:2008 Health informatics - Electronic health record communication - Part 2: Archetype interchange specification (2008)
4. Onyshkevych, B.: Template Design for Information Extraction. In: Proc. of the TIPSTER Text Program: Phase I, Virginia, USA, pp. 141–145 (September 1993), available in the ACL Anthology http://www.aclweb.org/anthology/X93-1015
5. Sowa, J.: Conceptual Information Processing in Mind and Machines, Reading, MA (1984)
6. Chambers, N., Jurafsky, D.: Template-Based Information Extraction without the Templates. In: Proc. of the 49th ACL Ann. Meeting, Oregon, pp. 976–986 (June 2011)
7. Boytcheva, S.: Structured Information Extraction from Medical Texts in Bulgarian. In: Proc. of the SINUS Workshop Semantic Technologies in the Humanities, Sozopol, Bulgaria, June 7-8 (2012); to appear in a Special Issue of the Journal Cybernetics and Information Technologies
8. Nikolova, I.: Unified Extraction of Health Condition Descriptions. In: Proc. of the NAACL HLT 2012 Student Research Workshop, Montreal, Canada, June 3-8, pp. 23–28 (2012), http://www.aclweb.org/anthology-new/N/N12/N12-2005.pdf

Using Conceptual Structures in the Design of Computer-Based Assessment Software

Uta Priss, Nils Jensen, and Oliver Rod

Zentrum für erfolgreiches Lehren und Lernen
Ostfalia University of Applied Sciences
Wolfenbüttel, Germany
www.upriss.org.uk, {n.jensen,ol.rod}@ostfalia.de

Abstract. This paper discusses the use of conceptual structures in the design of computer-based assessment (CBA) tools for e-assessment of programming exercises. In STEM (science, technology, engineering and maths) subjects, universities often observe high dropout and failure rates among the first year students. There are a number of research initiatives that investigate the use of interactive teaching methods and e-learning technologies for improving STEM education. This paper presents a conceptual model of programming exercises and discusses more generally how conceptual structures can be employed for the implementation of CBA tools.

1 Introduction

STEM (science, technology, engineering and maths) subjects are notoriously difficult to learn and teach as demonstrated by high dropout and failure rates among first year university students. There are a number of reasons for the difficulty of such subjects. Researchers in Physics Education Research (for example, Hestenes et al. (1992)) have observed that students often have misconceptions which are not easily overcome by traditional lecturing methods even if these include exercises and demonstrations. Hestenes et al. (1992) explain that misconceptions are commonsense beliefs which can be regarded as reasonable hypotheses grounded in everyday experience. Unfortunately, commonsense belief are not always correct. For example, Newtonian physics includes many concepts that are contradictory to commonsense beliefs and in fact counter-intuitive. Students find it very difficult to overcome such misconceptions. Even though they may be able to apply Newtonian concepts in calculations by following an algorithm (which is frequently sufficient for passing exams), if asked to provide conceptual explanations students will often revert to non-Newtonian, incorrect concepts. Furthermore, if students are passing an exam only by applying memorised facts and algorithms, they will forget the subject matter quickly after the end of the semester. On the other hand, as soon as students achieve a conceptual understanding of a subject matter they will often retain such knowledge for 20 years and more (Conway et al., 1992).

H.D. Pfeiffer et al. (Eds.): ICCS 2013, LNAI 7735, pp. 121–134, 2013.

A second problem is what can be called a "teacher's dilemma". McDermott (2001) observes that at least in USA, the people who teach physics are not at all like the undergraduate students they teach because teachers have achieved a masters or doctoral level of understanding of the subjects whereas undergraduate students have often no ambition or interest to progress in the subject any further than required. By definition, teachers are usually not people who have ever dropped out of university or experienced learning difficulties but instead usually have been comfortable with the learning styles presented by traditional university teaching. Teacher training attempts to help prospective teachers develop an understanding of the students' conceptual models. For example, Prediger (2010) discusses the diagnostic competences that maths teachers need to develop in order to be able to listen to students and to analyse and understand their thinking. Tall (1977) argues that learning of mathematics involves cognitive conflicts. The acquisition of new concepts by a maths student is not a continuous process but includes conceptual jumps and states of confusion and emotional upset. A teacher must be able to detect occurrences of conflict in the mind of a learner and select an appropriate approach for conflict resolution amongst many different possible approaches. A further potential challenge to be overcome are teachers' attitudes towards trying new teaching methods (Pundak et al. 2009). Interestingly, changing a teacher's pre-existing belief about teaching methods may not be any easier than it is for students to overcome their misconceptions.

While traditional lecturing styles seem to be less than optimal in STEM subjects, interactive engagement methods (Hake, 1998) appear to be more successful. Hake defines "interactive engagement methods as those designed at least in part to promote conceptual understanding through interactive engagement of students in heads-on (always) and hands-on (usually) activities which yield immediate feedback through discussion with peers and/or instructors". An example is Mazur's (1996) peer instruction which uses cycles consisting of questions that are voted on by the students, peer discussion, group discussion, debriefing and then again the original questions. Apparently, while students might find it difficult to learn from a teacher's explanations, they find it easier to understand complex concepts and resolve cognitive conflicts when they can discuss these with other students (peers) who tend to be at a similar level of conceptual development as they are. Thus peer instruction is a means of overcoming the teacher's dilemma. The teacher becomes more of a facilitator or coach than an authoritarian source of information. A theoretical foundation for this approach to teaching is a constructivist model of learning (e.g., Ben-Ari, 1998).

It would be of interest to replace the currently prevailing constructivist model of learning with a Peircean pragmatist model. Levy (2007) observes that Peirce already discussed a "teacher's dilemma" because "in order to learn you must desire to learn, and in so desiring not be satisfied with what you already incline to think" (CP.1.135)[1]. But a teacher needs to be reasonably convinced of the

---

[1] The usual manner of citing Peirce' papers is adopted where CP refers to the Collected Papers of Charles Sanders Peirce followed by volume and paragraph numbers.

truthfulness of the subject matter to be able to teach. Thus the state of teaching (a state of belief) is in contrast with the state of learning which is a state of doubt. According to Levy, Peirce's solution to the dilemma is that a teacher must be willing to learn while teaching and that the learning process must be a cooperation between teacher and student. Therefore, it can be argued that Peirce anticipated interactive engagement teaching. A more in depth analysis of the relevance of Peirce's work for education would be of interest (in particular with respect to a pragmatic instead of a constructivist philosophy). While that is beyond this paper, the ICCS community might be a suitable audience for such research.

The core application area of this paper within STEM education is teaching programming to computer science students. In particular we are interested in how conceptual structures can be used to support tools for teaching programming such as computer based assessment (CBA) tools. With "conceptual structures" we are referring to tools and technologies commonly used in the ICCS community, for example, conceptual graphs and formal concept analysis. Section 2 introduces CBA tools. Section 3 describes a conceptual model of programming exercises. Section 4 discusses more generally how conceptual structures can be used to support CBA tools. The paper ends with a short concluding section.

2   Computer-Based Assessment Software

A large body of literature exists on the topic of STEM education. Our particular interest is the teaching of programming languages in computing or similar formalisms in mathematics. In this domain, computer-based assessment (CBA) software has been developed which allows students to submit code that is automatically evaluated (e.g. Pears et al. (2007), Rongas et al. (2004)). CBA software is more narrow in scope than virtual learning environments or course management systems which usually provide access to lecture materials, timetables and communication tools. If virtual learning environments provide automatically evaluated assessments at all, these are of a simpler, more static nature such as multiple-choice or fill-in the blanks tests. CBA tools without graded assessments are quite popular as add-ons to on-line tutorials which contain pastebins for sourcecode execution[2]. Advantages of using CBA tools in university courses are according to Pears et al. (2007) the fact that even students in large classes can be provided with detailed feedback in a timely manner. Automatic assessment is often seen as more fair and objective than assessment by tutors. Since CBA tools are employed in practicals, not lectures, they are more likely to be used with interactive engagement methods. Drawbacks of CBA tools are that exercises need to be specified very carefully to avoid misunderstandings. Furthermore automatic evaluation can miss problems. A student's work could receive full marks although it is written poorly and contains errors that were not anticipated by the

---

[2] For example the Tryit editor at `www.w3schools.com` or the SQL tutorials at `sqlzoo.net`

designers of the exercise. Unrestricted access to instant feedback can encourage students to employ a trial-and-error approach to programming.

CBA tools should be deployed with a suitable pedagogical method as can be found in the literature, for example, by Leron & Dubinsky (1995) since the 1990s. Without a sound pedagogical method or without being embedded into an interactive engagement style of teaching, CBA tools may not provide any benefits. If used correctly, CBA tools save time because the tools provide automated feedback and can be used by large numbers of students simultaneously - only limited by the size of the computer labs. Ideally the time lecturers save by not having to provide feedback on simple mistakes which are automatically detected by the CBA tool, lecturers should spend on helping students with conceptually challenging problems (or misconceptions) that require more in depth discussion (Priss et al. (2012b)).

Creating exercises for a CBA tool is more labour-intensive than creating other exercises because CBA exercises need to be specified very precisely so that they cannot be misinterpreted by students and they need to be tested before being used. Furthermore, the algorithms used for automatically evaluating the student-submitted code need to be provided usually either using software testing methods or intelligent tutoring techniques. CBA tools are only labour-saving if tools are provided that assist lecturers in the creation of exercises and the exercises can be reused. Thus in addition to the software required for the CBA tools themselves, one needs authoring tools and an infrastructure for the storage, retrieval and exchange of exercises. These are the areas where we see conceptual structures as potentially very useful. Although there are already many existing e-learning tools and standards for exchanging exercises available, as Rey-Lopez et al. (2008) observe these existing tools are not suited for the more detailed and content-rich exercises used for teaching programming. The problem of exchanging programming exercises and integrating CBA tools with other e-learning tools is according to Rey-Lopez et al. still an unsolved problem.

3   A Conceptual Model of Programming Exercises

In order to improve authoring tools for CBA software and to support exchanging exercises across tools and users, a solid understanding of the conceptual structure of programming exercises is beneficial. This section discusses a conceptual model of programming exercises developed using the Protege[3] editor. The only reason for using Protege was because it has a sophisticated, stable user interface and many graphical output options. The functionality used was classes, is-a relations and attributes (or slots) with value restrictions which are provided by many kinds of conceptual structures tools. Thus the discussion in this section is not meant to focus on the technology used but instead on the conceptual model that was derived.

---

[3] http://protege.stanford.edu/

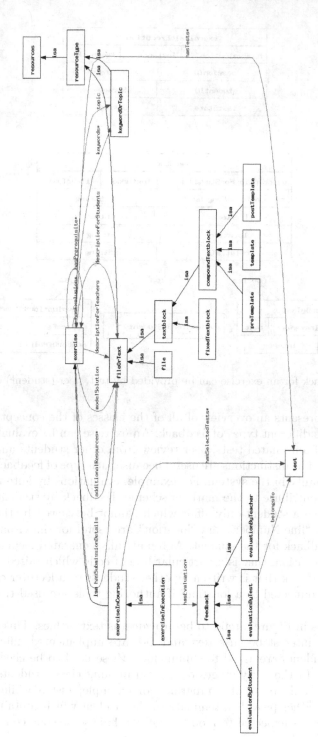

Fig. 1. Overview of the model

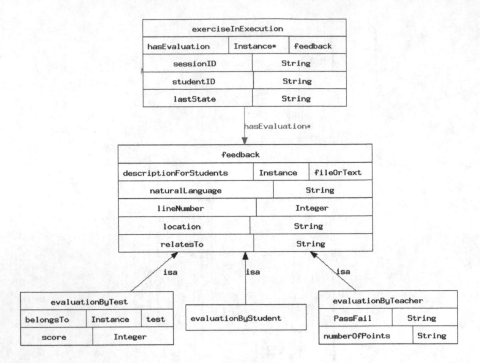

Fig. 2. Feedback for an exercise can be provided by test, peer (student) or teacher

Figure 1 represents an overview of all of the classes of the conceptual model. Figure 2 shows different types of feedback. An exercise can be evaluated by any combination of automated tests, peer review from other students and feedback from lecturers. The distinctions are useful because each type of feedback has a different functionality in the system. For example, evaluations by automated tests and teachers contribute to the marking scheme. Feedback by students is sometimes considered a student-only affair which cannot be viewed by the teachers. The attributes "line Number" and "location" are useful for the visual presentation of the feedback for the student. Automatically generated feedback usually has a precise location, that is a particular line of code which raised an error or a warning. Feedback that is written by other students or a lecturer can only be economically connected to a location, if authoring tools are used that allow to annotate code.

Resources as in Figure 3 tend to be provided as text or files. This distinction is of technical interest because text and files are implemented differently. For example, if students are asked to submit files, these need to be checked for file size and type. In the early stages of a programming class, students are often asked to write only parts of a program, for example, just a while loop. The CBA tool can either provide a template to the student which contains the code that the student is expected to modify or it can hide some code completely and

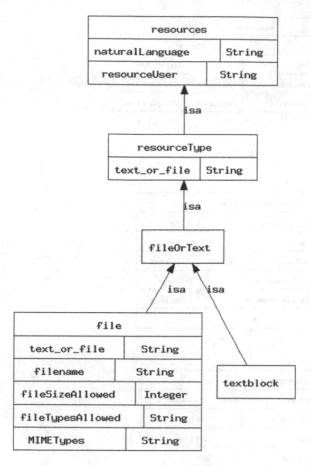

Fig. 3. Resources that can be up- or downloaded can be textblocks or files

automatically attach it before or after a student-submitted code snippet. Each resource has a "resource User" attribute which determines who has access to the resource, in particular whether the students are allowed to see the resource.

Figure 4 shows that a programming exercise exists on three levels: a general description independently of when and where the exercise is used; an "exercise in course" which has additional attributes about deadlines and about the actually selected tests from all available tests; and an "exercise in execution" which contains attributes about the student-submitted code, its evaluation results and session and state information so that a student can return to an exercise which has not been completed.

Exercises themselves are part of an ordered set: each exercise can have some prerequisites which the students need to pass beforehand. This is particularly useful if the CBA tool has intelligent tutor functionality. In order to avoid

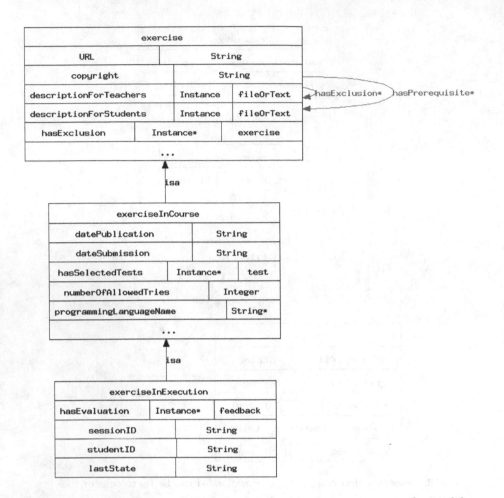

Fig. 4. An exercise exists on three levels: abstract, in a course or submitted by a student

plagiarism, it is helpful to have a larger question bank from which randomised questions are selected so that not every student receives the same questions (Russell & Cummings, 2005). This means that there needs to be an equivalence relation on the exercises ("has Exclusion") which shows which exercises are of similar difficulty and content and can be used as alternatives.

Last but not least, Figure 5 provides examples of available tests. Many current CBA tools use standard software engineering tests (unit, style checking and code coverage tests) for evaluating student-submitted code. CBA tools often incorporate standard testing software for such purposes so that the lecturers need not learn new technologies for writing their tests. Some CBA tools support writing blackbox tests which analyse the in- and output of a program.

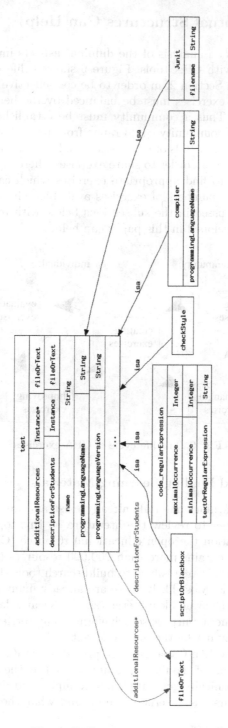

Fig. 5. Different types of tests

4   How Conceptual Structures Can Help

This section provides an analysis of the different aspects involved in preparing and using exercises with CBA tools. Figure 6 shows a life cycle of CBA exercises. As explained in Section 2, in order to be cost-effective the additional cost required for creating exercises must be balanced by the benefit of reusing and sharing of exercises. Thus a community must be established that shares and reuses exercises. This community could range from just a few lecturers within a department to lecturers in 100s of universities as, for example, the user group of the Lon-Capa[4] software. In order to share exercises, there must be a mechanism that allows lecturers to find appropriate exercises which can be a challenging task if there are large numbers of exercises available. The individual stages of the life cycle are discussed in the subsections below with reference to how the conceptual model developed in this paper can help.

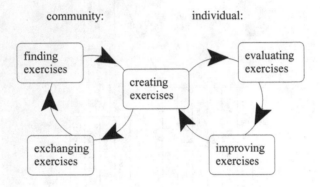

Fig. 6. The life cycle of CBA exercises

4.1   Searching and Finding Learning Materials

Large amounts of learning materials already exist and are available for reuse. It is beyond this paper to review the literature on this topic in any detail. It might suffice to mention the open educational resource (OER)[5] efforts or the fact that Lon-Capa contains more than 200,000 resources (Kortemeyer, 2006). A major challenge in this area is how to build search tools that help lecturers to find relevant materials. Most likely there are already many duplicate or similar documents amongst the available materials simply because lecturers do not know what is available. General purpose search engines only retrieve the most popular documents which may not be the most relevant.

As an example, the Lon-Capa software provides metadata for its more than 100,000 exercises. Some of the metadata are created by the authors of the exercises. Some are dynamically generated, for example, information about in how many and which courses an exercise is used and what the student results are

---

[4] http://www.lon-capa.org
[5] http://en.wikipedia.org/wiki/Open_educational_resources

for each exercise. A review of the existing metadata recently revealed[6] that the manually created metadata are entirely useless for search purposes because they are inconsistent and often missing (supporting what has been known in the library and information science community for decades). In fact in the future, Lon-Capa may drastically reduce the collection of manually created metadata and focus on the automatically generated data instead.

Currently most of the Lon-Capa exercises are of a more static nature and not programming exercises. We argue that the metadata of programming exercises according to the conceptual model in this paper is also manually created, but of a different type than the already existing metadata in Lon-Capa. The information recorded in the conceptual model is essential to the functioning of a programming exercise. For example, it is necessary to specify what programming language and what tests are to be used and how the marks are calculated. Once an exercise has been created this information is precise and unambiguous. This metadata is different from metadata, such as subject headings or keywords, which are more subjective, optional and debatable.

We argue that a conceptual model of programming exercises as developed in this paper can improve retrieval of programming exercises. In contrast to manually created metadata which tend to be inconsistent, our model structures only the essential data of the exercises which is thus rendered more accessible for searching. Other search details can be obtained from automatically collected metadata (such as the degree of difficulty of an exercise from the metadata about student results). Typical, re-occurring examples of programming exercises (such as the "Towers of Hanoi" or "Fibonacci numbers") can be found by searching within the full text of the exercises. Thus, an automatic exploitation of structured data, automatically generated metadata and the full text of the exercise with standard data mining methods is possible.

4.2   Exchanging Exercises

In order for exercises to be exchanged, a standard format needs to be defined which represents the data and metadata of the exercise in a structured manner. An XML representation of the conceptual model developed in this paper could be an example of such a format. As mentioned before, Rey-Lopez et al. (2008) observe that existing standards for learning materials are not suitable for representing the greater amount of detail required for programming exercises. Automatically-assessed programming exercises not only need a description of the content of the exercise but also of the technical requirements, for example, as to which programming language, which versions of the tools, and which testing technologies are used. Furthermore, if the exercises are to be exchanged in a manner that does not require extensive amount of manual editing for importing exercises, then there need to be means for automatically detecting version differences and to convert into formats required by a specific tool.

---

[6] G. Kortemeyer, personal communication, August 31, 2012.

There is currently an effort to create an exchange format for programming exercises undertaken by a working group as part of the eCULT project[7] which we are part of. Because that work is on-going and yet unpublished we cannot discuss more details about the format in this paper. Our contribution to the working group is based on the conceptual model developed in this paper. But because not all members of the working group are familiar with conceptual structures, the exchange format is represented in XML and not conceptually. Our conceptual structures model is somewhat more detailed and abstract than the format developed by the group and represents our view of the topic.

4.3   Creating Exercises

Creating programming exercises consists of creating the content and the technical implementation. With respect to developing appropriate content, Priss et al. (2012a) discuss in detail how conceptual structures (in the form of Formal Concept Analysis (FCA)) can be used for modelling conceptual difficulties in learning processes in mathematics. A conceptual model as represented in this paper provides structures for authoring tools for programming exercises that can be used in implementations. Programming exercises need a detailed specification of testing tools, versions and the tests themselves. Some details are repetitive and could be supplied semi-automatically; other details are specific for each exercises and need to be manually supplied. In our experience it takes about 2 hours to convert an existing programming exercise into one that is usable with a CBA tool. Using the conceptual model developed in this paper, it would be possible to design templates that would shorten the time required for writing exercises.

4.4   Evaluating and Improving Exercises

As mentioned in Section 2, the quality and precision of CBA exercises must be higher than that of manually-assessed exercise. If a lecturer makes a mistake in the wording of a manually-assessed exercise, this mistake can be rectified when the exercise is marked, for example, by adjusting the marking scheme to reflect that slightly different interpretations of the exercise are acceptable. If a CBA exercise is ambiguously worded and thus provides misleading feedback to students then the labour-saving effect of the exercise is lost because the lecturer needs to contact every individual student to provide additional information to remove the ambiguity. The resulting confusion could easily destroy any pedagogical benefit of using a CBA tool. If the CBA tool is used for an exam, the exercise may need to be manually-assessed after all. If the error is detected too late, the whole assessment may become worthless; or if the error is not detected at all, students will receive unjustified marks. Therefore exercises need to be well-tested before they are used with larger groups of students. During and at the end of a semester, the performance of the exercises needs to be evaluated, for example, by

---

[7] http://www.ecult-niedersachsen.de/

statistical analysis of the points students achieved for each exercise. High failure rates for an exercise could indicate that there is a problem with the wording of the exercise or it could be that the exercise highlights a misconception which the students have that must be addressed by other learning materials. Based on the evaluation, exercises (and supporting learning materials) should then be improved before they are used again.

Evaluation and improvement of exercises can only be performed by individual lecturers who use the exercises in their course. But the improvements of an exercise then need to be shared again with the community. The Lon-Capa software has essentially solved these problems by establishing mechanisms for creating and maintaining copies of exercises that have been modified and for communicating changes to other current users of an exercise. Furthermore, Lon-Capa provides mechanisms for alerting authors of exercises to potential problems detected with an exercise. A conceptual model could further assist by providing additional semi-automated checks, for example, if the version of a programming language is changed for one exercises it could automatically be checked whether other exercises might require a similar change.

5   Conclusion

This paper argues that teaching STEM topics is difficult by nature but pedagogical methods and tools exist that lead to improved teaching success. With respect to teaching programming languages, CBA tools can be beneficial if they are employed with a suitable interactive engagement style of teaching. The creation, maintenance, exchange and retrieval of programming exercises is labour-intensive but can be supported by conceptual structures. An example of a conceptual model for programming exercise is presented in this paper. The model is currently being used to guide our involvement in a working group for creating an exchange format for programming exercises and as a design aid in our implementation of a CBA tool which is further described by Priss et al. (2012b).

Acknowledgements. This work has been partially funded by the German Federal Ministry of Education and Research (BMBF) under grant number 01PL11066H. The sole responsibility for the content of this paper lies with the authors. We would also like to thank the other members of the eCULT working group on creating an exchange format for programming exercises: Sebastian Becker, Stefan Bisitz, Helmar Gust, Sven Strickroth. We have been careful not to use any materials from those discussions in this paper but it is likely that the conceptual model developed in this paper has been influenced to some degree by discussions of that group.

References

1. Ben-Ari, M.: Constructivism in computer science education. SIGCSE Bull. 30(1), 257–261 (1998)
2. Conway, M.A., Cohen, G., Stanhope, N.: Very long-term memory for knowledge acquired at school and university. Applied Cognitive Psychology 6, 467–482 (1992)

3. Hake, R.R.: Interactive-engagement versus traditional methods: A six-thousand-student survey of mechanics test data for introductory physics courses. American Journal of Physics 66(1), 64–74 (1998)
4. Hestenes, D., Wells, M., Swackhamer, G.: Force Concept Inventory. Phys. Teach. 30, 141–158 (1992)
5. Kortemeyer, G.: The Evolving Growth of LON-CAPA. Campus Technology (March 10, 2006), http://campustechnology.com/articles/2006/10/the-evolving-growth-of-loncapa.aspx
6. Leron, U., Dubinsky, E.: An Abstract Algebra Story. The American Mathematical Monthly 102(3), 227–242 (1995)
7. Levy, R.: Peirce's Theory of Learning. Educational Theory 2, 151–176 (2007)
8. Mazur, E.: Peer Instruction: A User's Manual. Prentice-Hall, New Jersey (1996)
9. McDermott, L.C.: Oersted Medal Lecture 2001: Physics Education Research-The Key to Student Learning. American Journal of Physics 69(11), 1127–1137 (2001)
10. Pears, A., Seidman, S., Malmi, L., Mannila, L., Adams, E., Bennedsen, J., Devlin, M., Paterson, J.: A Survey of Literature on the Teaching of Introductory Programming. SIGCSE Bull. 39(4), 204–223 (2007)
11. Prediger, S.: How to develop mathematics-for-teaching and for understanding: the case of meanings of the equal sign. J. Math. Teacher Educ. 13, 73–93 (2010)
12. Priss, U., Riegler, P., Jensen, N.: Using FCA for Modelling Conceptual Difficulties in Learning Processes. In: Domenach, Ignatov, Poelmans (eds.) Contributions to the 10th International Conference on Formal Concept Analysis (ICFCA 2012), pp. 161–173 (2012a)
13. Priss, U., Jensen, N., Rod, O.: Software for E-Assessment of Programming Exercises. In: Goltz, et al. (eds.) Informatik 2012, Proceedings of the 42. Jahrestagung der Gesellschaft für Informatik, GI-Edition, Lecture Notes in Informatics, p. 208, pp. 1786–1791 (2012b)
14. Pundak, D., Herscovitz, O., Shacham, M., Wiser-Biton, R.: Instructors' Attitudes toward Active Learning. Interdisciplinary Journal of E-Learning and Learning Objects 5, 215–232 (2009)
15. Rey-Lopez, M., Brusilovsky, P., Meccawy, M., Diaz-Redondo, R., Fernandez-Vilas, A., Ashman, H.: Resolving the Problem of Intelligent Learning Content in Learning Management Systems. International Journal on E-Learning 7(3), 363–381 (2008)
16. Rongas, T., Kaarna, A., Kalviainen, H.: Classification of Computerized Learning Tools for Introductory Programming Courses: Learning Approach. In: Proceedings of the IEEE International Conference on Advanced Learning Technologies (ICALT 2004), pp. 678–680. IEEE Computer Society (2004)
17. Russell, G., Cummings, A.: Online Assessment and Checking of SQL: Detecting and Preventing Plagiarism. In: 3rd Workshop on Teaching Learning and Assessment in Databases (TLAD 2005). HEA-ICS, pp. 46–50 (2005)
18. Tall, D.: Cognitive Conflict and the Learning of Mathematics. In: First Conference of The International Group for the Psychology of Mathematics Education at Utrecht, Netherlands (1977)

Modeling Ontological Structures
with Type Classes in Coq

Richard Dapoigny and Patrick Barlatier

LISTIC/Polytech'Annecy-Chambéry
University of Savoie, P.O. Box 80439, 74944 Annecy-le-vieux cedex, France
richard.dapoigny@univ-savoie.fr

Abstract. In the domain of ontology design as well as in Conceptual Modeling, representing universals is a challenging problem. Most approaches which have addressed this problem rely either on Description Logics (DLs) or on First Order Logic (FOL), but many difficulties remain especially about expressiveness. In mathematical logic and program checking, type theories have proved to be appealing but so far, they have not been applied in the formalization of ontologies. To bridge this gap, we present here the main capabilities of a theory for representing ontological structures in a dependently-typed framework which relies both on a constructive logic and on a functional type system. The usability of the theory is demonstrated with the Coq language which defines in a precise way what ontological primitives such as classes, relations, properties and meta-properties, are in terms of type classes.

1 Introduction

On the one hand, many researchers are striving for an ontologically well-founded representation language e.g., by adding new operators to Description Logics (DL) while preserving decidability, while on the other hand, mathematical and logical theories built on the Curry-Howard isomorphism have promoted the well-typedness as a foundational paradigm (above intuitionistic logic) leading to highly powerful theorem provers. Most modeling languages that have been proposed so far to express ontological constraints (or rules) are based on very simple meta-conceptualization as underlined in [21]. These languages offer appropriate structuring mechanisms such as classes, relationships and subsumption (subclass relations). However, in the representation of a formula, some structures have meaning whereas other do not make sense. This aspect requires "suitable ontological distinctions" understood as meta-properties of ontological structures as pointed out in [16] (e.g., the principles of identity or rigidity). In addition, the distinction of ontological meta-level categories such as types, kinds, roles, relations, etc., further make accurate and explicit the real-world semantics of the terms that are involved in domain representations. Not only an ontology is committed to represent knowledge of reality in a way that is independent of the different uses one can make of it, but it is intended to provide a certified and coherent map of a domain. All these constraints can be fulfilled within a highly expressive language built on a solid logical background. For that purpose in this paper, we propose a two-layered theory including a higher-order dependent type theory as a lower layer and an ontological layer as upper layer. This theory

H.D. Pfeiffer et al. (Eds.): ICCS 2013, LNAI 7735, pp. 135–152, 2013.

referred to, as K-DTT (Knowledge-based Dependent Type Theory) [2] is derived from [12] for the modeling of contexts. The logic in the lower layer operates on (names of) types whose meaning is constrained in the upper (ontological) layer.

Dependent types are based on the notion of indexed families of types and provide a high expressiveness since they can represent subset types, relations or constraints as typed structures. They will be exploited for representing knowledge in an elegant and secure way. This last aspect is analyzed in [8] where the authors investigate typing applied to reasoning languages of the Semantic Web and point out that dependent types ensure normalization. For example, type theory enjoys the property of subject reduction which ensures that no illegal term will appear during the execution of a well-typed query in a well-typed program. Alternatively, in [14], the authors have shown the ability of the type-theoretical approach to cope with scalability on the SUMO foundational ontology. Therefore, we introduce a a simple, coherent and decidable theory called K-DTT (Knowledge-based Dependent Type Theory) departing from the existing ones such as usual first-order logic theories. We will demonstrate with code fragments written in the support language Coq that the theory is able to satisfy most of all the constraints inherent in an expressive conceptual model.

2   Motivations for a Type-Theoretical Framework

In the conceptualization of information systems, Mylopoulos [32] has pointed out that the related language should be able to "formally represent the relevant knowledge". This assertion means that the conceptual language should (i) be expressive enough to represent the "relevant knowledge" and (ii) offer deduction capabilities to provide a valid model. It follows that there are three possible ways for increasing both the expressiveness and the soundness of a conceptual language (i) controlling the semantics of the conceptual language with a formal ontology (ii) using an expressive language e.g., an Object-Oriented language or (iii) using existing logic-based approaches such as Conceptual Graphs (CGs) or Description Logics. Let us review the basic features of these approaches.

Some authors such as [19] claim that a foundational ontology should allow to evaluate the "ontological correctness of a conceptual model" and to develop guidelines telling how the constructs of a conceptual modeling language should be used (e.g., association inclusion, specialization and redefinition). The author suggests that ontology adequacy should be a measure of the distance between the models produced by a modeling language and the real-world situations they are supposed to represent. To fulfill these constraints, the author has proposed an ontologically well-founded modeling language whose formal semantics is defined in a logical system as expressively as possible [20].

Object-Oriented languages [5], are the most significant formalisms for representing knowledge. They stem from frames which are seen as data structures that glue pieces of information together in their slots. In other words, OO languages are a computational implementation of frames. Classes correspond to frame descriptions while objects are identified to frame instances after filling in the slots. OO models offer two salient properties (i) the analogy between software models and physical models and (ii) the reusability of their components. The design of the program appears to be isomorphic to the components which result from the analysis of a given application with e.g., the UML tool. The central paradigm of OO languages is the notion of class. Classes encapsulate data (fields) and their properties (methods) in a single structure. classes are instantiated into objects which are designed to represent anything. Classes can be arranged into hierarchies using inheritance. A class can inherit behavior (i.e., data and properties) from another class called its superclass or parent class. Subsumption (i.e., polymorphism), is the ability to use a subclass where an object of its superclass is expected. However, many if not most, knowledgeable computing professionals recognize that the object-oriented paradigm is not the best one for every problem. In particular, despite some tentatives to add some logic to OO languages (e.g., F-logic [26]), the major weakness is that it lacks an expressive logical background for reasoning. For example, F-logic is very expressive but is generally undecidable.

The logical formalism of Conceptual Graphs (CGs)[43,7] originates in semantic networks and in the existential graphs of C.S. Peirce. It includes classes, relations, individuals and quantifiers with the purpose of providing a form humanly readable and computationally tractable. A CG has direct translation to the language of first order predicate logic, from which it takes its semantics. It results that CGs have the same expressing power as predicate logic. Labeled graphs with entities and relationships between them describe knowledge. CGs offer a significant advantage over concurrent formalisms, they are easily interpreted by end-users provided that they are not too complex. The model-theoretic semantics for the CGs is also specified in the ISO standard for Common Logic (CL). Common Logic includes the usual predicate-calculus notation for first-order logic. While CL semantics may represent entities of any type, it lacks ability for relating such entities to the internal structure of CL sentences. Reasoning in CGs relies on six canonical formation rules at the semantic level [43]. One can extend semantic operations with combinations of the rules, i.e., projection and maximal join. Graph operations such as projections (i.e., graph homomorphisms) are a major form of reasoning. A fundamental problem in simple CGs, i.e., deduction, starts with two simple CGs as input and searches whether a projection from the first to the second exists. The problem can be extended to a set of simple CGs (i.e., a knowledge base) and a simple CG representing a query (query answering). In querying simple conceptual graphs with negation, it has been shown that in the case of incomplete knowledge intuitionistic logic can be very attractive for capturing an answer to the query [31]. Finally, while CGs are efficient for representing natural language semantics, they do not support modality [44] and contexts are not taken into consideration when reasoning with CGs.

Alternatively, Description logics (DLs) refers to a family of knowledge representation formalisms that represent the relevant concepts of the domain (its terminology) together with properties of objects and individuals occurring in the domain. Semantically they are fragments of predicate logic, but their language is formed so that it would be enough for practical modeling purposes and also so that the logic would have good computational properties such as decidability. Very expressive DLs are likely to meet inference problems of high complexity, or to become undecidable.

We follow the idea of using an ontology for controlling the semantics of a conceptual model and will show in this paper how a well-founded model for the semantics, i.e., an ontologically correct conceptual model can be designed. Alternatively, the conceptual model must also be syntactically correct w.r.t. a set of rules and here we speak of a well-formed model (this part has widely been explored in the literature and will not be addressed here). The analysis of ontological correctness boils down to design specifications. A specification has the same status as axioms of a mathematical theory, i.e., they can be proved. More precisely, one can prove that a specification is consistent (it does not include a contradiction), just as one can prove that the axioms of a theory are consistent. For that purpose, a strong theoretical framework together with a core foundational ontology are required. This ontology will rely on the classical dichotomy between universals and particulars. While some approaches exist, the selected framework should emphasize expressiveness while assuming a strong formal theory.

Most philosophers have only some knowledge about First-Order Logic (FOL) and as a consequence, most claims about universals, particulars, properties and relations have a FOL-based logical bias. If now we rather adopt a constructive logic, then the picture is different and if we associate this logic to a rich type system rooted in the lambda-calculus, then a different but coherent picture can be drawn. Therefore, we exploit here the Knowledge-based Dependent Type Theory (K-DTT) theory already introduced in [2]. The interesting point is that available tools exist (e.g., the Coq theorem prover) making more exploitable the theoretical picture. We will demonstrate with code fragments, written in the support language Coq, that the theory is able to satisfy most of all the constraints inherent in an expressive conceptual model. Using a higher-order polymorphic type theory provides a lot of benefits. First, higher-order is useful (i) to permit instances of categorization types to be types themselves, (ii) to abstract away from level distinctions and (iii) to directly support quantification over sets and general concepts. Second, the typed framework enjoys (i) the reduction of the search space by restricting the domains/ranges of functions, predicates and variables to subsets of the universe of discourse, (ii) a structured knowledge representation facilitating both assertions and class-hierarchies and (iii) the detection of type errors with well-typed formulas. Finally, using dependent types is crucial to offer a high expressiveness and to enforce semantic conditions. The approach of [24] combining an order-sorted logic with the ontological property classification is a first step in this direction. But we can do more by including a type system with a strong proof theory. Using an unified theory providing high expressiveness together with the ability to constrain semantics will give the knowledge engineer the tools to produce models with certain guaranteed properties in terms of ontological transparency, well-foundedness and re-usability. These aspects are possible since properties are treated on a par with meta-properties (see section 6.2). In addition, there are available theorem provers (e.g., Coq) which can be used to check the well-formedness of user-defined typed structures.

We assume a layered structure including a logical level subsumed by an ontological level. The lowest level is an intensional type theory based on previous works [9,36,46] giving rise to a computational theory and, at the highest level, to knowledge structures whose semantics relies on a hierarchy of concepts (e.g., the DOLCE hierarchy of particular categories) and the so-called meta-properties (as e.g., in Ontoclean). As a consequence, all the ontological classes introduced in the following must have corresponding structures satisfying typing mechanisms provided at the logical level. Unlike most representation languages, K-DTT provide constructs able to distinguish among terms having similar logical structure but different ontological meaning [18].

3   K-DTT: The Type-Theoretical Layer

The Type-Theoretical layer of K-DTT is both rooted in a constructive logic and on a typed λ-calculus using dependent types. Dependent type theories allow a type to be predicated on a value which makes them much more flexible and expressive than conventional type systems [30]. The constructive logic pre-supposes a logic centered on the concept of proof rather than truth and follows the Curry-Howard isomorphism [22] in which proving is "equivalent" to computing (or querying a database). Reasoning in K-DTT consists either in reducing types to their normal form or finding proofs for reduced types. In K-DTT, the type of a type is called a universe. Universes are partially ordered and are organized into an infinite hierarchy of predicative type universes $Type_i$ for data types together with an impredicative[1] universe noted $Prop$ for logic. This hierarchy follows a kind of cumulativity: $Prop \subseteq Type_0 \subseteq Type_1 \subseteq \ldots$. A universe is seen as a type that is closed under the type-forming operations of the calculus. Using the Curry-Howard isomorphism, terms of the type-theoretical layer can represent data structures as well as properties of these structures and proofs of these properties.

Definition 1. *Let Γ be a valid environment.*
A term T is called a type in Γ if $\Gamma \vdash T : U$ for some universe U.
A term M is called a proof object in Γ if $\Gamma \vdash M : T$ for some type T.

The underlying theory, i.e., the Calculus of Constructions with inductive types and universes (see e.g., [3]), has given rise to the Coq language[2] [10] which has recently promoted very powerful primitives such as Types Classes (TCs) [41,45] for describing data structures. TCs in Coq are a lot like type classes in Haskell, however Coq allows us to do better by specifying the rules inside TCs. Coq both combines a higher-order logic and a richly-typed functional programming language. All logical judgments in Coq are typing judgments such as $x : T$, where x is a variable and T, a term. The type-checker checks the correctness of proofs, that is, it checks using proof search that a data structure complies to its specification. The proof engine also provides an interactive proof assistant to build proofs using specific programs called tactics. The language of the Coq theorem prover consists in a sequence of declarations and definitions. A declaration associates a name with a specification. Specifications can be either logical propositions which reside in the universe $Prop$, mathematical collections which are in Set or abstract types which belong to a universe $Type_i$ with $i \in N$. The theory includes dependent types generalizing function spaces and Cartesian products.

Lemma 1. *[10] Let A, B, two types defined in the current context Γ such that $\Gamma \vdash s : \phi(x : A, B[x])$ where ϕ denotes a dependent type, then the universe of $\phi(x : A, B[x])$ is the maximum universe among the universes of A and B w.r.t. coercions.*

Dependent types give new power to TCs while types and values are unified. TCs have a structure derived from record types with fields, but they are more powerful by allowing parametric arguments, inheritance and multiple fields. For example if one wants to represent reflexive relations, we can introduce the `Reflexive` TC with:

---

[1] Impredicativity is a kind of conceptual circularity.

[2] The Coq language has reached a state where it is well usable as a research tool.

```
Class Reflexive {A} {R : relation A} :=
                reflexivity : forall x, R x x.
```

where { ... } denotes implicit arguments, A stands for any type and R : A → A → Prop is a dependent type expressing mathematical relations. Notice that (i) the implicit type of variable x is automatically resolved in Coq to x : A and (ii) relation refers to the basic Coq library and complies with the above definition of R.

4   K-DTT: The Ontological Layer

To explain how to represent an ontology with K-DTT, we take the example of the DOLCE taxonomy of particulars [29,15] which does not classify universals and leaves room for conceptual choices about universal structures. The hierarchical taxonomy of particular categories will serve as a backbone, referred to as DOLCE backbone. The DOLCE backbone plus ontological commitments on appropriate structures expressed within type theory will form the ontological layer of the K-DTT theory. It can be used to express knowledge as long as the added features respect the core structures together with their logical constraints (see e.g., [11]). All data structures will be expressed either with operational TCs having a single field and returning a value in $Type_i$ or predicate classes returning a truth value in $Prop$.

4.1   Representing Ontological Classes

We follow the position adopted by most formal ontologies in computer science, i.e., that universals are general entities which are further refined in subcategories (e.g., relations) and that particulars are specific entities which exemplify universals but which cannot have themselves instances. K-DTT objects are equipped with meanings using the Curry-Howard isomorphism and assuming that any type (or typed structure) corresponds to a universal. Notice that "type" here is a mathematical notion and is not itself an object for ontological modeling. The fundamental ontological distinction between universals and particulars can be informally understood using the typing relation ":". Possible worlds which is a way of characterizing the distinction between descriptions (i.e., intensions) and particulars corresponding to the descriptions (i.e., extensions) is implicitly accounted for in K-DTT respectively with typed structures and sets of proof objects[3]. Properties and relations which correspond to predicates in a logical language are usually considered as universals. Terms of the ontological layer of K-DTT are built from the category $Universal$, (the highest universe in the lower layer) which includes the four basic categories (sub-universes in the lower layer), C, Rel, P and Rol which stand respectively for concept types, relation types, property types and role types (see fig 1). The last category is not discussed here but more details can be found in [2].

The universe of concepts includes the set of universes for the foundational ontology $\{PT, AB, R, TR, T, PR, \ldots, STV, ST, PRO, \ldots, \}$ which refers to the DOLCE taxonomy. It classifies categories of particulars such as APO which stands for Agentive Physical Object. Each of these universes is closed under the type formation rule

---

[3] Proof objects are e.g., the result of queries on a database related to the ontology.

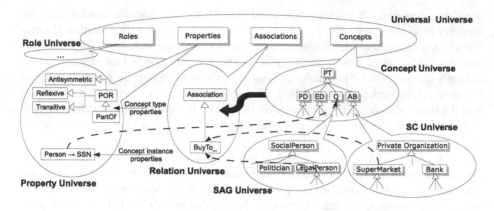

Fig. 1. Ontological categories in K-DTT

(lemma 1) and includes itself a set of sub-universes which can be defined in a domain ontology. All universes are partially ordered by subsumption formalized here by co-ercions [38] (e.g., ED is subsumed by PT, ...). In TCs, inheritance is implemented with (i) implicit arguments and (ii) the :> operator. Concepts can be either primitive or compound. Primitive concepts are in line with the existence of "natural types", i.e., they can be identified as types in isolation (see e.g., [42]). Each category of particulars which belong to the DOLCE taxonomy is described with a TC where their position in the hierarchy is computed w.r.t. the coercion rules. All terms of the ontological layer must be well-formed.

Definition 2. *(Well-formed concepts) A term T is well-formed if for some formal ontology $\mathcal{O}$ providing the environment, we have either $\mathcal{O} \vdash T : C$ or $\mathcal{O} \vdash T : R$ or $\mathcal{O} \vdash T : P$ for some $C \in \mathcal{C}, R \in \mathcal{R}el$ and $P \in \mathcal{P}$.*

In the following, we often forget the environment for the sake of clarity, while any assertion will be relative to an explicit environment (e.g., a foundational ontology). For example, the judgment $LegalPerson : SAG \vdash x : LegalPerson$ where SAG denotes a Social AGent asserts that the variable x belongs to the concept $LegalPerson$ provided that it is a well-formed term, i.e., that $LegalPerson$ belongs to the universe SAG. A value for the variable x which can be e.g., $JohnDoe$, is called a proof object because it is both considered as an object (from the programming language side) and as a proof (from the logical side).

The relation of instantiation (:) between a universal and its instance corresponds here to type inhabitation. However, when this relation connects a type and its universe, it behaves like a Grothendieck universe [6] w.r.t. lemma 1. Unlike FOL-based ontologies, it describes in a natural and simple way the relation between universals and particulars without the need to introduce specialized ad hoc formulations. For example, in [40], the proposed theory of Is_a and $Part\_of$ is based on a relation of instantiation between an instance and a class and requires axioms for governing its use. In K-DTT, the relation of instantiation is already part of the theory and does not require any further axioms.

The equality between terms in the ontological layer is ascribed to be coherent w.r.t. the Leibniz equality of the lower layer (two types are logically identical iff they have the same properties). It relates to the usual definition of the identity condition for an arbitrary property P, i.e., $P(x) \wedge P(y) \to (R(x,y) \leftrightarrow x = y)$ with a relation R satisfying this formula. This definition is carried out for any type in K-DTT since equality between types requires the Leibniz equality. The major reason is that identity can be uniquely characterized if the language is an higher-order language in which quantification over all properties is possible [34]. This property yields that Leibniz's Law, which is at the basis of identity in the lower layer of K-DTT is expressible in this language.

Relation types are rooted in the mathematical structure of a relation over a set A. The set is replaced by a general type $A : Type$ and rules can be defined over the basic TC $relation$ as described in section 3. It can be seen as a generalization of the work already presented in [11]. Specifications are a restriction of TCs with a first field for data structure and a second field for the property. Furthermore, fields of TCs can reside in any universe assuming the rule of lemma 1. Notice that proof objects for relation types are tuples and are closely related to lines of a table in a database. Universes of relation types are closed under type forming operations, i.e., Π-types and TCs constructs.

Property types reside in a sub-universe of $Universal$, which means that they are also universals. Properties of concept instances correspond to quality types (the DOLCE category Q) while properties of concept types correspond to meta-properties. Property types (concept or instance type properties) are divided into two sub-categories describing respectively mandatory properties (rigid properties) and possible properties (anti-rigid properties). This choice complies with the constraints of [17] arguing that anti-rigid properties cannot subsume rigid properties related to ontological distinctions in the current practice of conceptualization. While we agree with the introduction of three formal properties, i.e., identity, rigidity and dependence as proposed in [18], we do not share the view of properties as unary predicates since it relies on the restricted support of FOL (i.e., it has a logical bias).

4.2   Expressing Generalization

Generalization consists in deciding whether one category is "more general than" another one and is formalized by the subsumption relation "A subsumes B" which says that being a B logically implies being an A. The notion of subsumption has several readings, the more important ones being extensional and intensional [47,33]. There are some drawbacks to the extensional interpretation of subsumption because (i) determining whether the extension of one concept is included in the extension of another one is often undecidable and (ii) observing that two concepts have the same extension does not mean that they are identical.

In K-DTT, generalization and refinement are intensional and take advantage of two mechanisms (i) the simple mechanism of coercive subtyping [38,28] and (ii) TCs [41]. The hierarchy of generic concept types from DOLCE is isomorphic to a stratified hierarchy of universes in Coq (e.g., PD is (i) defined as a universe with the definition := Type (ii) ordered with the typing assertion : PT and (iii) explicitly coerced with the parameter u1). In such a way, subsumption hierarchies can be designed provided that the coherence between coercion paths is preserved. This coherence is automatically checked in Coq (see e.g., [38]). The following fragment details this mechanism on some DOLCE categories showing e.g., that every perdurant (PD) and every endurant (ED) are also particulars (PT).

```
Definition PT          : Concept       := Type.
Definition PD          : PT            := Type.
Definition ED          : PT            := Type.
Parameter u1 : PD->PT.  Coercion u1 : PD>->PT.
Parameter u2 : ED->PT.  Coercion u2 : ED>->PT.
                        . . .
```

The other mechanism creates inheritance hierarchies by refining TCs. Using the :> operator within a field of a TC means that any instance of the actual class is also an instance of the parent class. It results that TCs are a kind of bounded quantification where the subtyping relation needs not be internalized. When we represent higher and higher structures, TCs avoid the set of arguments growing as well. Predicate classes also support multiple inheritance which can be exploited in e.g., biomedical ontologies. Furthermore, they allow overlapping multiple inheritance which enable inherited structures to share components [45]. In such a way, very expressive hierarchies can be composed out of predicate classes and operational classes based on inheritance.

For example, we can add a TC for transitive relations to the operational TC of section 3 in a similar way. Then, a pre-order TC can inherit of these classes as follows:

```
Class Reflexive {A} {R : relation A} : Prop :=
           reflexivity : forall x, R x x.}
Class Transitive {A} {R : relation A} : Prop :=
           transitivity : forall x y z, R x y -> R y z -> R x z.}
Class PreOrder {A}{R:relation A} : Prop := {
           PreOrder_Reflexive   :> @Reflexive A R;
           PreOrder_Transitive  :> @Transitive A R }.
```

Here, the syntax :> declares each projection (i.e., each field) of the TC PreOrder as an instance of the respective TCs Reflexive and Transitive. It follows that each pre-order can be seen as a reflexive and as a transitive relation. This simple example highlights how multiple inheritance is implemented with implicit parameters.

5   Representing Relations

The strength of dependent types in type theory allows expressing relations as primitives of the language. The first consequence is that they are terms of the logic and can be involved in complex predicates. At the ontological level, relations are hierarchical (e.g., subsumption or part-of relations[4]) or non-hierarchical (e.g., domain relations). For the sake of simplicity, we only discuss here binary relations. Hierarchical relations are discussed in detail in section 6.2 since they require the specification of properties. Non-hierarchical relations denote tuples involving particulars and precisely correspond to instances of TCs having a first and a second field which detail their component types and a third one explaining how they are constructed and possibly other field(s) giving the additional properties they are subject to (see section 6.2). The basic TC BinaryRel defines the generic structure built out of a type A, another type B and a predicate type. For example, a domain relation type expressing persons which may suffer from a given

---

[4] We refer here to the part-of relation which is transitive by contrast with the partonomic relation which is usually not (see [11] for more explanations).

(or multiple) disease(s) inherits this basic structure through their implicit arguments. The specification `Class Person : APO` assumes that *Person* belongs to the universe *APO* which itself is a sub-universe of *Concept*. Implicit respective parameters A and B are automatically resolved[5] with *Person* and *Disease* giving rise to the relation type `SufferFrom`:

```
Definition Disease          : PRO              := Type.
Parameter u20 : Disease->PRO. Coercion u20: Disease>->PRO.
Class Person                : APO              := { }.

Class BinaryRel {A B:PT}    : Association   := {
        BinaryRel_arg1   : A;
        BinaryRel_arg2   : B;
        BinaryRel_rule   : A->B->Prop}.
Class SufferFrom            : Association   := {
        SufferFrom_struc :> @BinaryRel Person Disease}.
```

Instances of *Person* and *Disease* correspond respectively to a set of persons and a set of diseases these persons are subject to. Notice that *Disease* is defined as a universe since it cannot have direct instances. Each time an object having the type `@BinaryRel Person Disease` is introduced, Coq will try to construct an object of type `SufferFrom`. To clarify the meaning of relations in the typed framework, let us consider the appropriate table within a database (proof objects are items in the table):

Person	*Disease*
John_Doe	*Influenza* *AIDs*
Mike_Hammer	*Herpes_Simplex*
Henry_Mann	*Rhinovirus*
Franck_Burch	*Hepatitis_A* *nail_infection*

The relation type $SuffersFrom$ belongs to the universe of (ontological) binary relations. A tuple $\langle John\_Doe, AIDs \rangle$ is a proof for the type $SuffersFrom$. Here proof objects are (constructively) obtained with requests to the database rather than by mathematical computation as usual.

Let us denote Tab and DB the respective contexts of the table related to the relation and the whole database. The set of persons[6] $Obj_{Tab}(Person) = \{John\_Doe, Mike\_Hammer, Henry\_Mann, Franck\_Burch\}$ relative to the table does not rule out the possibility to have other persons involved in other tables. We assume that DB is a valid context which contains at least every component of Tab. Using the weakening rule for contexts with $DB \vdash Person : Type$ and $Tab \vdash Person : Type$, we can assert $DB \vdash x : Person \supset Tab \vdash x : Person$[7]. The same reasoning holds

---

[5] Instance resolution is part of the Coq unifier.

[6] More formally, the set of proof objects.

[7] The symbol $\supset$ denotes the logical implication in higher-order logic.

for the type *Disease*. Then using the extensionality of computational equality between inhabited types [9], it follows that $Obj_{Tab}(Person) \subset Obj_{DB}(Person)$ and $Obj_{Tab}(Disease) \subset Obj_{DB}(Di\ sease)$. Furthermore, the diseases undergone by a person is a subset of the diseases undergone by any person within the table.

These definitions for binary relations can be extended to n-ary relations but this aspect will not be discussed here. The TC `BinaryRel` is general and applies at any abstraction level which is less than *PT*. Then Coq checks for the coherence of subsumption paths, for instance the sequence: `[u20; u15; u14]` : `Disease >--> PD`, `[u20; u15; u14; u1]` : `Disease >--> PT`, `[u20; u15]` : `Disease >--> STV`, `[u20]` : `Disease >--> PRO]` assumes that `Disease` is a `Concept` while providing the sequence of the coercion hops between these two terms. Notice that Coq checks for the coherence of these paths by avoiding multiple paths between two terms. This example illustrates how the types of the arguments are controlled throughout the refinement process and highlights the benefit of TCs for capturing knowledge.

6 Representing Properties

6.1 Concept Instances Properties

Concept instance properties or quality types can be attributed to things or predicated about them and then it can be said that objects exemplify quality types. In OO modeling, qualities (attributes) are embedded within the scope of a class while in CGs and DLs they rather have a relational flavor. In DOLCE, these qualities belong to a finite set of quality types (e.g., color, size, shape, etc.) and inhere in specific individuals. It results that two particulars cannot have the same properties (due to Leibniz equality) but they can have the same qualities, and at any time, a quality cannot exist unless the entity it inheres in also exists. A quality type is close to the "relational moment type" [20] which is also existentially dependent on other particulars. The inherence relation is isomorphic with type dependence on values. In K-DTT, the idea is to consider qualities as represented by TCs and then to attach them to other concepts through dependent arguments. Let us consider how a moment universal can be captured in K-DTT. We only consider here (for the sake of simplicity) an intrinsic moment which uniquely depends on a single particular. For example a person which has an attribute SSN (Social Security Number) can be conceptualized with first introducing two TCs with their appropriate coercions as follows:

```
Class SSN            : Q          := { SSN_Quale : nat}.
Class Person         : APO        := { ... }.
Class HasSSN         : Prop       := {
            SSN_Attr : Person->SSN;
            SSN_mul  : SelectArity Person (card_0_1 Person)}.
```

where `SSN_Quale` is a natural number describing the SSN value. The TC `HasSSN` relates a person with its SSN through the type of the first field `Person->SSN`. The second field describe the arity of the SSN (we presuppose that a person has 0 or 1 SSN). Each arity has a lower and an upper value. Arity 1 is assumed if the field is a simple type declaration. Arity 0..1 can be represented with the predefined inductive type *option*. This type has two constructors, one referred to as *none* says that there are no elements having the type whereas *Some t* provides a term t having the type. Then, any arity value greater than 1, i.e., the arity $1..n$ is easily represented with the list type. The resulting encoding formalizes these assumptions:

```
Inductive Arity (A:Type)   : Type   :=
             card_1         : Arity A
           | card_0_1       : Arity A
           | card_0_n       : Arity A.
Definition SelectArity (A:Type)(x:Arity A) : Type :=
           match x with
             | card_1      => A
             | card_0_1    => option A
             | card_0_n    => list A
           end.
```

With these definitions, it becomes easy to understand the second field in the HasSSN TC. It means that the type *person* has an arity 0..1, that is he has 0 or 1 SSN. If the SSN value is 0, then the related person has no SSN while any positive integer will provide the person as output. In summary, to express that any concept has a quality, a predicate TC (i) formalizes the dependency of the concept over the quality in a first field and (ii) formalizes the arity which inheres in the link person-SSN.

For each instance of HasSSN, one must be able to (i) construct an object (e.g., "33") having the type SSN and (ii) a proof that the cardinality 0..1 has been selected.

```
Instance SSNPerson       : HasSSN        := {
          SSN_Attr John    := (Build_SSN 33);
          SSN_mul          := Multiplicity0_1 (Build_SSN 33) John }.
with:
Definition Multiplicity0_1 (s:SSN)(p:Person) : option Person :=
          match s with Build_SSN 0 => None
          | Build_SSN s => Some p end.
```

On the one hand the dependent type allows to filter out unexpected values as arguments and unexpected quality types for the structural part of the property seen as a predicate TC, while on the other hand, inductive types provide a suitable mechanism for arity checking. Rigid and anti-rigid properties are useful if one can reason about them in a meta-schema as claimed in [17,18]. While basic constructors of type theory can be used for that purpose (see e.g., [2]), reasoning can be made easier by supplying a supplementary field to type classes, i.e., Quality_rig. Then this information can be used in a process which automate the control of domain ontologies resulting in well-formed hierarchies.

6.2   Concept Type Properties

Properties of concept types or meta-properties are described with rules, that is with predicate classes. For example to introduce partial order relations (POR), a widely-used concept in reasoning, one will extend the previous definitions for Reflexive and Transitive type classes as follows:

```
Class Antisymmetric { A } {R : relation A} : Prop :=
          antisymmetry : forall x y, R x y -> R y x -> x = y.
```

```
Class Irreflexive A R : relation A :=
        irreflexivity : forall x, R x x -> False.
Class Asymmetric {A}{R:relation A} := {
        Asym_Irreflexive                :> @Irreflexive A R;
        Asym_Antisymmetric              :> @Antisymmetric A R }.
Class POR { A }{R:relation A} : Prop    := {
        POR_Reflexive                   :> @Reflexive A R;
        POR_Antisymmetric               :> @Antisymmetric A R;
        POR_Transitive                  :> @Transitive A R }.
```

First, the coercions over class instances (e.g., @Reflexive A R) in the *POR* TC express multiple inheritance diagrams and second, the required types A and relation A are general and can be applied to any kind of relation. It follows that such a kind of inheritance diagram can be reused in any ontology-based application and give rise to modular hierarchies of axiomatic structural properties.

7   Constructing Inheritance Hierarchies

Usual theories about part-whole relations do not consider categories of the entity types involved in a part-whole relation and subsumption between these relations presuppose that their arguments are identical. The K-DTT theory is able to control both the types of each argument if required and inheritance between the type classes representing distinct relation types. This property extends the expected expressiveness of the theory beyond the usual power of ontology languages since it involves not only the predicates of relation types but also all arguments for these types. Let us consider mereotopological relations for endurants such as the has_3D property [25] which refers to both the endurant itself and the region (R) it occupies. The authors claim that the *contained_in* relation which is widely used in biology, involves both parthood and containment and can be captured with the first-order formula:

$$\forall x, y(contained\_in(x, y) \triangleq part\_of(x, y) \wedge R(x) \wedge R(y) \wedge$$
$$\exists z, w(has\_3D(z, x) \wedge has\_3D(w, y) \wedge ED(z) \wedge ED(w))$$

In K-DTT, this dual property is expressed using multiple inheritance. The inheritance diagram requires first the specification of a generic part-of relation. The inheritance with instances of type classes (i.e., :> @PartOf R rr;) avoids the introduction of a structural parthood in the taxonomy (see [25] for more details). Furthermore, simple inheritance with restriction on the type of argument will describe the involved_in relation, which relates two perdurants (see figure 2).

```
Definition ED   : PT                    := Type;
Definition AB   : PT                    := Type;
Definition R    : AB                    := Type;
Class PartOf {A:PT}{rr:relation A} : Association := {
        PO_prop         :> @POR A rr }.
Class ProperPartOf {A:PT}{r:relation A} : Association := {
        PPO_propAs      :> @Asymmetric A r;
        PPO_propTr      :> @Transitive A r}.
```

```
Class ContainedIn {rr: relation R} : Association := {
        CIpartof_struct :> @PartOf R rr;
        CIregion_struct :> @BinaryRel R ED}.
Class InvolvedIn {rt: relation PD} : Association := {
        IIpartof_struct :> @PartOf PD rt}.
```

Then, assuming for example that we declare variable `regions : relation R`, we can prove with a tactic that any relation of type `ContainedIn` is transitive. The fact that we can apply transitivity to the arguments which belong to the class `ContainedIn`, means that it is propagated along the TC hierarchy until it reaches the TC `Transitive`. Then, applying twice the unification yields the result. Notice that tactics may be registered and reused which makes designer's task more easy by abstracting away the logical part.

```
Goal forall c:@ContainedIn regions, forall x y z:R, regions x y->
                                 regions y z -> regions x z.
Proof.
intros.
eapply transitivity.
eassumption.
assumption.
Qed.
```

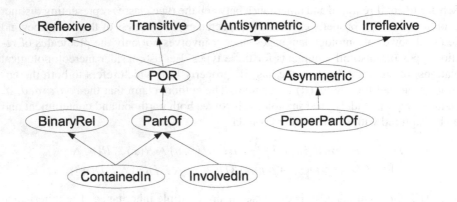

Fig. 2. An inheritance diagram in K-DTT

In a similar way, wider inheritance hierarchies can be investigated for describing expressive knowledge hierarchies in which type checking allow for proving well-formed ontologies. Automated proof-search could be improved by specifying ad'hoc requests on a database using parametrized requests (this aspect will be soon investigated).

8   Related Works

A subset of the CG theory called Prolog+CG is a Java implementation of Prolog where CGs are first-class datatypes on a par with terms [23]. It includes also Object oriented extensions. This approach is based on typed hierarchy of concepts which is a lattice (The Sowa ontology) while our approach stems from the DOLCE foundational ontology and Ontoclean for the meta reasoning. Both approaches are using types but in Prolog+CG, typing requires multiple programming layers since Prolog does not naturally work with types. Instead, the K-DTT approach has a native typing system. While Prolog+CG has a Prolog core, nothing is said about negation whose handling is problematic. Due to the intuitionistic framework of K-DTT based on proof construction rather than discovering truth values, this problem is avoided. For that purpose, Coq has a lot of built-in tactics, and many more are available in libraries. Typically in Prolog, one expects the result(s) of the resolution and obtains a truth value while in Coq, one has to say what type has to be proved and obtains more information since a type is identified with the set of its proof objects. The creation of instances requires a primitive goal (CreateInstance) while Coq has a built-in notion of instanciation (:). The higher order capabilities of the type-theoretical layer are a crucial advantage for meta-reasoning.

Description logics are widely described and many tools exist which allow developers of to export their ontologies into a DL formalism. However, in DL-based biomedical ontologies, analysis have been investigated to check how their terminology complies with a basic set of ontological principles [4]. The authors have pointed out some inaccuracies of subsumption links, incomplete description and other conceptual ambiguities. Alternatively, in spatial reasoning [1], some limitations of OWL (and therefore of DL) are identified such as the difficulty of the language to represent the constraints which govern the coordinates of spatial objects in containment hierarchies. The comparison of datatype properties of individuals is the result of the lack of variables in DLs. Moreover, since DLs are conceptually oriented, they lack a rule-based reasoning mechanism such as the one found in logic programming (e.g., Horn clauses). Therefore, various approaches have been proposed for integrating logic programming and description logics such as [13,27,37]. Nevertheless, these reasoning techniques suffer from some limitations such as the restriction of DLs to "safe rules" and complicate the reasoning process by adding a translation mechanism between DLs and logic programming. More precisely, using these languages for practical applications raises several challenges [35]. The restriction to binary predicates in both SWRL and OWL is a first difficulty leading to violate safe rules for expressing the higher-order nature of the rules that have to be constructed. The main advantage of the present approach holds in the unified framework in which numerical values can be manipulated, rules can be applied and proved and multiple inheritance can be exploited while offering an expressive higher-order language.

9 Conclusion

The K-DTT theory is an attempt at constraining the semantics of knowledge representation based on expressive typed structures. While the logical part of the representation languages is neutral as concerns ontological choices, typing is not (e.g., the DOLCE backbone). K-DTT is a unifying theory both sufficiently expressive and logically founded together with a logic which supports different abstraction levels. We have (i) introduced the basic features of K-DTT and (ii) illustrated how modeling ontological knowledge can be checked in the Coq theorem prover. It is demonstrated that type classes can model several non-trivial aspects of classes such as meta-level properties and multiple inheritance. TCs unify the two representations of relations, i.e., the logical view in which relations are predicates and the conceptual modeling view where a relation is seen as a set of tuples.

On the one hand, the present theory is more expressive than usual predicate logic in which it is neither possible to apply a function symbol to a proposition, nor to bind a variable except with a quantifier. In addition, the language of K-DTT is richer than FOL-based languages in allowing proofs to appear as parts of the propositions so that the propositions can express properties of proofs (and not only of individuals like in FOL). On the other hand, the distinction between usual Object Oriented programming and type theory relies on the ability of their representative computational structures to correctly express the semantics of ontological components. While their expressiveness is comparable, many aspects of object-oriented programming can be preserved in type theory since it unifies functional programming, component based programming, meta-programming (MDA), and logical verification (see [39] for more details). Further works include the realization of an interface between Coq and databases to collect proof objects with requests and a tool for proving well-formed ontologies.

References

1. Alia, I., Abdelmoty, A.I., Smart, P.D., Jones, C.B., Fu, G., Finch, D.: A critical evaluation of ontology languages for geographic information retrieval on the Internet. Journal of Visual Languages & Computing 16(4), 331–358 (2005)
2. Barlatier, P., Dapoigny, R.: A Type-Theoretical Approach for Ontologies: the Case of Roles. Applied Ontology 73, 311–356 (in press, 2012)
3. Bertot, Y., Castéran, P.: Interactive Theorem Proving and Program Development. Coq'Art: The Calculus of Inductive Constructions. Texts in Theoretical Computer Science. An EATCS series. Springer (2004)
4. Bodenreider, O., Smith, B., Kumar, A., Burgun, A.: Investigating subsumption in SNOMED CT: An exploration into large description logic-based biomedical terminologies. Artificial Intelligence in Medicine 39, 183–195 (2007)
5. Booch, G.: Object-Oriented Design with Applications. Benjamin Cummings, Redwood City (1991)
6. Bourbaki, N.: Univers, Séminaire de Géométrie Algébrique du Bois Marie Théorie des topos et cohomologie étale des schémas (SGA 4), 1. Lecture notes in mathematics, vol. 269, pp. 185–217. Springer (1972)
7. Chein, M., Mugnier, M.L., Simonet, G.: Nested graphs: a graph-based knowledge representation model with FOL semantics. In: Procs. of KR 1998, pp. 524–534. Morgan Kaufmann (1998)
8. Cirstea, H., Coquery, E., Drabent, W., Fages, F., Kirchner, C., Maluszynski, J., Wack, B.: Types for Web Rule Languages: a preliminary study. Technical report A04-R-560, PROTHEO - INRIA Lorraine - LORIA (2004)
9. Coquand, T., Huet, G.: The calculus of constructions. Information and Computation 76(2-3), 95–120 (1988)
10. Coq Development Team, The Coq Reference Manual, Version 8.3., INRIA, France (2010)
11. Dapoigny, R., Barlatier, P.: Towards Ontological Correctness of Part-whole Relations with Dependent Types. In: Procs. of the Sixth Int. Conference (FOIS 2010), pp. 45–58 (2010a)

12. Dapoigny, R., Barlatier, P.: Modeling Contexts with Dependent Types. Fundamenta Informaticae 104(4), 293–327 (2010b)
13. Eiter, T., Lukasiewicz, T., Schindlauer, R., Tompits, H.: Combining answer set programming with description logics for the semantic web. In: Proc. of Ninth Int. Conf. on the Principles of Knowledge Representation and Reasoning (KR 2004), pp. 141–151. AAAI Press (2004)
14. Angelov, K., Enache, R.: Typeful Ontologies with Direct Multilingual Verbalization. In: Rosner, M., Fuchs, N.E. (eds.) CNL 2010. LNCS, vol. 7175, pp. 1–20. Springer, Heidelberg (2012)
15. Gangemi, A., Guarino, N., Masolo, C., Oltramari, A., Schneider, L.: Sweetening Ontologies with DOLCE. In: Gómez-Pérez, A., Benjamins, V.R. (eds.) EKAW 2002. LNCS (LNAI), vol. 2473, pp. 166–181. Springer, Heidelberg (2002)
16. Guarino, N.: The Ontological Level. In: Casati, R., Smith, B., White, G. (eds.) Philosophy and the Cognitive Science, pp. 443–456. Holder-Pivhler-Tempsky (1994)
17. Guarino, N., Welty, C.: An Overview of OntoClean. In: Handbook on Ontologies, pp. 151–172 (2004)
18. Guarino, N.: The Ontological Level: Revisiting 30 Years of Knowledge Representation. In: Borgida, A.T., Chaudhri, V.K., Giorgini, P., Yu, E.S. (eds.) Conceptual Modeling: Foundations and Applications. LNCS, vol. 5600, pp. 52–67. Springer, Heidelberg (2009)
19. Guizzardi, G., Herre, H., Wagner, G.: On the General Ontological Foundations of Conceptual Modeling. In: Spaccapietra, S., March, S.T., Kambayashi, Y. (eds.) ER 2002. LNCS, vol. 2503, pp. 65–78. Springer, Heidelberg (2002)
20. Guizzardi, G.: Ontological Foundations for Structural Conceptual Models. University of Twente (Centre for Telematics and Information Technology) (2005)
21. Guizzardi, G., Masolo, C., Borgo, S.: In Defense of a Trope-Based Ontology for Conceptual Modeling: An Example with the Foundations of Attributes, Weak Entities and Datatypes. In: Embley, D.W., Olivé, A., Ram, S. (eds.) ER 2006. LNCS, vol. 4215, pp. 112–125. Springer, Heidelberg (2006)
22. Howard, W.A.: To H.B. Curry: Essays on Combinatory Logic, Lambda Calculus and Formalism. The formulae-as-types notion of construction, pp. 479–490. Academic Press (1980)
23. Kabbaj, A., Janta-Polczynski, M.: From PROLOG++ to PROLOG+CG: A CG Object-Oriented Logic Programming Language, B. In: Ganter, B., Mineau, G.W. (eds.) ICCS 2000. LNCS (LNAI), vol. 1867, pp. 540–554. Springer, Heidelberg (2000)
24. Kaneiwa, K., Mizoguchi, R.: Ontological Knowledge Base Reasoning with Sort-Hierarchy and Rigidity. In: Procs. of KR 2004, pp. 278–288. AAAI Press (2004)
25. Keet, C.M., Artale, A.: Representing and reasoning over a taxonomy of part-whole relations. Applied Ontology 3(1-2), 91–110 (2008)
26. Kifer, M., Lausen, G., Wu, J.: Logical foundations of object-oriented and frame-based languages. Journal of the ACM 42, 741–843 (1995)
27. Krötzsch, M., et al.: How to reason with OWL in a logic programming system. In: Procs. of RuleML 2006 (2006)
28. Luo, Z.: Coercive subtyping. Journal of Logic and Computation 9(1), 105–130 (1999)
29. Masolo, C., Borgo, S., Gangemi, A., Guarino, N., Oltramari, A.: Ontology Library (D18). Laboratory for Applied Ontology-ISTC-CNR (2003)
30. McKinna, J.: Why dependent types matter. In: Procs. of the 33rd ACM SIGPLAN-SIGACT Symposium on Principles of Programming Languages, vol. 41(1), p. 1 (2006)
31. Mugnier, M.L., Leclère, M.: On querying simple conceptual graphs with negation. Data & Knowledge engineering 60(3), 468–493 (2007)
32. Mylopoulos, J., Borgida, A., Jarke, M., Koubarakis, M.: Telos: Representing Knowledge About Information Systems. ACM Trans. on Information Systems 8(4), 325–362 (1990)

33. Napoli, A.: Subsumption and classification-based reasoning in object-based representations. In: Procs. of the 10th European Conference on Artificial Intelligence (ECAI 1992), pp. 425–429. John Wiley & Sons Ltd. (1992)
34. Noonan, H.: Identity. In: Zalta, E.N. (ed.) The Stanford Encyclopedia of Philosophy (2011), http://plato.stanford.edu/archives/win2011/entries/identity/
35. Pires, L.F., van Sinderen, M., Munthe-Kaas, E., Prokaev, S.M.H., Plas, D.J.: Techniques for describing and manipulating context information, Freeband/A MUSE D3.5v2.0, Lucent Technologies (2005)
36. Paulin-Mohring, C.: Inductive Definitions in the System Coq - Rules and Properties. In: Bezem, M., Groote, J.F. (eds.) TLCA 1993. LNCS, vol. 664, pp. 328–345. Springer, Heidelberg (1993)
37. Rosati, R.: DL+log: Tight integration of description logics and disjunctive datalog. In: Proc. of Tenth Int. Conf. on Principles of Knowledge Representation and Reasoning (KR 2006), pp. 68–78. AAAI Press (2006)
38. Saibi, A.: Typing algorithm in type theory with inheritance. In: Procs. of the 24th ACM SIGPLAN-SIGACT Symposium on Principles of Programming Languages (POPL 1997), pp. 292–301. ACM Press (1997)
39. Setzer, A.: Object-Oriented Programming in Dependent Type Theory. In: Trends in Functional Programming, Intellect, vol. 7, pp. 91–108 (2007)
40. Smith, B., Rosse, C.: The Role of Foundational Relations in the Alignment of Biomedical Ontologies. In: Fieschi, M., et al. (eds.) MEDINFO 2004. IOS Press, Amsterdam (2004)
41. Sozeau, M., Oury, N.: First-Class Type Classes. In: Mohamed, O.A., Muñoz, C., Tahar, S. (eds.) TPHOLs 2008. LNCS, vol. 5170, pp. 278–293. Springer, Heidelberg (2008)
42. Sowa, J.F.: Using a lexicon of canonical graphs in a semantic interpreter. Relational models of the lexicon, pp. 113–137. Cambridge University Press (1988)
43. Sowa, J.F.: Knowledge Representation: Logical, Philosophical, and Computational Foundations. Brooks Cole Publishing Co., Pacific Grove (2000)
44. Sowa, J.F.: Conceptual Graphs. In: van Harmelen, F., Lifschitz, V., Porter, B. (eds.) Handbook of Knowledge Representation, ch. 5, pp. 213–237. Elsevier (2008)
45. Spitters, B., van der Weegen, E.: Type classes for mathematics in type theory. Mathematical Structures in Computer Science 21(4), 795–825 (2011)
46. Werner, B.: On the strength of proof-irrelevant type theories. Logical Methods in Computer Science 4(3) (2008)
47. Woods, W.A.: Understanding Subsumption and Taxonomy: a Framework for progress. In: Sowa, J. (ed.) Principles of Semantic Networks, pp. 45–94. Morgan Kaufmann (1991)

Parse Thicket Representation for Multi-sentence Search

Boris A. Galitsky[1], Sergei O. Kuznetsov[2], and Daniel Usikov[3]

[1] eBay Inc San Jose CA USA
boris.galitsky@ebay.com
[2] Higher School of Economics, Moscow Russia
skuznetsov@hse.ru
[3] Dept. of Physics University of Maryland MD USA
usikov@hotmail.com

Abstract. We develop a graph representation and learning technique for parse structures for sentences and paragraphs of text. This technique is used to improve relevance answering complex questions where an answer is included in multiple sentences. We introduce Parse Thicket as a sum of syntactic parse trees augmented by a number of arcs for inter-sentence word-word relations such as coreference and taxonomic. These arcs are also derived from other sources, including Rhetoric Structure theory, and respective indexing rules are introduced, which identify inter-sentence relations and joins phrases connected by these relations in the search index. Generalization of syntactic parse trees (as a similarity measure between sentences) is defined as a set of maximum common sub-trees for two parse trees. Generalization of a pair of parse thickets to measure relevance of a question and an answer, distributed in multiple sentences, is defined as a set of maximal common sub-parse thickets. The proposed approach is evaluated in the product search domain of eBay.com, where user query includes product names, features and expressions for user needs, and the query keywords occur in different sentences of text. We demonstrate that search relevance is improved by single sentence-level generalization, and further increased by parse thicket generalization. The proposed approach is evaluated in the product search domain of eBay.com, where user query includes product names, features and expressions for user needs, and the query keywords occur in different sentences of text.

Keywords: learning taxonomy, learning syntactic parse tree, syntactic generalization, search relevance.

1   Introduction

The task of answering complex questions, where desired information is distributed through multiple sentences in a document, becomes the bottleneck of modern search engines. The demand for access to different types of information have led to a renewed interest in answering questions posed in ordinary human language and seeking exact, specific and complete answer. After having made substantial achievements in fact-finding and list questions, natural language processing (NLP)

H.D. Pfeiffer et al. (Eds.): ICCS 2013, LNAI 7735, pp. 153–172, 2013.
© Springer-Verlag Berlin Heidelberg 2013

community turned their attention to more complex information needs that cannot be answered by simply extracting named entities (persons, organization, locations, dates, etc.) from single sentences in documents [4]. Complex questions often seek multiple different types of information simultaneously, located in multiple sentences, and do not presuppose that one single sentence could meet all of its information seeking expectations. To systematically analyze how keywords from query occur in multiple sentences in a document, one needs to explore coreferences and other relations between words within a sentence and between sentences.

Modern search engines attempt to find the occurrence of query keywords in a single sentence in a candidate search results [11]. If it is not possible or has a low search engine score, multiple sentences within one document are used. However, modern search engines have no means to determine if the found occurrences of the query keywords in multiple sentences are related to each other, to the same entity, and, being in different sentences, are all related to the query term.

In this study we attempt to systematically extract semantic features from paragraphs of text using a graph-based learning, assuming that an adequate parse trees for individual sentences are available. In our earlier studies [8,9] we applied graph learning to parse trees at the sentence level, and here we proceed to learning the structure of paragraphs, relying on *parse thickets*. Parse thicket is defined as a sum of parse trees with additional arcs between nodes for words in different sentences. We have defined the least general generalization of parse trees (we call it *syntactic generalization*), and in this study we extend it to the level of paragraphs. We propose parse thicket matching algorithm and apply it to re-rank multi-sentence answers to complex questions. Computing generalization of a pair of paragraph, we performed a pair-wise generalization for each sentence in paragraphs. This approach ignores the richness of coreference information, and in the current study we develop graph learning means specifically oriented to represent paragraphs of text as respective parse thickets with nodes interconnected by arcs for a number of relations including coreference. We consider a number of discourse-related theories such as Rhetoric Structure and Speech Acts as source of arcs to augment the parse thicket. These arcs will connect edges for words within as well as between parse trees for sentences.

Machine learning at the paragraph level is required for text classification problems, where handling the meaning (via collection of keywords) at the sentence level is insufficient, and taking advantage of coreference information is necessary [6]. In this paper we will demonstrate how building adequate paragraph structure is necessary when a paragraph is indexed for search. We will consider two cases for text indexing, where establishing proper coreferences inside and between sentences links entities in an index for proper match with a question (Fig. 1):

Text for indexing1: ... Tuberculosis is usually a lung disease. It is cured by doctors specializing in pulmonology.

Text for indexing2: ... Tuberculosis is a lung disease... Pulmonology specialist Jones was awarded a prize for curing a special form of disease.

Question: Which specialist doctor should treat my tuberculosis?

Fig. 1. Multi-sentence indexing cases

In the first case, establishing coreference link *Tuberculosis → disease → is cured by doctors pulmonologists* helps to match these entities with the ones from the question. In the second case this portion of text does not serve as a relevant answer to the question, although it includes keywords from this question. Hence at indexing time, keywords should be chained not just by their occurrence in individual sentences, but additionally on the basis of coreferences. If words X and Y are connected by a coreference relation, an index needs to include the chain of words $X_0, X_1...X, Y_0, Y_1...Y$, where chains $X_0, X_1...X$ and $Y_0, Y_1... Y$ are already indexed (phrases including X and Y). Hence establishing coreference is important to extend index in a way to improve search recall. Notice that usually, keywords from different sentences can only be matched with a query keywords with a low score (high score is delivered by inter-sentence match).

Since our problem concerns with finding the best sentence that contains the answer to any given question, we need some mechanism that can measure how close the a candidate answer is to the question. This allows us to choose the final answer which is the one that matches the most closely to the question. To achieve this we need a representation of the sentences that allows us to capture useful information in order to accommodate the matching process. We also need an efficient matching process to work on the chosen representation.

The evaluation of matching mechanism in this study is associated with improvement of search relevance by checking syntactic similarity between query and sentences in search hits, obtained via a search engine API. This kind of syntactic similarity is important when a search query contains keywords which form a phrase, domain-specific expression, or an idiom, such as "shot to shot time", "high number of shots in a short amount of time". In terms of search implementation, this can be done in two steps:

1)   Keywords are formed from query in a conventional manner, and search hits are obtained taking into account statistical parameters of occurrences these words in documents, popularity of hits, page rank and others.
2)   Above hits are filtered with respect to syntactic similarity of the snapshots of search hits with search query. Parse thicket generalization comes into play here.

Hence we obtain the results of the conventional search and calculate the score of the generalization results for the query and each sentence and each search hit snapshot. Search results are then re-ranked and only the ones syntactically close to search query are assumed to be relevant and returned to a user.

2   Generalizing Portions of Text

To measure similarity of abstract entities expressed by logic formulas, a least-general generalization was proposed for a number of machine learning approaches, including explanation based learning and inductive logic programming. Least general generalization was originally introduced in [14]. Its realization within the predicate logic is opposite to the most general unification; therefore it is also called *anti-unification*. In this study, to measure similarity between portions of text such as paragraphs, sentences and phrases, we extend the notion of generalization from logic formulas to sets of syntactic parse trees of these portions of text. The purpose of an abstract generalization is to find commonality between portions of text at various semantic levels. Generalization operation occurs on the levels of Article/Paragraph/Sentence/Phrases (noun, verb and others)/Individual word.

At each level except the lowest one, individual words, the result of generalization of two expressions is a *set* of expressions. In such set, expressions for which there exist less general expressions are eliminated. Generalization of two sets of expressions is a set of sets which are the results of pair-wise generalization of these expressions.

We outline the algorithm for two sentences and then proceed to the specifics for particular levels (Fig. 2). The algorithm we present in this paper deals with paths of syntactic trees rather than sub-trees, because it is tightly connected with language phrases. We refer the reader to [8,9] for more details.

---

1) Obtain parsing tree for each sentence. For each word (tree node) we have the *word (lemma), part of speech* and *form of word* information. This information is contained in the node label. We also have an arc to the other node.

2) Split parse trees for sentences into sub-trees which are phrases for each type: *verb, noun, prepositional* and others; these sub-trees are overlapping. The sub-trees are coded so that information about occurrence in the full tree is retained.

3) All sub-trees are grouped by phrase types.

4) Extending the list of phrases by adding equivalence transformations

5) Generalize each pair of sub-trees for both sentences for each phrase type.

6) For each pair of sub-trees yield an alignment, and then generalize each node for this alignment. For the obtained set of trees (generalization results), calculate the score.

7) For each pair of sub-trees for phrases, select the set of generalizations with highest score (least general).

8) Form the sets of generalizations for each phrase types whose elements are sets of generalizations for this type.

9) Filtering the list of generalization results: for the list of generalization for each phrase type, exclude more general elements from lists of generalization for given pair of phrases.

Fig. 2. Sentence-level syntactic generalization algorithm

For a pair of phrases, generalization includes all *maximum* ordered sets of generalization nodes for words in phrases so that the order of words is retained. In the following example

To buy digital camera today, on Monday
Digital camera was a good buy today, first Monday of the month

Generalization is {*<JJ-digital, NN-camera>* ,*<NN- today, ADV-*, NN-Monday>*} where the generalization for noun phrases is followed by the generalization by adverbial phrase. Verb *buy* is excluded from both generalizations because it occurs in a different order in the above phrases. *Buy - digital - camera* is not a generalization phrase because *buy* occurs in different sequence with the other generalization nodes. Further details on sentence level generalization are available in [8].

2.1   Direct Paragraph-Paragraph Match

We build a model of generalizing paragraphs taking into account coreference and taxonomic relationship between words between sentences, as well as within sentences. We will provide a number of examples to introduce the representation via parse thicket.   We start with a simple example of how a discourse can be visualized by a forest.

> Lady Gaga has revealed that her next album will be released as an app.
> The singer confirmed that the album, called ARTPOP, will be a multimedia experience.
> She says she wants fans to "fully immerse" themselves in the project.
> Content will include extra music, videos, chat options and games.

To answer cross-sentence questions, we need to establish connections between the words of different sentences, taking into account that each consecutive sentence elaborate on the previous one.

Question "multimedia experience from lady gaga" (Fig. 3) will need the path in Parse Thiket *Lad_Gaga <possession>*→ *album <same entity relation>*→ *album <is-a relation>* → *multimedia_experience*. Question "Does Lady Gaga rely on games content" will involve the path "*Lad_Gaga <possession>*→ *album <has-a relation>*→ *content <has-a relation>*→ *games*.

Fig. 3. Parse Thicket- supported search for a concert at StubHub/eBay

To establish the semantic relationships above, we need to use multiple sources. Notice that we cannot rely on ontologies, only on syntactic information, searching in a horizontal domain. Within a sentence, we use its parse tree. *Same-entity relation* is based on anaphora resolution, and has-a relation is based on syntactic structure within a sentence. Between sentences, we use the elaboration assumption that each consecutive sentence elaborates on some entities from previous sentences. It turns out that Rhetoric Structure theory provides a systematic framework to do that.

We now show a paragraph which includes four sentences with the relations between the words. These relations can be established once taxonomy of domain entities is available [3]. In this Section we are interested in the structure of the paragraph, encoded by these relations. Below we will be representing and visualizing arcs for these relations together with edges of constituency parse tree. We use the relations *Same entity/Sub-entity (a partial case)/Super-entity (more general)/Sibling entity/New predicate for an entity*.

We can visualize information flow in a paragraph by just showing the structure of entities, without original sentences. Then it becomes clear how each sentence brings in a new form of constraint for the entities from the previous set of sentences in a paragraph. This structure is fairly important for answering a question: one needs to determine which level of specificity is best to answer it. The structure of relations must be taken into account indexing this paragraph for search in addition to keywords for each sentence. The best match between the parse thicket for a question (usually, a trivial parse tree) and the set of parse thickets for an answers does not only indicates the best answer, but also the most appropriate sentences within this answer according to desired specificity as expressed in the query.

Notice that the answer relevance to a question is measured by the cardinality of maximal common sub-thicket. In a conventional search engine, the closer the answer to the question, the higher the number of keywords common between the question and the answer (weighted according to TF*IDF model and according to distances between these keywords in the answer. Parse Thicket approach makes similarity measure more linguistically aware to the structure of text by means of forming maximum common sub-thicket.

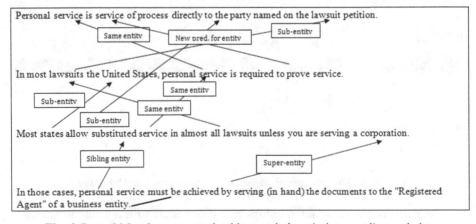

Fig. 4. Parse thicket for a paragraph with *super/sub-entity/new predicate* relations

This parse thicket (Fig.4) is helpful to answer a question 'How to serve a corporation in United States' where we need to link the second and the third sentences via the nodes *service* and *lawsuit*.

In our third example, we visualize the discourse structure of customer review (Fig.5). We now draw the detailed constituency parse thicket and augment it by all possible arcs we discover for relations between words, including coreferences and sub-entities. The text to be represented as parse thicket is as follows:

> After numerous attempts to bring my parents into the digital world, I think I have finally succeeded. I failed a few years ago with a Sony digital camera that they could not quite figure out how to use and have succeeded only modestly with regards to the computer and internet surfing. But, heck I decided to give it another try when they asked about a digital camera the other day.

3   Extending Parse Thickets with Rhetoric Structure-Based Arcs

We have demonstrated how to build parse thicket based on coreference arcs and similar/related-words arcs. In this section we attempt to treat computationally, with a unified framework, two approaches to textual discourse:

- Rhetoric structure theory (RST [12]);
- Speech Act theory;

Although both these theories have psychological observation as foundations and are mostly of a non-computational nature, we will build a specific computational framework for them. We will use these theories to find links between sentences to enhance indexing for search. For the concept structure based formalization of Speech Act Theory, we refer to our earlier paper [7]. We proposed a graph-based mechanism to represent a structure of a dialog using nodes for communicative actions and edges for temporal and other relationships between them. We used a vocabulary of communicative actions to

1) find their subjects,
2) add respective arcs to the parse thicket,
3) index combination of phrases as subjects of communicative actions

For RST, we introduce explicit indexing rules which will be applied to each paragraph and:

1) attempt to extract an RST relation,
2) build corresponding fragment of the parse thicket, and
3) index respective combination of formed phrases (noun, verb, prepositional), including words from different sentences.

People sometimes assume that whenever a text has some particular kind of discourse structure, there will be a signal indicating that structure. A typical case would be a conjunction such as 'but'.

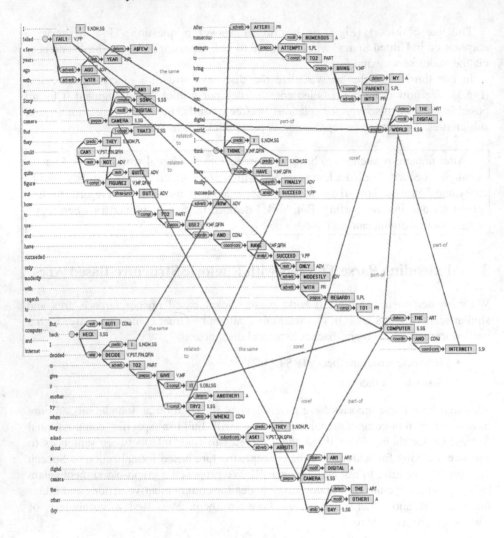

Fig. 5. Parse thicket for coreference, sub-entity and part-of relations

What structure is seen depends vitally on the words and sentences of the text are, but the relationship between words and text structure is extremely complex. Phrases and syntactic patterns can also be used to signal discourse structure. One expects discourse structure to be conveyed by signals. So the idea that discourse structure can be conveyed without signals is unexpected. Even more unexpected is the fact that for the discourse structure that RST represents, more than half is conveyed without explicit signals. On a relation-by-relation basis, it appears that every relation can be signaled in some contexts, and also that every relation can be conveyed without an explicit signal.

RST was originally developed as part of studies of computer-based text generation at Information Sciences Institute (part of University of Southern California) in about 1983 by Bill Mann, Sandy Thompson and Christian Matthiessen. The theory is designed to explain the coherence of texts, seen as a kind of function, linking parts of a text to each other. This coherence is explained by assigning a structure to the text, which slightly resembles a conventional sentence structure. We adjust this structure for the purpose of multi-sentence search ability.

We write *Syntactic template*

Index(Part-of- Syntactic template)

where *Syntactic template* indicates how to extract a particular RST relation from text, in syntactic generalization format, and

Index(Part-of- Syntactic template) is a set of expressions which will be indexed in addition to the original sequence of words. *Part-of- Syntactic template* is a set of sub-lists of *Syntactic template*.

For the purpose of search, we build syntactic templates to express RST Relational classes. We don't have to cover all RST relation, and we don't have to be precise in establishing them, unless relation type is matched with query term. We give examples of some relations and respective templates we use to detect an RST links, and specify respective indexing rules for how to add additional joined phrases to the search index.

- Consequence (N/S), Result, Cause, Cause-Result

Nonvolitional-cause: ImperMentalVB ...NP. Maybe... VP
----
index(NP, VP), {remember, recall, notice} ∈ ImperMentalVB.

In-response to NP ResponseVP
----
index(NP, ResponseVP)

NP AllowVP to ResultVP
----
index(NP, ResultVP), {allow, help, assist, give-ability}∈ ResultVP
- Manner, Means, Medium ('Medium demonstrates ... feature ... of the system')

MeansNP DemoVB NP to ToNP
----
index(NP, ToNP), {show, demonstrate, indicate} ∈ DemoVB
- Temporal-before, Temporal-same-time, Temporal-after

VP until UntilVP
----
index(NP, UntilNP)

For the following text, we build parse thicket for RST (Fig. 6) and SpActT (Fig. 7).

> Recently I tried to log into my account.
> I received an error message that my account had been locked.
> The site informed me to contact their appeal email address.
> I have done so several times; however:
> I get an email message back from Paypal stating that I cannot receive an answer of how to get into my account until I go to the site and login.
> Well, this is impossible because my account is locked.

For the latter ParseThicket, we have the following structure of communicative actions which form the inter-sentence arcs in Fig.7:

> *try [log-into-my-account]→ receive [error-message]*
>   /
> *Inform [to-contact...]-*
> *contact [their-appeal-email-address] → do [contact ... however]*
>   /
> *get [*email message back]
> *state [*I cannot]→ *receive []*
>   *answer [*how to ...]
>   *get* [get into my account until I go to the site and login]

We can now define a generalization operation on two parse thickets.

Given two parse thickets $C_x=(V_x, E_x)$ and $C_y=(V_y, E_y)$, generalization denoted $C_x \wedge C_y$ is defined as the set $\{G_1, G_2,...,G_k\}$ of all inclusion-maximal common subgraphs of C_x and C_y, such that each graph $G_i \in C_x \wedge C_y = (V_x, E_x)$ is characterized as follows:

1) v_i is a vertex in G_i iff v_i is a vertex in both C_x and C_y which corresponds to CAs of the same party (opponent or proponent;

2) (v_x, v_y) is a thick (resp. thin) arc in G_i iff (v_i, v_j) is a thick (resp. thin) arc in C_x and C_y;

3) (v_x, v_y) is a thick (resp. thin) arc in G_i iff (v_i, v_j) is a thick (resp. thin) arc in C_x and (v_i, v_j) is a thin (resp. thick) arc in C_y

4) G_i contains at least one thick arc (v_i, v_j).

Note that when (v_i, v_j) is of the same type (thin or thick) in both C_x and C_y, then that type is adopted for (v_i, v_j) in G_i.

Condition 3) specifies that a thin arc (v_i, v_j). is adopted as an arc in G_i whenever there are arcs (v_i, v_j) in C_x and C_y of different types (thin arcs are seen thus as a weaker generalization of both thick and thin arcs).

By applying this definition of generalization we are now able to provide a criterion for accepting/rejecting an answer by generalizing it with the question and earlier approved/rejected answers. We outline a nearest neighbor approach to relating a new answer to the class of relevant/irrelevant answer classes, on the basis of its similarity with previous question-answer pairs in the training dataset.

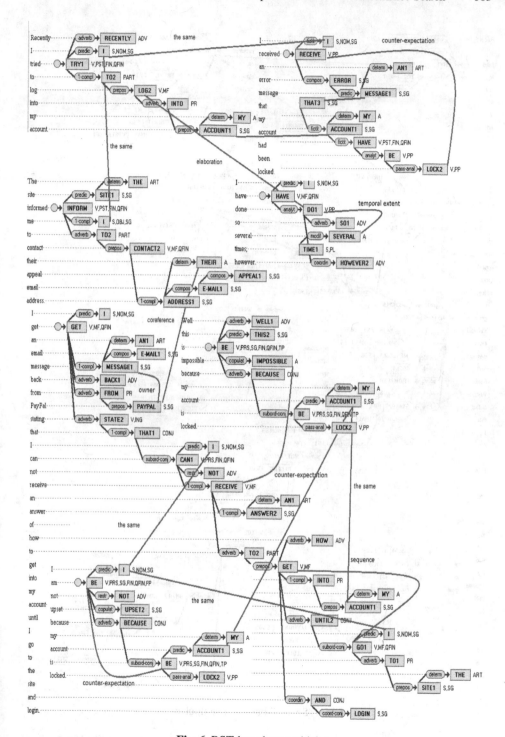

Fig. 6. RST-based parse thicket

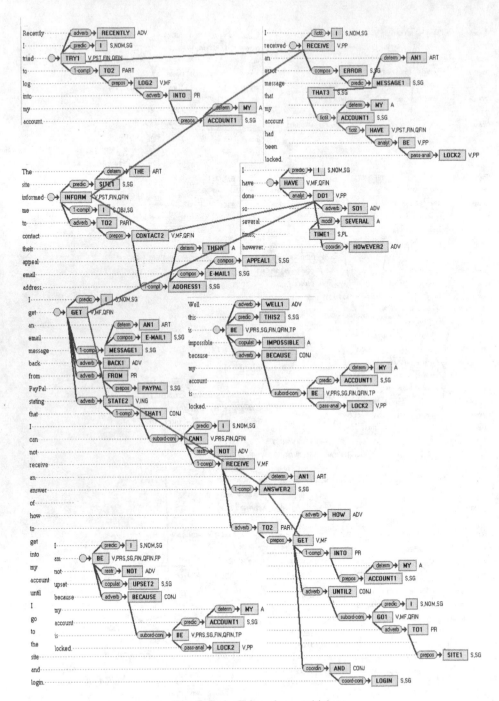

Fig. 7. SpActT-based parse thicket

Full parse thicket for the same paragraph we used for RST is depicted in Fig.7.

4   Evaluation of Parse Thicket Generalization

Syntactic generalization and parse thicket-based search has been implemented for entertainment-related domain at eBay's site StubHub.com. The query includes the desired performer, reference to a particular performance, as well as associated sentiments and feelings. Naturally, all such search criteria occur in different sentences, so the indexing system needs to find inter-sentence relations to verify that performers, events and user feelings are all properly related to each other.

The notion of query is rather broad in our case, including a posting in a blog, Facebook wall posting, or an email expressing an event attendance intent. The system is designed to answer complex queries about all products and associated sentiments, not just entertainment events. Queries are expected to include multiple sentences, where it is essential to track similarity between a query and abstract to improve user experience in search. In particular, the search is oriented to opinions data in linked aggregated form from various sources. To search for an opinion, a user specifies a product class, a name of particular products, a set of its features, specific concerns, needs or interests. A search can be narrowed down to a particular source, otherwise multiple sources of opinions (review portals, vendor-owned reviews, forums and blogs available for indexing) are combined. Search phrase may combine multiple sentences, for example: "*I am a beginner user of digital camera. I want to take pictures of my kids and pets. Sometimes I take it outdoors, so it should be waterproof to resist rain*". Obviously, this kind of specific opinion request can hardly be represented by keywords like 'beginner digital camera kids pets waterproof rain'.

We conducted evaluation of relevance of syntactic generalization – enabled search engine, based on Yahoo and Bing search engine APIs. For an individual query, the relevance was estimated as a percentage of correct hits among the first ten, using the values: {correct, marginally correct, incorrect} (compare with (Resnik, and Lin 2010)). Accuracy of a single search session is calculated as the percentage of correct search results plus half of the percentage of marginally correct search results. Accuracy of a particular search setting (query type and search engine type) is calculated, averaging through 40 search sessions.

For our evaluation, we use customers' queries to eBay entertainment and product-related domains, from simple questions referring to a particular product, a particular user need, as well as a multi-sentence forum-style request to share a recommendation. In our evaluation we split the totality of queries into noun-phrase class, verb-phrase class, how-to class, and also independently split in accordance to query length (from 3 keywords to multiple sentences). The evaluation was conducted by the authors.

To compare the relevance values between search settings, we used first 100 search results obtained for a query by Yahoo and Bing APIs, and then re-sorted them according to the score of the given search setting (syntactic generalization score and taxonomy-based score). To evaluate the performance of a hybrid system, we used the weighted sum of these two scores (the weights were optimized in an earlier search sessions).

Table 1 shows the search relevance evaluation results for single-sentence answers. The third and fourth columns show baseline Yahoo and Bing searches. The fifth column shows relevance of re-ranked search, and the last column shows relevance improvement compared with the baseline, the averaged Yahoo and Bing relevance.

Table 1. Evaluation of single-sentence search

Query	phrase sub-type	Relevancy of baseline Yahoo search, %, averaging over 20 searches	Relevancy of baseline Bing search, %, averaging over 20 searches	Relevancy of re-sorting by generalization, %, averaging over 40	Relevancy re-improvement: sorted relevance /(averaged for Bing & Yahoo)
3-4 word phrases	noun phrase	86.7	85.4	87.1	1.012
	verb phrase	83.4	82.9	79.9	0.961
	how-to expression	76.7	78.2	79.5	1.026
	Average	82.3	82.2	82.2	0.999
5-10 word phrases	noun phrase	84.1	84.9	87.3	1.033
	verb phrase	83.5	82.7	86.1	1.036
	how-to expression	82.0	82.9	82.1	0.996
	Average	83.2	83.5	85.2	1.022
2-3 sentences	one verb one noun phrases	68.8	67.6	69.1	1.013
	both verb phrases	66.3	67.1	71.2	1.067
	one sent of how-to type	66.1	68.3	73.2	1.089
	Average	67.1	67.7	71.2	1.056

We observe that using syntactic generalization improves the relevance of search in cases where query is relatively complex. For shorter sentences there is a slight drop in accuracy (-0.1%), for medium-length queries of 5-10 keywords we get 2% improvement, and 5.6% improvement for multi-sentence query. As the absolute performance of search naturally drops when queries become more complex, relative contribution of syntactic generalization increases.

We did not find a significant correlation between a query type, phrase type, and search performance with and without syntactic generalization for these types of phrases. Verb phrases in questions did well for multi-sentence queries perhaps because the role of verbs for such queries is more significant than for simpler queries where verbs can be frequently ignored.

Modern search engines attempt to find the occurrence of query keywords in a single sentence in a candidate search results. If it is not possible or has a low search engine score, multiple sentences within one document are used. However, modern search engines have no means to determine if the found occurrences of query keywords:

Are related to each other / Are related to the same entity /Being in different sentences, all related to the query term.

---

microvision laser projector fits in the palm of my hand

Amazon.com: MicroVision SHOWWX Classic **Laser** Pico **Projector ...**
www.amazon.com › ... › Computers & Accessories › Video Projectors
Microvision SHOWWX+ HDMI **Laser** Pico **Projector** (AA0123600-020) ... projected images that are always in focus from a device that **fits** right in your pocket. ... Its sleek contours feel great in your **hand** and slim enough to put in your pocket. ... I received **my** ShowWx **Laser** Pico **Projector** from Device Plus in Madrid, Spain ...

Amazon.com: Microvision SHOWWX+ HDMI **Laser** Pico **Projector ...**
www.amazon.com › ... › Camera & Photo › Projectors › Video Projectors
Amazon.com: **Microvision** SHOWWX+ HDMI **Laser** Pico **Projector** ... Lightweight, portable, **fits** in your pocket, doesn't include RCA cables .... underneath your iphone, so you can easily hold both devices in one **hand**. ... **My** review is based on the product and not on a comparison to other more ..... For BlackBerry and **Palm** ...

MicroVision SHOWWX+ **Laser** Pocket **Projector** - Amazon
www.amazon.com/MicroVision...Laser...Projector/.../B004EKM5B4
This review is from: **MicroVision** SHOWWX+ **Laser** Pocket **Projector** (Office Product) .... I was excited about the technology and couldn't wait to get **my hands** on one, ... The ShowWX+ requires you to choose 4:3 or 16:9 whichever **fits** best the current .... a pillow, somebodies back, a corner of a room, the **palm** of your **hand**.....it ...

Immersive Gaming part 1- iPhone 3GS plus **Microvision** ShowWX ...

www.youtube.com/watch?v=isn0m7eTHUI
7 May 2010 - 2 min - Uploaded by soundkite
... 3GS and my **laser** pico **projector** the ShowWX from **Microvision**, my entire ... for completely immersive ...

Fig. 8. Search results for the query requiring PARSE THICKET to provide relevance answers

---

To illustrate this statement, we search Google for   'microvision laser projector which fits in the palm of my hand'.

The expected/desired answer is as follows:

---

http://www.popularmechanics.com/technology/gadgets/4244056
I also saw another projection technology yesterday that looked pretty close to production. A company called **Microvision** produces a tiny, portable **projector** that uses red, green and blue **lasers** and a single tiny micromirror to project an image much the way old cathode ray tube televisions did, by scanning lines to create 60 frames per second. The whole **projector fits in the palm of your hand**, but a lot of that size comes from the battery; if you had an external power source, such as a USB, this thing could be as small as your thumb.

Read more: Innovative Projectors Will Fit in Your Palm, Cellphone: Buzzword @ CES 2008 (With Video) - Popular Mechanics

First few hits obtained by Google are shown in Fig 8. We observe all of the above tree problems in each of the search result. All answers are indeed about a 'Microvison projector', but user needs is represented by neither search result snippet. And if the keywords from the user need part of the query occur, they are not related to the main entity, 'Microvison projector'.

One can see that 'Microvision' (company name), 'laser' (type of product), 'fits in the palm of my hand' (user need) are most likely occur in different sentences, so matching of parse thicket is required to find a document with relevant information. Parse thicket approach would work best at indexing time, however in this paper we evaluate search relevance improvement in horizontal domain, re-ranking search engine API results.

4.1   Evaluation of Multi-sentence Search

To conduct a multi-sentence search evaluation, we also use Yahoo and Bing search APIs as for the single-sentence answers. We selected queries from eBay product searches which included reference to a product and a number of user need. Frequently expressions for these needs occurred in multiple sentences in product reviews, shopping forums and blogs. We also automatically filtered out the cases which gave satisfactory one-sentence answers to build multi-sentence parse thicket-based evaluation set.

Discovering trivial (in terms of search relevance) links between different sequences (such as coreferences) is not as important for search as finding more implicit links provided by text discourse theories. We separately measure search relevance when parse thicket is RST-based and SpActT-based. Since these theories are the main sources for establishing non-trivial links between sentences, we limit ourselves to measuring the contributions of these sources of links. Our hybrid approach includes both these sources for links.

We now conduct specific evaluation where answers are distributed through two or more sentences. If it is not the case, we exclude a query from our evaluation set. We consider all cases of questions (phrase, one, two, and three sentences) and all cases of search results occurrences (compound sentence, two, and three sentences) and measured how parse thicket improved the search relevance, compared to original search results ranking averaged for yahoo and Bing.

One can see that even the simplest cases of short query and a single compound sentence gives more than 5% improvement. Parse thicket - based relevance improvement stays within 7-9% as query complexity increases by a few keywords, and then increases to 9-11% as query becomes one-two sentences. For the same query complexity, naturally, search accuracy decreases when more sentences are required for answering this query. However, contribution of the parse thicket does not vary significantly with the number of sentences the answer occurs in.

Table 2. Search improvement results for parse thicket approach

Query	Answer	Relevancy of baseline Yahoo search, %, averaging over 20 searches	Relevancy of baseline Bing search, %, averaging over 20 searches	Relevancy of re-sorting by pair-wise sentence generalization, %, averaging over 40 searches	Relevancy of re-sorting by forest generalization based on RST, %, averaging over 20 searches	Relevancy of re-sorting by forest generalization based on SpActT, %, averaging over 20 searches	Relevancy of re-sorting by hybrid RST+SpActT forest generalization, %, averaging over 40 searches	Relevancy improvement for parse thicket approach, comp. to pair-wise generalization
3-4 word phrases	1 comp. sentence	81.7	82.4	86.6	88.0	87.2	91.3	1.054
	2 sent	79.2	79.9	82.6	86.2	84.9	89.7	1.086
	3 sent	76.7	75.0	79.1	85.4	86.2	88.9	1.124
	Average	79.2	79.1	82.8	86.5	86.1	90.0	1.087
5-10 word phrases	1 comp. sentence	78.2	77.7	83.2	87.2	84.5	88.3	1.061
	2 sent	76.3	75.8	80.3	82.4	83.2	87.9	1.095
	3 sent	74.2	74.9	77.4	81.3	80.9	82.5	1.066
	Average	76.2	76.1	80.3	83.6	82.9	86.2	1.074
1 sentence	1 comp. sent	77.3	76.9	81.1	85.9	86.2	88.9	1.096
	2 sent	74.5	73.8	78.	82.5	83.1	86.3	1.101
	3 sent	71.3	72.2	76.5	80.7	81.2	83.2	1.088
	Average	74.4	74.3	78.7	83.0	83.5	86.1	1.095
2 sentences	1 comp. sent	75.7	76.2	82.2	87.0	83.2	83.4	1.015
	2 sent	73.1	71.0	76.8	82.4	81.9	82.1	1.069
	3 sent	69.8	72.3	75.2	80.1	79.6	83.3	1.108
	Average	72.9	73.2	78.1	83.2	81.6	82.9	1.062
3 sentences	1 sentence	73.6	74.2	78.7	85.4	83.1	85.9	1.091
	2 sentences	73.8	71.7	76.3	84.3	83.2	84.2	1.104
	3 sentences	67.4	69.1	74.9	79.8	81.0	84.3	1.126
	Average	71.6	71.7	76.6	83.2	82.4	84.8	1.107
Average for all Query and Answer type								1.085

Notice that there is a noticeable improvement of accuracy in the comparable cases between Tables 1 and 2. While single-sentence syntactic match gives 5.6% improvement, multi-sentences parse thickets provides 8.7% for the comparable query complexity (5.4% for single-sentence answer) and up to 10% for the cases with more complex answers. One can see that parse thicket improves single sentence syntactic generalization by at least 2%. On average through the cases of Table 2, parse thickets outperforms single sentence syntactic generalization by 6.7%, whereas RST on its own gives 4.6% and SpActT-4.0% improvement respectively. Hybrid RST + SpActT gives 2.1% improvement over RST-only and 2.7% over SpActT only. We conclude that RST links compliment SpActT links to properly establish relations between entities in sentences for the purpose of search.

5   Related Work and Conclusions

Usually, classical approaches to semantic inference rely on complex logical representations. However, practical applications usually adopt shallower lexical or lexical-syntactic representations, but lack a principled inference framework. [2] proposed a generic semantic inference framework that operates directly on syntactic trees. New trees are inferred by applying entailment rules, which provide a unified representation for varying types of inferences. The current work deals with syntactic tree transformation in the graph learning framework, treating various phrasings for the same meaning in a more unified and automated manner.

The set of semantic problems addressed in this paper is of a much higher semantic level compared to SRL, therefore more sensitive tree matching algorithm is required for such semantic level. In terms of this study, semantic level of classification classes is much higher than the level of semantic role labeling or semantic entailment. SLR does not aim to produce complete formal meanings, in contrast to our approach. Unlike [19] who uses edit distance for finding optimal dependency tree matching, we use maximal set of common sub-graphs which obeys logical properties of least general generalization and is therefore better suited to ascend to semantic level (of logical forms representation). This study operates on the level of paragraphs instead of sentences and our previous studies [7, 8]. Also, we apply re-ranking to search engine results and not a raw index. Lexical chain formalism can be considered as a special case of parse thicket. Paper [5] considered keywords as condensed versions of documents and short forms of their summaries. The authors treat the problem of automatic extraction of keywords from documents as a supervised learning task.

In this study we introduced the notion of syntactic generalization to learn from parse trees for a pair of sentences, and extended it to learning augmented parse thickets for two paragraphs. Parse thicket is intended to represent syntactic structure of text as well as a number of semantic relations for the purpose of indexing for search. To accomplish this, parse thicket includes relations between words in different sentences, such that these relations are essential to match queries with portions of texts to serve as an answer.

We considered the following sources of relations between words in sentences:

Coreferences, Taxonomic relations such as sub-entity, partial case, predicate for subject etc.; Rhetoric structure relation and Speech acts. Since the first and second source of relations has been explored in details, we focus our evaluation on the contribution of third and fourth sources. We demonstrated that search relevance can be improved, if search results are subject to confirmation by sentence-level syntactic generalization, if answer occurs in a single sentence, and by parse thicket generalization, if answer occurs in multiple sentences.

Traditionally, machine learning of linguistic structures is limited to keyword forms and frequencies. At the same time, most theories of discourse are not computational, they model a particular set of relations between consecutive states. In this work we attempted to achieve the best of both worlds: learn complete parse tree information augmented with an adjustment of discourse theory allowing computational treatment.

Graphs have been used extensively to formalize ranking of NL texts [18]. Graph-based ranking algorithms are a way of deciding the importance of a vertex within a graph, based on global information recursively drawn from the entire graph. Using semantic information for query ranking has been proposed in [1]. Moreover, relying on matching of parse trees of a question and an answer has been the subject of [13]. However, we believe the current study leads the way in multi-sentence relevance improvement, relying on learning parse trees and linguistic theories of the nature of the coherence of texts.

Acknowledgments. The second author was supported by the project "Mathematical Models, Algorithms, and Software Tools for Intelligent Analysis of Structural and Textual Data" within the framework of the Basic Research Program at the National Research University Higher School of Economics, Moscow, Russia.

References

1. Alcman-Meza, B., Halaschek, C., Arpinar, I., Sheth, A.: A Context-Aware Semantic Association Ranking. In: Proc. First Int'l Workshop Semantic Web and Databases (SWDB 2003), pp. 33–50 (2003)
2. Bar-Haim, R., Dagan, I., Greental, I., Shnarch, E.: Semantic Inference at the Lexical-Syntactic Level AAAI (2005)
3. Bhogal, J., Macfarlane, A., Smith, P.: A review of ontology based query expansion. Information Processing & Management 43(4), 866–886 (2007)
4. Chali, Y., Hasan, S.A., Joty, S.R.: Improving graph-based random walks for complex question answering using syntactic, shallow semantic and extended string subsequence kernels. Inf. Process. Manage. 47(6), 843–855 (2011)
5. Ercan, G., Cicekli, I.: Using lexical chains for keyword extraction. Information Processing & Management 43(6), 1705–1714 (2007)
6. Galitsky, B.: Natural Language Question Answering System: Technique of Semantic Headers. Advanced Knowledge International, Australia (2003)
7. Galitsky, B., González, M.P., Chesñevar, C.I.: A novel approach for classifying customer complaints through graphs similarities in argumentative dialogue. Decision Support Systems 46(3), 717–729 (2009)

8. Galitsky, B., Dobrocsi, G., de la Rosa, J.L.: Inferring semantic properties of sentences mining syntactic parse trees. Data & Knowledge Engineering 81-82, 21–45 (2012)
9. Galitsky, B., Dobrocsi, G., de la Rosa, J.L., Kuznetsov, S.O.: Using Generalization of Syntactic Parse Trees for Taxonomy Capture on the Web. In: 19th International Conference on Conceptual Structures, ICCS 2011, pp. 104–117 (2011)
10. Kapoor, S., Ramesh, H.: Algorithms for Enumerating All Spanning Trees of Undirected and Weighted Graphs. SIAM J. Computing 24, 247–265 (1995)
11. Kim, J.-J., Pezik, P., Rebholz-Schuhmann, D.: MedEvi: Retrieving textual evidence of relations between biomedical concepts from Medline. Bioinformatics 24(11), 1410–1412 (2008)
12. Mann, W.C., Christian, M.I., Matthiessen, M., Thompson, S.A.: Rhetorical Structure Theory and Text Analysis. In: Mann, W.C., Thompson, S.A. (eds.), pp. 39–78. John Benjamins, Amsterdam (1992)
13. Moschitti, A.: Efficient Convolution Kernels for Dependency and Constituent Syntactic Trees. In: Fürnkranz, J., Scheffer, T., Spiliopoulou, M. (eds.) ECML 2006. LNCS (LNAI), vol. 4212, pp. 318–329. Springer, Heidelberg (2006)
14. Plotkin, G.D.: A note on inductive generalization. In: Meltzer, B., Michie, D. (eds.) Machine Intelligence, vol. 5, pp. 153–163. Elsevier North-Holland, New York (1970)
15. Punyakanok, V., Roth, D., Yih, W.: The Necessity of Syntactic Parsing for Semantic Role Labeling. In: IJCAI (2005)
16. OpenNLP (2012), http://incubator.apache.org/opennlp/documentation/manual/opennlp.html
17. Marcu, D.: From Discourse Structures to Text Summaries. In: Mani, I., Maybury, M. (eds.) Proceedings of ACL Workshop on Intelligent Scalable Text Summarization, Madrid, Spain, pp. 82–88 (1997)
18. Mihalcea, R., Tarau, P.: TextRank: Bringing Order into Texts. In: Empirical Methods in NLP (2004)
19. Punyakanok, V., Roth, D., Yih, W.: Mapping dependencies trees: an application to question answering. In: Proceedings of AI & Math., Florida, USA (2004)

FCA-Based Models and a Prototype Data Analysis System for Crowdsourcing Platforms

Dmitry I. Ignatov[1], Alexandra Yu. Kaminskaya[1,2], Anastasya A. Bezzubtseva[1,2], Andrey V. Konstantinov[1], and Jonas Poelmans[1,3]

[1] National Research University Higher School of Economics, Russia, 101000, Moscow, Myasnitskaya str., 20
dignatov@hse.ru
http://www.hse.ru
[2] Witology
http://www.witology.com
[3] KU Leuven, Belgium

Abstract. This paper considers a data analysis system for collaborative platforms which was developed by the joint research team of the National Research University Higher School of Economics and the Witology company. Our focus is on describing the methodology and results of the first experiments. The developed system is based on several modern models and methods for analysing of object-attribute and unstructured data (texts) such as Formal Concept Analysis, multimodal clustering, association rule mining, and keyword and collocation extraction from texts.

Keywords: collaborative and crowdsourcing platforms, Data Mining, Formal Concept Analysis, multimodal clustering.

1 Introduction and Related Work

The success of modern collaborative technologies is marked by the appearance of many novel platforms for holding distributed brainstorming or carrying out so called "public examination". There are a lot of such crowdsourcing companies in the USA (Spigit [1], BrightIdea [2], InnoCentive [3] etc.) and Europe (Imaginatik [4]). There is also the Kaggle platform [5] which is most beneficial for data practitioners and companies that want to select the best solutions for their data mining problems. In 2011 Russian companies launched business in that area as well. The two most representative examples of such Russian companies are Witology [6] and Wikivote [7]. The reality as yet is far away from technological breakthrough, though some all-Russian projects have already been finished successfully (for example, Sberbank-21, National Entrepreneurial Initiative-2012 [8] etc.). The core of such crowdsourcing systems is a socio-semantic network [9,10,11,12], which data requires new approaches to analyze. This paper is devoted to the new methodological base for the analysis of data generated by collaborative systems, which uses modern data mining and artificial intelligence models and methods. As a rule, while participating in a project, users of such

H.D. Pfeiffer et al. (Eds.): ICCS 2013, LNAI 7735, pp. 173–192, 2013.

crowdsourcing platforms [13] discuss and solve one common problem, propose their ideas and evaluate ideas of each other as experts. Finally, as a result of the discussion and ranking of users and their ideas we get the best ideas and users (their generators). For deeper understanding of users's behavior, developing adequate ranking criteria and performing complex dynamic and statistic analyses, special means are needed. Traditional methods of clustering, community detection and text mining need to be adapted or even fully redesigned. Moreover, these methods require ingenuity for their effective and efficient use (finding nontrivial results). We briefly describe models of data used in crowdsourcing projects in terms of Formal Concept Analysis (FCA) [14]. Furthermore, we present the collaborative platform data analysis system CrowDM (Crowd Data Mining), its architecture and methods underlying the key steps of data analysis.

The remainder of the paper is organized as follows. Section 2 contains descriptions of the Witology crowdsourcing methodology and Sberbank-21 project. In section 3 we describe some key notions from FCA, our data and methods. In section 4 we discuss the analysis scheme of the developed system. In section 5 we present the results of our first experiments with the Sberbank-21 data. Section 6 concludes our paper and describes some possible directions for future research.

2   Witology Crowdsourcing Methodology and Projects

One proverb says "Two heads are better than one", but crowdsourcing projects may take several thousands of heads. The term "crowdsourcing" is a portmanteau of "crowd" and "outsourcing", coined by Jeff Howe in 2006 [13]. There is no general definition of crowdsourcing, but it takes some specific features. Crowdsourcing is a process, both online and offline, that includes task solving by a distributed and large group of people who are usually from different organisations, and not necessarily paid by money for their work.

We shortly describe the methodology of the Witology crowdsourcing company, Witodology, considering as an example its Sberbank-21 project. Note that the company clearly says that Witodology is based on the notion of socio-sematic networks [11,12]. In 2011, from October till November, Witology and Sberbank launched one of the first successful crowdsourcing projects in Russia: Sberbank-21. The Russian company Sberbank is the largest and oldest Russian bank which history started in 1841; its name can be formally translated into English as "Savings Bank of the Russian Federation". The project was devoted to the theme "Office of Sberbank in 2012 (Office SB-2012)". The main project topics include "Office of SB-21 for private clients", "Office SB-21 for individual entrepreneurs", "Office SB-21 for small businesses", "Internal filling of "physical" SB-21 office", and "Internal filling of "virtual" SB-21 office". The goal was formulated as "a selection of the best well-founded and innovative solutions for the formulated tasks, which include format and content of the proposed solutions as well as reasons for their appearance and adoption". During the preliminary test, 450 experts were selected out of 5198 people. Amongst them 33% were women and 67% were man, 21% were Sberbank employees and 79% of them were either clients or other interested persons. The main stages of the project were "Solution's generation", "Selection of similar solutions", "Generation of counter-solutions", "Total voting", "Solution's improvements", "Solution's stock", "Final improvements" and finally "Solution's review".

In total, 222 experts proposed 1581 solutions for 15 tasks which were grouped into 5 topics. After "Selection of similar solutions" by participants, 24 574 analytical operations including comparison, clustering, and filtering of ideas were performed. As a result of this stage, 589 solutions were selected. The stage "Total voting" resulted in the selection of 182 solutions. After the stage "Solution's stock" 75 solutions were left. From 15 remaining solutions after "Solution's review" 3 solutions were nominated as the best ones.

The first stage "Solution's generation" is performed individually by each user. A key difference between traditional brainstorming and the "Solution's generation" stage is that nobody can see or listen to ideas of other participants. The main similarity is the absence of criticism which was moved to later stages. In the "Selection of similar solutions" phase participants are selecting similar ideas (solutions) and their aggregated opinions are transformed to clusters of similar ideas.

For Sberbank-21 projects all the proposed solutions were divided into 15 clusters (tasks), three per topic: (Sberbank and private client: interface in 2021?, Sberbank-21 service for every 21 years old person?, Unique service of 2021 for private clients?), (Sberbank and entrepreneur: interface in 2021?, Service in 2021 for startupers?, Unique service in 2021 for entrepreneurs?), (Sberbank and small businesses: interface in 2021?, Service in 2021 for new businesses?, Unique service for small businesses?), (What will disappear in the "physical" office of SB-21?, What will change in the "physical" office SB-21?, What will appear in the "physical" office of SB-21?) (What will disappear in the online office of SB-21?, What will change in the online office of SB-21?, What will appear in the online office of SB-21?).

Counter-solutions generation includes criticism (pros and cons) and evaluation of proposed ideas by communication between an author and experts. During this stage an idea's author can invite other experts to his team taking into account their contribution to discussion and criticism. Total voting is performed by evaluation of each proposed idea by all users in terms of their attitude and quality levels of the solution (marks are integers between -3 and 3). Two stages, i.e. "Solution's improvements" and "Final improvements", involve active collaboration by experts and authors who improve their solutions together.

The system calculates 10 user's ratings based on their activity; among them are "Popularity","Social capital", "Performance", "Gamer", "Actor", "Judge", "Commenter", "Importance", "Influence", and "Reputation". For texts the company uses the following rates: "Significance", "Influence", "Popularity", "Quality", "Attitude" and also "Reputation".

Solution's stock is one of the most interesting game stages of the project when all participants with a positive reputation rate accumulated on the previous stages take money in internal currency "wito" and can perform stock trade.

The solutions with the highest price become winners. Finally, during the review of solutions, experts with high reputation make their final evaluations based on several criteria in -3 to 3 scale: "Solution efficiency", "Solution originality", "Solution performability", and "Return on investment".

We have to tell the reader about some best solutions but cannot go into details because of non-disclosure requirements. For example, for the task "What will disappear in the "physical" SB-21 office?" the best solution said that "You shouldn't to fill in the same documents several times". For the task "What will change in the SB-21 "physical" office" the solution was "Changing the access mode for a safe deposit box to biometric data for corporate clients and Near Based Communication (NFC) chips for private clients". And the answer for the question "What will appear in the SB-21 "physical" office?" was "Videowall", a sort of interface for communication with a distant operator and making regular financial transactions.

3   Mathematical Models and Methods

At the initial stage of collaborative platform data analysis two data types were identified: data without using keywords (links, evaluations, user actions) and data with keywords (all user-generated content). These two data types totally correspond with two components of a socio-semantic network. For the analysis of the 1st type of data (with keywords) we suggest to apply Social Network Analysis (SNA) methods, clustering (biclustering and triclustering [15,16,17], spectral clustering), FCA (concept lattices, implications, association rules) and its extensions for multimodal data, triadic, for instance [18]; recommender systems [19,20,21,22] and statistical methods of data analysis [23] (the analysis of distributions and average values).

3.1   Formal Concept Analysis and OA-biclustering

Methods described in this paper are mainly from the multimodal clustering block at the analysis scheme (see fig. 2). The protagonists of crowdsourcing projects (and corresponding collaborative platforms) are platform users (project participants). We consider them as *objects* for analysis. More than that, each object can (or cannot) possess a certain set of *attributes*. The user's attributes can be: topics which the user discussed, ideas which he generated or voted for, or even other users. The main instrument for analysis of such object-attribute data is FCA [14]. Let us give formal definitions. *The formal context in FCA is a triple* $\mathbb{K} = (G, M, I)$, *where* G *is a set of objects*, M *is a set of attributes*, and the relation $I \subseteq G \times M$ shows which object possesses which attribute. For any $A \subseteq G$ and $B \subseteq M$ one can define *Galois operators*:

$$A' = \{m \in M \mid gIm \text{ for all } g \in A\}, \tag{1}$$
$$B' = \{g \in G \mid gIm \text{ for all } m \in B\}.$$

The operator $''$ (applying the operator $'$ twice) is a *closure operator*: it is idempotent $(A'''' = A'')$, monotonous $(A \subseteq B$ implies $A'' \subseteq B'')$ and extensive $(A \subseteq A'')$. The set of objects $A \subseteq G$ such that $A'' = A$ is called closed. The same properties hold for closed attribute sets, i.e. subsets of the set M. A couple (A, B) such that $A \subset G$, $B \subset M$, $A' = B$ and $B' = A$, is called *formal concept* of a context K. The sets A and B are closed and called *extent* and *intent* of a formal concept (A, B) respectively. For the set of objects A the set of their common attributes A' describes the similarity of objects of the set A, and the closed set A'' is a cluster of similar objects (with the set of common attributes A'). The relation "to be more general concept" is defined as follows: $(A, B) \geq (C, D)$ iff $A \subseteq C$. We denote by $\mathfrak{B}(G, M, I)$ the set of all concepts of a formal context $\mathbb{K} = (G, M, I)$. The concepts of a formal context $\mathbb{K} = (G, M, I)$ ordered by extensions inclusion form a lattice, which is called a *concept lattice*. For its visualization a *line diagram* (Hasse diagram) can be used, i.e. the cover graph of the relation "to be a more general concept".

To represent datasets with numerical (e.g., age, word frequency, number of comments) and categorical (e.g., gender, job) attributes there are many-valued contexts. A *many-valued context* (G, M, W, I) consists of sets G, M and W and a ternary relation $I \subseteq G \times M \times W$ for which it holds that

$$(g, m, w) \in I \text{ and } (g, m, v) \in I \Rightarrow w = v.$$

The elements of G are still called objects, those of M (many-valued) attributes and the elements of W attribute values. Sometimes we write $m(g) = w$ to show that the object g has the value w of the attribute m.

We can transform the many-valued context into a one-valued one by means of **conceptual scaling** [14].

In the worst case (Boolean lattice) the number of concepts is equal to $2^{\{\min |G|, |M|\}}$, thus, for large contexts, FCA can be used only if the data is sparse. Moreover, one can use different ways of reducing the number of formal concepts (choosing concepts by stability [24] index or extent size). The alternative approach is a relaxation of the definition of formal concept as maximal rectangle in object-attribute matrix which elements belong to the incidence relation. One of such relaxations is the notion of object-attribute bicluster [16]. If $(g, m) \in I$, then (m', g') is called *object-attribute bicluster* with the density $\rho(m', g') = |I \cap (m' \times g')| / (|m'| \cdot |g'|)$.

The main features of OA-biclusters are listed below:

1. For any bicluster $(A, B) \subseteq 2^G \times 2^M$ it is true that $0 \leq \rho(A, B) \leq 1$.
2. OA-bicluster (m', g') is a formal concept iff $\rho = 1$.
3. If (m', g') is a bicluster, then $(g'', g') \leq (m', m'')$.

Let $(A, B) \subseteq 2^G \times 2^M$ be a bicluster and $\rho_{\min}$ be a non-negative real number such that $0 \leq \rho_{\min} \leq 1$, then (A, B) is called *dense*, if it fits the constraint $\rho(A, B) \geq \rho_{\min}$. The above mentioned properties show that OA-biclusters differ from formal concepts since unit density is not required. Graphically it means that not all the cells of a bicluster must be filled by a cross (see fig. 1). Besides

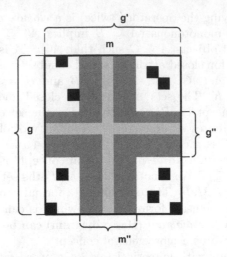

Fig. 1. OA-bicluster

formal lattice construction and visualization by means of Hasse diagrams one can use implications and association rules for detecting attribute dependencies in data. Then, using the obtained results, it is easy to form recommendations (for example, offering users the most interesting discussions for them). Furthermore, structural analysis can be performed and then used for finding communities. Statistical methods are helpful for frequency analysis of the different users' activities. Almost all of the above mentioned methods can be applied to data containing users' keywords (in this case they become attributes of a user).

3.2   Triadic FCA and OAC-triclustering

To deal with three-way data within FCA, an extension to Triadic Concept Analysis (TCA) was proposed by Lehman and Wille [25,26]. In [18] the author introduced the TRIAS algorithm for mining all frequent triconcepts from 3-dimensional data and applied it to the popular Bibsonomy (users-tags-papers) dataset. Voutsadakis [27] extended triadic concept analysis to n-dimensional contexts.

There exist some known difficulties in mining binary data, such as a lack of fault tolerance, an explosion of the number of patterns leading to large computational complexity and to many small patterns that appear to be false positive observations. In triadic or n-ary contexts these problems are seriously aggravated. To cope with these issues, several techniques have been introduced for faster selection of interesting patterns. For example, there is an extended box clustering approach [28] and triadic concept factors [29]. Another approach, called constraint-based mining, also scales up to n-ary relations and is discussed in [30] and [31]. In [17] we also proposed a new triclustering approach for mining so-called (dense) OAC-triclusters, where OAC stands for Object Attribute

Condition. This algorithm has a better theoretical time complexity than existing exact algorithms like TRIAS and is therefore better suited for very large datasets. Moreover, during experimentations with the bibsonomy dataset, we found the number of triclusters generated by our algorithm to be significantly lower than the number of triconcepts extracted by TRIAS. Manual validation of the extracted tricommunities revealed that the majority of them was meaningful.

A triadic context $\mathbb{K} = (G, M, B, Y)$ consists of sets G (objects), M (attributes), and B (conditions), and ternary relation $Y \subseteq G \times M \times B$. An incidence $(g, m, b) \in Y$ shows that the object g has the attribute m under condition b.

For convenience, a triadic context is denoted by (X_1, X_2, X_3, Y). A triadic context $\mathbb{K} = (X_1, X_2, X_3, Y)$ gives rise to the following diadic contexts

$$
\begin{aligned}
\mathbb{K}^{(1)} &= (X_1, X_2 \times X_3, Y^{(1)}), \\
\mathbb{K}^{(2)} &= (X_2, X_1 \times X_3, Y^{(2)}), \\
\mathbb{K}^{(3)} &= (X_3, X_1 \times X_2, Y^{(3)}),
\end{aligned}
\tag{2}
$$

where $gY^{(1)}(m, b) :\Leftrightarrow mY^{(1)}(g, b) :\Leftrightarrow bY^{(1)}(g, m) :\Leftrightarrow (g, m, b) \in Y$. The derivation operators (primes or concept-forming operators) induced by $\mathbb{K}^{(i)}$ are denoted by $(.)^{(i)}$. For each induced dyadic context we have two kinds of derivation operators. That is, for $\{i, j, k\} = \{1, 2, 3\}$ with $j < k$ and for $Z \subseteq X_i$ and $W \subseteq X_j \times X_k$, the (i)-derivation operators are defined by:

$$
\begin{aligned}
Z \mapsto Z^{(i)} &= \{(x_j, x_k) \in X_j \times X_k | x_i, x_j, x_k \text{ are related by Y for all } x_i \in Z\} \\
W \mapsto W^{(i)} &= \{x_i \in X_i | x_i, x_j, x_k \text{ are related by Y for all } (x_j, x_k) \in W\}.
\end{aligned}
\tag{3}
$$

Formally, a triadic concept of a triadic context $\mathbb{K} = (X_1, X_2, X_3, Y)$ is a triple (A_1, A_2, A_3) of $A_1 \subseteq X_1, A_2 \subseteq X_2, A_3 \subseteq X_3$, such that for every $\{i, j, k\} = \{1, 2, 3\}$ with $j < k$ we have $A_i^{(i)} = (A_j \times A_k)$. For a certain triadic concept (A_1, A_2, A_3), the components A_1, A_2, and A_3 are called the extent, the intent, and the modus of (A_1, A_2, A_3). It is important to note that for interpretation of $\mathbb{K} = (X_1, X_2, X_3, Y)$ as a three-dimensional cross table, according to our definition, under suitable permutations of rows, columns, and layers of the cross table, the triadic concept (A_1, A_2, A_3) is interpreted as a maximal cuboid full of crosses. The set of all triadic concepts of $\mathbb{K} = (X_1, X_2, X_3, Y)$ is called the concept trilattice and is denoted by $\mathfrak{T}(X_1, X_2, X_3, Y)$.

To simplify notation, we denote by $(.)'$ all prime operators, as it is usually done in FCA. For our purposes consider a triadic context $\mathbb{K} = (G, M, B, Y)$ and introduce primes, double primes and box operators for particular elements of G, M, B, respectively. In what follows, we write g' instead of $\{g\}'$ for 1-set $g \in G$ and similarly for $m \in M$ and $b \in B$: m' and b'.

We do not use double primes, because of their rigid structure; they do not tolerate exceptions like some amount of missing pairs. To allow missing pairs in the operators results we introduce box operators:

$$
\begin{aligned}
g^{\Box} &= \{ g_i \mid (g_i, b_i) \in m' \text{ or } (g_i, m_i) \in b' \} \\
m^{\Box} &= \{ m_i \mid (m_i, b_i) \in g' \text{ or } (g_i, m_i) \in b' \} \\
b^{\Box} &= \{ b_i \mid (g_i, b_i) \in m' \text{ or } (m_i, b_i) \in g' \}.
\end{aligned}
\tag{4}
$$

Table 1. Prime and double prime operators of 1-sets

Prime operators of 1-sets	Their double prime counterparts
$m' = \{ (g,b) \mid (g,m,b) \in Y \}$	$m'' = \{ \tilde{m} \mid (g,b) \in m' \quad and \quad (g,\tilde{m},b) \in Y \}$
$g' = \{ (m,b) \mid (g,m,b) \in Y \}$	$g'' = \{ \tilde{g} \mid (m,b) \in g' \quad and \quad (\tilde{g},m,b) \in Y \}$
$b' = \{ (g,m) \mid (g,m,b) \in Y \}$	$b'' = \{ \tilde{b} \mid (g,m) \in b' \quad and \quad (g,m,\tilde{b}) \in Y \}$

Let $\mathbb{K} = (G, M, B, Y)$ be a triadic context. For a certain triple $(g, m, b) \in Y$, the triple $T = (g^{\square}, m^{\square}, b^{\square})$ is called a *OAC-tricluster* based on box operators.

The density of a certain tricluster (A, B, C) of a triadic context $\mathbb{K} = (G, M, B, Y)$ is given by the fraction of all triples of Y in the tricluster, that is $\rho(A, B, C) = \frac{|I \cap A \times B \times C|}{|A||B||C|}$.

The tricluster $T = (A, B, C)$ is called *dense* if its density is greater than a predefined minimal threshold, i.e. $\rho(T) \geq \rho_{min}$. For a given triadic context $\mathbb{K} = (G, M, B, Y)$ we denote by $\mathbf{T}(G, M, B, Y)$ the set of all its (dense) triclusters.

The main features of OAC-triclusters are listed below:

1. For every triconcept (A, B, C) of a triadic context $\mathbb{K} = (G, M, B, Y)$ with nonempty sets A, B, and C we have $\rho(A, B, C) = 1$.
2. For every triclucter (A, B, C) of a triadic context $\mathbb{K} = (G, M, B, Y)$ with nonempty sets A, B, and C we have $0 \leq \rho(A, B, C) \leq 1$.

Proposition 1. *Let* $\mathbb{K} = (G, M, B, Y)$ *be a triadic context and* $\rho_{min} = 0$. *For every* $T_c = (A_c, B_c, C_c) \in \mathfrak{T}(G, M, B, Y)$ *there exists a tricluster* $T = (A, B, C) \in \mathbf{T}(G, M, B, Y)$ *such that* $A_c \subseteq A, B_c \subseteq B, C_c \subseteq C$.

In the table 2 we have $3^3 = 27$ formal triconcepts, 24 with $\rho = 1$ and 3 void triconcepts with $\rho = 0$ (they have either emptyset of users or ideas or tags). Although the data is small, we have 27 patterns to analyze (maximal number of triconcepts for the context size $3 \times 3 \times 3$); this is due to the data being the power set triadic context. We can conclude that users u_1, u_2, and u_3 share almost the same sets of tags and resources. So, they are very similar in terms of $(term, idea)$ shared pairs and it is convenient to reduce the number of patterns describing this data from 27 to 1. The tricluster $T = (\{u_1, u_2, u_3\}, \{t_1, t_2, t_3\}, \{i_1, i_2, i_3\})$ with $\rho = 0.89$ is exactly such a reduced pattern, but its density is slightly less than 1. Each of the triconcepts from $\mathfrak{T} = \{(\emptyset, \{t_1, t_2, t_3\}, \{i_1, i_2, i_3\}), (\{u_1\}, \{t_2, t_3\}, \{i_1, i_2, i_3\}), \ldots$ $(\{u_1, u_2, u_3\}, \{t_1, t_2\}, \{i_3\})\}$ is contained, w.r.t. component-wise set inclusion, in T.

Table 2. A toy example with Witology data for users $\{u_1, u_2, u_3\}$ describing ideas $\{i_1, i_2, i_3\}$ by terms $\{t_1, t_2, t_3\}$

	t_1	t_2	t_3
u_1		×	×
u_2	×	×	×
u_3	×	×	×

i_1

	t_1	t_2	t_3
u_1	×	×	×
u_2	×		×
u_3	×	×	×

i_2

	i_1	i_2	i_3
u_1	×	×	×
u_2	×	×	×
u_3	×	×	

i_3

3.3 Socio-semantic Networks for Crowdsourcing

One of the possible models for crowdsourcing platforms is the so called *socio-semantic network* [12]. A *social network* is usually modeled as a weighted multi-graph

$$G = \{V, E_1, \ldots, E_k; \pi, \delta_1, \ldots, \delta_k\},$$

where

- V represents members of the network or crowdsourcing platform,
- $E_1, \ldots, E_k \subset V \times V$ denote different relations between the members, e.g. being a friend, follower, relative, co-worker etc.
- $\pi : V \to \Pi$ is a *user profile* function, which stores personal information about the network members.
- $\delta_i : E_i \to \Delta_i$ ($i \in \{1, \ldots, k\}$) keeps parameters and details of the corresponding relation.

The model of the content has a very similar definition. It is a multi-graph

$$C = \{T, R_1, \ldots, R_m; \theta, \gamma_1, \ldots, \gamma_m\},$$

where

- T stands for the set of all elements of the generated content, e.g. posts, comments, evaluations, tags etc.
- $R_1, \ldots, R_m \subset T \times T$ denote different relations on the content, e.g. being a reply on, have the same subject, etc.
- $\theta : T \to \Theta$ stores parameters of the content;
- $\gamma_i : R_i \to \Gamma_i$ ($i \in \{1, \ldots, k\}$), similarly, keeps parameters and details of the corresponding relation.

The basic connections between the social graph and the content are defined by the authorship relation $A \subset V \times T$.

One can also consider other kinds of connections between users and generated content items, but usually all of them could be modeled via introducing a new type of content. For example, the relation *John is interested in post "Announcement"* could be modeled by introducing a new content node *interest evidence*, which points to "Announcement" (use the corresponding relation R_i here) and is authored by John.

Since we deal with binary relations between users and users, users and items, and items and items, it is easy to turn the socio-semantic based models to FCA-language and possibly, by so doing, obtain some benefits for finding communities, groups of interests and making recommendations. For visualising socio-sematic networks refer to [32].

3.4   FCA-Based Models for Crowdsourcing Data

From socio-semantic networks we move on to formal concept analysis. It is easy to show that all key crowdsourcing platform data can be described in FCA terms by means of formal contexts (single-valued, multi-valued or triadic).

1. The data below are described by a single-valued formal context $\mathbb{K} = (G, M, J)$.

Let $\mathbb{K}_P = (U, I, P)$ be a formal context, where U is a set of users, I is a set of ideas, and $P \subseteq U \times I$ shows which user proposed which idea. Two other contexts, $\mathbb{K}_C = (U, I, C)$ and $\mathbb{K}_E = (U, I, E)$, describe binary relations of idea commenting and idea evaluation respectively.

The user-to-user relationships can also be represented by means of a single-valued formal context $\mathbb{K} = (U, U, J \subseteq U \times U)$, where $u_1 J u_2$ can designate, for example, that user u_1 commented some idea proposed by u_2. Relationships between content items can be modelled in the same way, e.g. $\mathbb{K} = (T, T, J \subseteq T \times T)$, where $t_1 J t_2$ shows that t_1 and t_2 occurred together is some text (idea or comment).

2. A multi-valued context $\mathbb{K}^W = (G, M, W, J)$ can be useful for representing data with numeric attributes.

Let $\mathbb{K}^F = (U, K, F, J)$ be a multi-valued context, where U is a set of users, K is a set of keywords, F is a set of keyword frequency values, $J \subseteq U \times K \times F$ shows how many times a particular user u applied a keyword k in an idea description or while discussing some ideas. The context $\mathbb{K}^F$ can be reduced to a plain context by means of (plain) scaling.

The commenting and evaluation relations can be described through multi-valued contexts in case we count each comment or evaluation for a certain topic. E.g, the multi-valued context $\mathbb{K}^V = (U, I, V = \{-3, -2, -1, 0, 1, 2, 3\}, J)$ describes which mark a particular user u assign to an idea i, where V contains values of possible marks; it can be written as $u(i) = w$, where $v \in V$.

3. A triadic context $\mathbb{K}_B = (G, M, B, Y \subseteq G \times M \times B)$ can be used for data containing tags as descriptors.

Consider the formal context $\mathbb{K}_T = (U, I, T, Y)$, where U is a set if users, I is a set of ideas, T is a set of tags (e.g. keywords and keyphrases), Y shows that a particular user u used keyword t in the description of an idea i.

It is worth to mention that all considered data can be sorted out into two groups: 1) data with keywords $\mathbb{K}^F, \mathbb{K}_T$ and 2) data without them $\mathbb{K}_P, \mathbb{K}_I, \mathbb{K}_E$.

The main advantage of such a representation is that FCA can be applied to community detection which is the main part of social network analysis. Social network analysis is a popular research field in which methods are developed for analysing 1-mode networks, like friend-to-friend, 2-mode [33,34,35], 3-mode

[18,36,17] and even multimodal dynamic networks [9,11,12]. Here we focused on the subfield of bicommunity and tricommunity identification.

As it was shown above, crowdsourcing data can be represented as bipartite or tripartite graphs. Standard techniques like "maximal bicliques search" return a huge number of patterns (in the worst case exponential w.r.t. the input size). Therefore we need some relaxation of the biclique notion and good interesting-ness measures for mining biclique communities.

It is widely known in the social network analysis community (see, e.g. [37,38,39,40,41]) that the notion of formal concept is (almost) the same thing as a biclique.

A concept-based bicluster (OA-bicluster) [16] is a scalable approximation of a formal concept (biclique). The advantages of concept-based biclustering are:

1. Less number of patterns to analyze;
2. Less computational time (polynomial vs exponential);
3. Manual tuning of bicluster (community) density threshold;
4. Tolerance to missing (object, attribute) pairs.

For analyzing three-mode network data like folksonomies [42] we also proposed a triclustering technique [17]. The reader can refer to [43] to see how that approach was empirically validated on real online social network data.

Thus every formal concept or OA-bicluster of contexts from paragraph 1 or 2 (after scaling) can be considered as a bicommunity of users sharing similar interests or behaving similarly and every triconcept or tricluster of contexts from paragraph 3 can be interpreted as a tricommunity of users, their ideas and keywords they used. These patterns are crucial for team building and recommendation of relevant topics and persons for discussions. According to practitioners in the field, exploiting these patterns can make crowdsourcing work more comfortable and increase user's activity.

3.5   FCA-Based Recommender Model

Two kinds of recommendations seem to be potentially useful for crowdsourcing. The first one is a recommendation of like-minded persons to a particular user, and the second one is able to find antagonists, users which discussed the same topics as a target one, but with opposite marks.

1. Recommendations of like-minded persons and interesting ideas

Let $\mathbb{K}_P = (U, I, P)$ be a context which describes idea proposals. Consider a target user $u_0 \in U$, then every formal concept $(A, B) \in \mathfrak{B}_P(U, I, P)$ containing u_0 in its extent provides potentially interesting ideas to the target user in its intent and prospective like-minded persons in $A \setminus \{u_0\}$.

Consider the set $\mathfrak{R}(u_0) = \{(A, B) | (A, B) \in \mathfrak{B}_P(U, I, P) \text{ and } u_0 \in A\}$ of all concepts containing a target user u_0. Then the score of each idea or user to recommend to u_0 can be calculated as follows $score(i, u_0) = \frac{|\{u | u \in A, (A,B) \in \mathfrak{R}(u_0) \text{ and } i \in u'\}|}{|\{u | u \in A \text{ and } (A,B) \in \mathfrak{R}(u_0)\}|}$ or $score(u, u_0) = \frac{|\{A | u \in A \text{ and } (A,B) \in \mathfrak{R}(u_0)\}|}{|\mathfrak{R}(u_0)|}$ respectively. As a result we have a set of ranked recommendations $R(u_0) = \{(i, score(i)) | i \in B \text{ and } (A, B) \in \mathfrak{R}\}$. One can select the topmost N of recommendations from R ordered by their score.

2. Recommendations of antagonists

Consider two evaluation contexts: the multi-valued context $\mathbb{K}^W = (U, I, W = \{-3, -2, -1, 0, 1, 2, 3\}, J)$ and binary context $\mathbb{K}_E = (U, I, E)$. Then consider (X, Y) from $\mathfrak{R}(u_0) = \{(A, B) | (A, B) \in \mathfrak{B}_P(U, I, P) \text{ and } u_0 \in A\}$. Set X contains people that evaluated the same set of topics Y, but we cannot say that all of them are like-minded persons w.r.t relation E. However, we can introduce a distance measure, which shows for every pair of users from X how distant they are in marks of ideas evaluation:

$$d_{(X,Y)}(u_1, u_2) = \sum_{\substack{u_1, u_2 \in X \\ i \in Y}} |i(u_1) - i(u_2)|. \tag{5}$$

As a result we again have a set of ranked recommendations $R_{(X,Y)}(u_0) = \{(u, d(i)) | u \in U \text{ and } (A, B) \in \mathfrak{R}\}$. The topmost pairs from $R_d(u_0)$ with the highest distance contain antagonists, that is persons with the opposite views on most of the topics which u_0 evaluated. To aggregate $R_{(X,Y)}(u_0)$ for different (X, Y) from $\mathfrak{R}(u_0)$ into a final ranking we can calculate

$$d_{(u_0, u)} = \max\{d_{(X,Y)}(u_0, u) | (X, Y) \in \mathfrak{R}(u_0) \text{ and } u_0, u \in X\}. \tag{6}$$

The proposed models need to be tuned and validated, and also assume several variations such as using biclusters instead of formal concepts and other ways of final distance calculation. An additional possible recommender model can exploit triadic data structures for more diverse recommendations from the different sets of a triadic context.

3.6 Keywords and Keyphrases Extraction

We consider *Keywords (keyphrases)* as a set of the most significant words (phrases) in a text document that can provide a compact description for the content and style of this document. In the remainder of this paper we do not always differentiate between keywords and keyphrases, assuming that a keyword is a particular case of a keyphrase. In our project two similar problems of keyword and keyphrase extraction arise:

1. Keywords and keyphrases of the whole Witology forum;
2. Keywords and keyphrases of one user, topic etc.

In the first case we concentrate on finding syntactically well associated keywords (keyphrases). In the second case specific words and phrases of a certain user or topic are the subject of interest. Hence, we have to use two different methods for each keyword (keyphras) extraction problem. The first one is solved by using any statistical measure of association, such as Pointwise Mutual Information (PMI), T-Score or Chi-Square [44]. To solve the second problem we may use TF-IDF or Mutual Information (MI) measures that reflect how important the word or phrase is for the given subset of texts. All the above mentioned measures define the weight of a specific word or phrase in the text. The words and phrases of the highest weight then can be considered as keywords and keyphrases. We are more interested in the quality of extracted keywords and keyphrases than in the way we obtain them. To tokenize texts we use a basic principle of word separation: there should be either a space or a punctuation mark between two words. A hyphen between two sequences of symbols makes them one word. To lemmatize words we use the Russian AOT lemmatizer [45], which is far from being ideal, but it is the only freely available one (even for commercial usage) for processing Russian texts. To normalize bi- and tri-grams we use one of our Python scripts that normalizes phrases according to their formal grammatical patterns. We are going to use formal contexts based on sets of extracted keyphrases and people who use them, the occurrence of keyphrases in texts and so on. By analogy, keyphrases, texts and users all together form a tricontext for further analysis. Moreover, keyphrases are an essential part of a socio-semantic network model, where they are used for semantic representation of the network's nodes.

4   Analysis Scheme

The data analysis scheme of CrowDM, which is developed now by the project and educational team of Witology and NRU HSE is presented in figure 2. As it was mentioned before, after downloading data from a platform database, we obtain formal contexts and text collections. In turn, the latter become formal contexts as well after keyword extraction. After that, the resulting contexts are analyzed. The FCA and multimodal clustering blocks of CrowDM were implemented by N. Romashkin and K. Blinkin in Python for the project.

5   First Experiments Results

We performed different experiments with the following methods: formal concepts, iceberg-lattices and stability indices, biclustering, triclustering, implications, association rules, power law analysis, and SNA methods.

For carrying out experiments we constructed formal concepts where objects are users of the platform and attributes are ideas which users proposed within one of 5 project topics ("Sberbank and private client"). We selected only the ideas that reached the end or almost the end of the project. An object "user" has an attribute "idea" if this user somehow contributed to the discussion of this idea, i.e. he is an author of the idea, commented on the idea and evaluated the idea or comments which were added to the idea. Thus, the extracted formal concepts (U, I), where U is a set of users, I is a set of ideas, correspond to so called epistemic communities (communities of interests), i.e. the set of users U who are interested in the ideas of I. Figure 3 displays the diagram of the obtained upper part of a certain concept lattice.

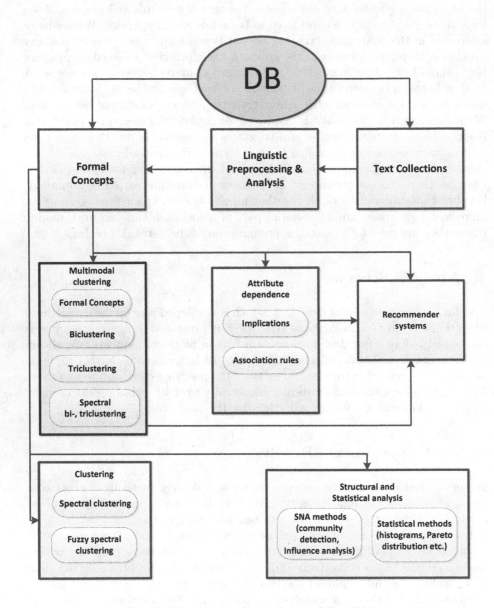

Fig. 2. The data analysis scheme of CrowDM

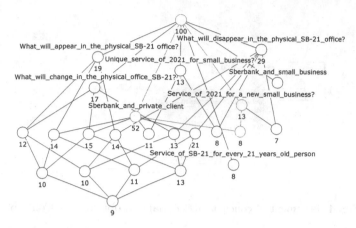

Fig. 3. The order filter (iceberg-lattice) diagram of the concept lattice for a certain users-tasks context with 24 concepts and *minsupp* = 7. The diagram is obtained by CrowDM system.

Each node of the diagram coincides with one formal concept (in total the lattice contains 198 concepts). A node cam be marked by the label of an object (or the count of objects in a formal concept extent) or an attribute if this object (moving bottom-up by diagram) or attribute (moving top-down) first appeared in this node. It is obvious that the obtained diagram is too awkward to be analyzed as a static image. Usually in such cases one can use order filters or diagrams of the sets of stable concepts or iceberg-lattices for visualization. We will showcase how to read a concept lattice using the lattice fragment in figure 4. Some first experiments were carried out using the program Concept Explorer (ConExp) which was developed for applying FCA algorithms to object-attribute data [46]. Later we applied our own data analysis system CrowDM. The system is able to build the formal concepts and biclusters for a given context. Clicking on a lattice node, one can see the objects and attributes corresponding to the concept which this node represents. Objects are accumulated from below (in the given example the set of objects contains User45 and User22), attributes come from above (we have only one attribute, "Microcredits from 1000 to 5000"). This means that User45 and User22 together took part in the discussion of the given idea and nobody else discussed it.

We demonstrate the results of applying biclustering algorithms on the same data below.

Let us explain the figure 5. During experiments we used the system for gene expression data analysis BicAT [15]. Rows correspond to users, columns are ideas of a given topic ("Sberbank and private client"), in the discussion of which users participated. The color of the cell of the corresponding row and column intersection depicts the contribution intensity of a given user to a given idea. The contribution is a weighted sum of the number of comments and evaluations to that idea and takes into account the fact whether this user is an author of this

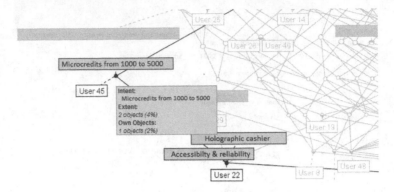

Fig. 4. Fragment of concept lattice diagram obtained by ConExp

idea. The lightest cells coincide with zero contribution, the brightest ones (fig. 5, top left cell) show the maximum contribution. After data discretization (0 – zero contribution, 1 – otherwise) we applied the BiMax algorithm which found some biclusters (see fig. 5 for example). Since one of the important crowdsourcing project problems is the search for people with similar ideas, the presented bicluster with 6 users is most interesting. The majority of the other found biclusters contained less than 4-5 users (we constrained the number of ideas in a bicluster to be strictly greater than 2).

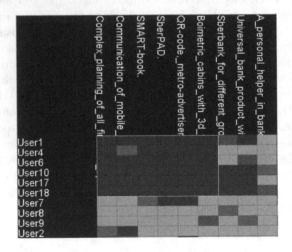

Fig. 5. Bicluster with a large number of users

Then, to gain a better understanding of the evaluation process in the project, the evaluation distribution was plotted in several ways. One of them is presented in fig. 6; it shows the cumulative number of users, who made more than a certain amount of evaluations during the entire project.

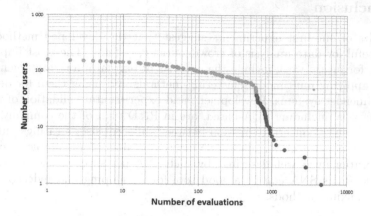

Fig. 6. Evaluation distribution

The horizontal axis displays the amount of submitted evaluations. The vertical axis represents the number of users, who made more than a fixed amount of evaluations. For instance, there is only one participant, who produced more than 5000 evaluations, and one more person, who made more than 3000 but less than 5000 evaluations. Thus, the rightmost dot on the X-axis shows the first participant (the y-coordinate is 1), and the next dot shows both of them (the y-coordinate is 2). The total number of users, who have once evaluated something, is 167. The set of graph points is explicitly split into two parts: the long gentle line (from $x = 0$ to 544 inclusive) and the steep tail. The fact, that both lines seem almost straight in logarithmic scales, indicates that the evaluation activity on the project might follow a Pareto distribution. It is reasonable to seek the individual distribution functions for the main and the tail parts of the sample, as testing the whole sample for goodness of fit to a Pareto distribution results in strong rejection of the null hypothesis ($H0$: "The sample follows a Pareto distribution").

We perform further analysis with two subsamples of the initial sample by means of Matlab tools from [23]. This analysis implies useful consequences according to the well-known "80:20" rule:

$$W = P^{(\alpha-2)/(\alpha-1)},$$

which means that the fraction W of the wealth is in the hands of the richest P of the population. In our case, 70% of users make 80% of all evaluations ($\alpha = 3.48$, p-value$= 0.78$ for $x \in [614, 5020]$). Thus, the situation is quite wealthy and there is no serious disproportions in evaluations. But if one finds strong disproportion in evaluation or commenting activity like 20% of users make 80% of actions, it implies that facilitators, who is responsible for monitoring and control in the crowdsourcing system, should involve more inactive users into the process. We also built a typology of Witology platform users [47].

6 Conclusion

The results of our first experiments suggest that the developed methodology will be useful for data analysis of crowdsourcing systems. The most important directions for future work include the analysis of textual information generated by users, applying multimodal clustering methods and using them for developing recommender systems. Development and experimental validation of recommender models, including FCA-based, are in R&D plan of the company. Some interesting results concern regression between different user's ratings and various measures of actor importance in SNA were obtained; for example, one of the Witology ratings was well-described by such SNA measures as user's centrality and degree. Thus SNA can be a good tool for developing and redesigning the Witology ranking methods.

Acknowledgments. The main part of this work was performed by the project and educational group "Algorithms of Data Mining for Innovative Projects Internet Forum". Further work was supported by the Basic Research Program at the National Research University Higher School of Economics in 2012 and performed in the Laboratory of Intelligent Systems and Structural Analysis. We also would like to thank Oleg Anshakov, Sergei Kuznetsov and Rostislav Yavorsky for their valuable comments and the rest of our group: Nikita Romashkin, Konstantin Blinkin, Ekaterina Chernyak, Olga Chugunova, Daria Goncharova, Daniil Nedumov, Fedor Strok.

References

1. Spigit company, http://spigit.com/
2. Brightidea company, http://www.brightidea.com/
3. Innocentive comp., http://www.innocentive.com/
4. Imaginatik company, http://www.imaginatik.com/
5. Kaggle, http://www.kaggle.com
6. Witology company, http://witology.com/
7. Wikivote company, http://www.wikivote.ru/
8. Sberbank-21, national entrepreneurial initiative-2012, http://sberbank21.ru/
9. Roth, C.: Generalized preferential attachment: Towards realistic socio-semantic network models. In: ISWC 4th Intl Semantic Web Conference, Workshop on Semantic Network Analysis. CEUR-WS Series, Galway, Ireland, vol. 171, pp. 29–42 (2005) ISSN 1613-0073
10. Cointet, J.-P., Roth, C.: Socio-semantic dynamics in a blog network. In: CSE (4), pp. 114–121. IEEE Computer Society (2009)
11. Roth, C., Cointet, J.P.: Social and semantic coevolution in knowledge networks. Social Networks 32, 16–29 (2010)
12. Yavorsky, R.: Research Challenges of Dynamic Socio-Semantic Networks. In: Ignatov, D., Poelmans, J., Kuznetsov, S. (eds.) CDUD 2011 - Concept Discovery in Unstructured Data, CEUR Workshop Proceedings, vol. 757, pp. 119–122 (2011)
13. Howe, J.: The rise of crowdsourcing. Wired (2006)
14. Ganter, B., Wille, R.: Formal Concept Analysis: Mathematical Foundations, 1st edn. Springer-Verlag New York, Inc., Secaucus (1999)

15. Barkow, S., Bleuler, S., Prelic, A., Zimmermann, P., Zitzler, E.: Bicat: a biclustering analysis toolbox. Bioinformatics 22(10), 1282–1283 (2006)
16. Ignatov, D.I., Kaminskaya, A.Y., Kuznetsov, S., Magizov, R.A.: Method of Biclusterzation Based on Object and Attribute Closures. In: Proc. of 8th International Conference on Intellectualization of Information Processing (IIP 2011), Cyprus, Paphos, October 17-24, pp. 140–143. MAKS Press (2010) (in Russian)
17. Ignatov, D.I., Kuznetsov, S.O., Magizov, R.A., Zhukov, L.E.: From Triconcepts to Triclusters. In: Kuznetsov, S.O., Ślęzak, D., Hepting, D.H., Mirkin, B.G. (eds.) RSFDGrC 2011. LNCS, vol. 6743, pp. 257–264. Springer, Heidelberg (2011)
18. Jäschke, R., Hotho, A., Schmitz, C., Ganter, B., Stumme, G.: TRIAS–An Algorithm for Mining Iceberg Tri-Lattices. In: Proceedings of the Sixth International Conference on Data Mining, ICDM 2006, pp. 907–911. IEEE Computer Society, Washington, DC (2006)
19. Ignatov, D.I., Kuznetsov, S.O.: Concept-based Recommendations for Internet Advertisement. In: Belohlavek, R., Kuznetsov, S.O. (eds.) Proc. CLA 2008. CEUR WS, vol. 433, pp. 157–166. Palacký University, Olomouc (2008)
20. Ignatov, D., Poelmans, J., Zaharchuk, V.: Recommender System Based on Algorithm of Bicluster Analysis RecBi. In: Ignatov, D., Poelmans, J., Kuznetsov, S. (eds.) CDUD 2011 - Concept Discovery in Unstructured Data. CEUR Workshop Proceedings, pp. 122–126 (2011)
21. Ignatov, D.I., Poelmans, J., Dedene, G., Viaene, S.: A New Cross-Validation Technique to Evaluate Quality of Recommender Systems. In: Kundu, M.K., Mitra, S., Mazumdar, D., Pal, S.K. (eds.) PerMIn 2012. LNCS, vol. 7143, pp. 195–202. Springer, Heidelberg (2012)
22. Ignatov, D.I., Konstantinov, A.V., Nikolenko, S.I., Poelmans, J., Zaharchuk, V.: Online recommender system for radio station hosting. In: [48], pp. 1–12
23. Clauset, A., Shalizi, C.R., Newman, M.E.J.: Power-law distributions in empirical data. SIAM Rev. 51(4), 661–703 (2009)
24. Kuznetsov, S.O.: On stability of a formal concept. Ann. Math. Artif. Intell. 49(1-4), 101–115 (2007)
25. Lehmann, F., Wille, R.: A Triadic Approach to Formal Concept Analysis. In: Ellis, G., Rich, W., Levinson, R., Sowa, J.F. (eds.) ICCS 1995. LNCS, vol. 954, pp. 32–43. Springer, Heidelberg (1995)
26. Wille, R.: The basic theorem of triadic concept analysis. Order 12, 149–158 (1995)
27. Voutsadakis, G.: Polyadic concept analysis. Order 19(3), 295–304 (2002)
28. Mirkin, B.G., Kramarenko, A.V.: Approximate Bicluster and Tricluster Boxes in the Analysis of Binary Data. In: Kuznetsov, S.O., Ślęzak, D., Hepting, D.H., Mirkin, B.G. (eds.) RSFDGrC 2011. LNCS, vol. 6743, pp. 248–256. Springer, Heidelberg (2011)
29. Belohlavek, R., Vychodil, V.: Factorizing Three-Way Binary Data with Triadic Formal Concepts. In: Setchi, R., Jordanov, I., Howlett, R.J., Jain, L.C. (eds.) KES 2010, Part I. LNCS, vol. 6276, pp. 471–480. Springer, Heidelberg (2010)
30. Cerf, L., Besson, J., Robardet, C., Boulicaut, J.F.: Data peeler: Contraint-based closed pattern mining in n-ary relations. In: SDM, pp. 37–48. SIAM (2008)
31. Cerf, L., Besson, J., Robardet, C., Boulicaut, J.F.: Closed patterns meet n-ary relations. ACM Trans. Knowl. Discov. Data 3, 3:1–3:36 (2009)
32. Drutsa, A., Yavorskiy, K.: Socio-semantic network data visualization. In: Tagiew, R., Ignatov, D.I., Neznanov, A.A., Poelmans, J. (eds.) EEML 2012 - Experimental Economics and Machine Learning. CEUR Workshop Proceedings, vol. 757 (2012)
33. Latapy, M., Magnien, C., Vecchio, N.D.: Basic notions for the analysis of large two-mode networks. Social Networks 30(1), 31–48 (2008)

34. Liu, X., Murata, T.: Evaluating community structure in bipartite networks. In: Elmagarmid, A.K., Agrawal, D. (eds.) SocialCom/PASSAT, pp. 576–581. IEEE Computer Society (2010)
35. Opsahl, T.: Triadic closure in two-mode networks: Redefining the global and local clustering coefficients. Social Networks 34 (2011) (in press)
36. Murata, T.: Detecting communities from tripartite networks. In: Rappa, M., Jones, P., Freire, J., Chakrabarti, S. (eds.) WWW, pp. 1159–1160. ACM (2010)
37. Freeman, L.C., White, D.R.: Using galois lattices to represent network data. Sociological Methodology 23, 127–146 (1993)
38. Freeman, L.C.: Cliques, galois lattices, and the structure of human social groups. Social Networks 18, 173–187 (1996)
39. Duquenne, V.: Lattice analysis and the representation of handicap associations. Social Networks 18(3), 217–230 (1996)
40. White, D.R.: Statistical entailments and the galois lattice. Social Networks 18(3), 201–215 (1996)
41. Roth, C., Obiedkov, S., Kourie, D.: Towards Concise Representation for Taxonomies of Epistemic Communities. In: Yahia, S.B., Nguifo, E.M., Belohlavek, R. (eds.) CLA 2006. LNCS (LNAI), vol. 4923, pp. 240–255. Springer, Heidelberg (2008)
42. Vander Wal, T.: Folksonomy Coinage and Definition (2007), http://vanderwal.net/folksonomy.html (accessed on March 12, 2012)
43. Gnatyshak, D., Ignatov, D.I., Semenov, A., Poelmans, J.: Gaining insight in social networks with biclustering and triclustering. In: [48], pp. 162–171
44. Manning, C.D., Schütze, H.: Foundations of statistical natural language processing. MIT Press, Cambridge (1999)
45. Russian project on automatic text processing, http://www.aot.ru
46. Grigoriev, P.A., Yevtushenko, S.A.: Elements of an Agile Discovery Environment. In: Grieser, G., Tanaka, Y., Yamamoto, A. (eds.) DS 2003. LNCS (LNAI), vol. 2843, pp. 311–319. Springer, Heidelberg (2003)
47. Bezzubtseva, A., Ignatov, D.I.: A New Typology of Collaboration Platform Users. In: Tagiew, R., Ignatov, D.I., Neznanov, A.A., Poelmans, J. (eds.) EEML 2012 - Experimental Economics and Machine Learning. CEUR Workshop Proceedings, vol. 757, pp. 9–19 (2012)
48. Aseeva, N., Babkin, E., Kozyrev, O. (eds.): BIR 2012. LNBIP, vol. 128. Springer, Heidelberg (2012)

Toward a Peircean Theory of Human Learning: Revealing the Misconception of *Belief Revision*

Mary Keeler and Uta Priss

Ostfalia University of Applied Sciences
Wolfenbüttel, Germany
mkeeler@uw.edu
www.upriss.org.uk

Abstract. *Belief Revision* was conceived to model how humans *do* think, and has found application in machine learning. This paper argues that Peirce's theory of inquiry conceives how we *must* think, if we want to keep improving our knowledge. Distinguishing between these two views, psychological (empirical) and pragmatic (normative), is crucial to our improvement of human learning methodology, especially as we develop interactive engagement methods for learning STEM concepts. Examining efforts to model *Belief Revision* in AI can reveal the limitations of this conceptualization for human learning, due to its misconception of Peirce's pragmatic theory of inquiry.

1   Introduction

In AI research, "Belief Revision" refers to the process of changing beliefs to accommodate new information, as part of the knowledge representation challenge to model learning by machine cognition. This research originates in early pragmatist philosophy, especially C.S. Peirce's essay, "The Fixation of Belief" [1877] and Dewey's learning theory. AI researchers have adopted Peirce's term "abductive reasoning" to explain belief revision as "guessing right" or "inference to the best explanation." Peirce considered abduction (or retroduction) to be a valid form of logical reasoning, but logical empiricists in cognitive science consider it to belong properly to psychology. Although he struggled to distinguish his logic of reasoning (as normative science) from psychology (as empirical science), the significance of this distinction remains unexamined, leaving cognitive research fundamentally confused about belief and reasoning in the process of human learning.

If, as Peirce explains, "the essence of belief is the establishment of a habit, and different beliefs are distinguished by the different modes of action to which they give rise" ["How to Make Our Ideas Clear" (1878)], and yet the purpose of logical reasoning (as inquiry, or learning) is to find the truth (which is an ideal limit we never reach, but must hope will keep our reasoning effectively progressing), then belief and reasoning are at odds in purpose. In fact, we might say that belief must be suspended during logical reasoning, so that it does not "block the way of inquiry," in Peirce's terms. Perhaps inquiry suspends belief, and belief suspends inquiry? If so, we might clarify the confusion in cognitive research by distinguishing belief from reasoning to enable more effective study of human learning. This effort begins to clear the way toward improvement upon the prevailing constructivist theory of learning, especially

H.D. Pfeiffer et al. (Eds.): ICCS 2013, LNAI 7735, pp. 193–209, 2013.

for teaching STEM subjects, by replacing Dewey's version of pragmatism with Peirce's pragmatic theory of inquiry—to advance the *how* toward the *why* of learning. The following sections respond to basic questions: What is the difference between belief and reasoning? What role can belief have in learning? Can we learn to use beliefs effectively in learning?

2   Belief Revision Theory: From Philosophy to AI

Tracing the history of *Belief Revision* (BR) development reveals the theoretical misunderstanding among philosophers, cognitive and computer scientists, linguists, and economists that we should expect in interdisciplinary research. Many in AI who have developed learning systems that incorporate some version of BR have ignored conceptual difficulties identified by philosophers. Even researchers who explicitly address conceptual problems, such as P. Thagard, have increased the confusion by simplifying theoretical fundamentals in their early work. The complexity of issues in that history is beyond this paper's scope, but we briefly cover its origins and evolution.

Although *Frontiers in Belief Revision* [Williams and Rott 2001] informs us that BR theory first came into focus in the work of the philosophers W. Harper [1976; 1977] and I. Levi [1977; 1980; 1991], we find evidence in [Doyle] that it began to take shape in the work of H. Kyburg [1961] who, in his delineation of the fundamentals from antiquity, erroneously conflates Peirce's with Dewey's pragmatism.

> There are two fundamentally distinct ways of thinking about thinking about the world. One is contemplative; the other is oriented toward action. One seeks pure knowledge; the other is pragmatic. One leads to hedged claims; the other leads to categorical claims in a hedged way. Both approaches to thinking about the world have ancient roots: Socrates, seeking wisdom; Alexander, the man of action. Both are represented in contemporary philosophy: Carnap wanted to associate with each statement of our language its appropriate degree of confirmation, relative to what we know; Peirce and Dewey took the impetus for deliberation to be the necessity to choose an action, and the outcome to be the act. Both approaches are represented in artificial intelligence: the probabilists taking the correct representation of our trans-evidential conclusions about the world to be hedged statements (the probability of rain tomorrow is .67), and the logicists taking the representation to be categorical statements (it will rain tomorrow), appropriately hedged in a non-monotonic logic—the conclusions can be withdrawn in the face of new evidence. [Kyburg 1994: 1-2]

In 1985, C. Alchourrón, P. Gärdenfors, and D. Makinson introduced their model (AGM) based on their earlier work (1978-82) [see Gärdenfors 1992, 2011], to address a critical problem. In simple terms, their work responds to: "how do you update a database of knowledge in the light of new information? What if the new information is in conflict with something that was previously held to be true? An intelligent system should be able to accommodate all such cases" (using an established set of postulates). The AGM model represents beliefs as sentences in some formal language that does not capture all aspects of belief:

> The beliefs held by an agent are represented by a set of such belief-representing sentences. It is usually assumed that this set is closed under logical consequence, i.e., every sentence that

follows logically from this set is already in the set. This is clearly an unrealistic idealization, since it means that the agent is taken to be "logically omniscient." However, it is a useful idealization since it simplifies the logical treatment; indeed, it seems difficult to obtain an interesting formal treatment without it. In logic, logically closed sets are called "theories". In formal epistemology they are also called "corpora", "knowledge sets", or (more commonly) "belief sets." [*Stanford Encyclopedia of Philosophy* (*SEP*): plato.stanford.edu/]

Levi [1984, 1986] clarified the nature of this idealization, pragmatically (in Dewey's sense of "pragmatic"): "a belief set consists of the sentences that someone is *committed to believe*, not those that she actually believes in" [*SEP*]. According to Levi's analysis, we are logically committed to believe in all the consequences of our beliefs, but typically our performance does not live up to this logical commitment. The belief set (as the set of an agent's epistemic commitments) is therefore larger than the set of her actually held beliefs. C. Misak's review of Levi's analysis hints at the confusion in its philosophical derivation:

> Levi's approach to the revision of belief and the growth of knowledge belongs, as he says, in the pragmatist tradition. Such an approach to epistemology emphasizes the context of inquiry and is teleological and decision oriented. Revisions of knowledge—scientific or otherwise—are taken to be central, and they are value laden in the sense that they are always made relative to the aims the agent is committed to promoting and to the agent's existing corpus of belief. Levi's decision theory is erected on these pragmatist foundations, and it promises to go a long way in clarifying how we should conduct our inquiries. But with respect to the foundations, the relationship between Levi and his predecessor Peirce is one of coincidence rather than supersession. [Misak, 264]

The original core assumption of belief revision was *minimal change*: the knowledge before and after the change should be as similar as possible. In the case of update, this principle formalizes the assumption of *inertia*. In the case of revision, this principle enforces the assumption of preservation of much information as possible in the change. In 1990, Thagard surveyed and assessed accounts of the psychological functions of concepts, and suggested that "conceptual change can come in varying degrees, with the most extreme consisting of fundamental conceptual reorganizations. ... Understanding epistemic change requires appreciation of the complex ways in which concepts are structured and organized and of how this organization can affect belief revision" [255]. Researchers in AI began describing their work in terms of combining BR and *abduction* [see Santos 1991; Boutilier and Becher 1993; survey of methods in Walliser, Zwirn, and Zwirn 2004]. Thagard took his work beyond belief revision, in *Conceptual Revolutions* [1992], to the question of revolutionary conceptual change.

> Theorists in philosophy, psychology, and artificial intelligence have proposed different views of the nature of concepts. A rich account of concepts and conceptual change is needed to overcome the widely held view that the growth of scientific knowledge can be understood purely in terms of belief revision with no reference to conceptual change. Concepts serve many psychological functions, and can be understood as complex computational structures organized into kind-hierarchies and part-hierarchies. Such structures involve rules that can combine to explanations. [33]

Based on his earlier work, *Computational Philosophy of Science* [1988], Thagard advocated that techniques derived from AI could be used "to understand the structure and growth of scientific knowledge," and reciprocally, "The theory of revolutionary conceptual change developed is germane to central issues in cognitive psychology and artificial intelligence, as well as to the philosophy of science" [3]. Thagard explained what had become the accepted view of *abduction* in AI—at the core of confusion between psychological and logical views of inference (emphasis added to points that will be examined, below).

The problem of inference to explanatory hypotheses has a long history in philosophy and a much shorter one in psychology and artificial intelligence. Scientists and philosophers have long considered the evaluation of theories on the basis of their explanatory power. In the late nineteenth century, *C.S. Peirce discussed two forms of inference to explanatory hypotheses:* hypothesis, *which involved the acceptance of hypotheses, and* abduction, *which involved merely the initial formation of hypotheses* (Peirce 1931-1958; Thagard 1988). Researchers in artificial intelligence and some philosophers have used the term "abduction" to refer to both the formation and the evaluation of hypotheses. [62: Thagard lists Pople 1977, Peng and Reggia 1990, Josephson et al. 1987, and Hobbs, Stickel, Appelt 1990, and Martin, Charniak and McDermott 1973, 1986]

Thagard proposed a theory of explanatory coherence (TEC) as "central to the general theory of conceptual change in science," which would "account for a wide range of explanatory inferences," in terms of principles to encompass the considerations "that suffice to make the judgments of explanatory coherence." He demonstrated the sufficiency of these principles by implementing his theory in a connectionist computer program called ECHO, which was applied to "complex cases of scientific and legal reasoning" [62-63; and see Thagard 1978: "Best Explanation: Criteria for Theory Choice"; and Van Fraassen's "The Pragmatics of Explanation"].

We abbreviate Thagard's summary of BR history in the following outline comparing philosophical to cognitive science perspectives and warning of terminological confusion.

1. Contemporary analytic philosophers take sentences to be the objects of epistemological investigation.
2. Knowledge is something like true justified belief, so increasing knowledge entails adding to what is believed.
3. Epistemology primarily evaluates strategies for improving stocks of beliefs, construed as sentences or attitudes toward sentence-like propositions (e.g., Gärdenfors models an individual's epistemic state as a consistent set of sentences that can change by expansion and contraction) [1992, 19].
4. Cognitive psychologists pay less attention to BR and far more attention to "what is the nature of concepts?"
5. Cognitive researchers in AI often follow philosophers' analysis of BR, but also pay attention to how knowledge can be organized in conceptual structures, or *frames* (Minsky 1975; for reviews see Thagard 1984, 1988).
6. Even philosophers who take cognitive science seriously consider BR to be the center of epistemology and pay little attention to conceptual change (e.g., A. Goldman 1986). The central question for epistemology has been: "when are we justified in adding and deleting beliefs from the set of beliefs judged to be known?" Epistemology should also address another question: "what are concepts and how do they change?" Concepts are relevant to epistemology if the question of conceptual change is not identical to the question of belief revision. [Based on Thagard 1992, 20]

Thagard's thesis of non-identity between conceptual change and belief revision did not attract serious response from belief-revision theorists, which W. Park [2010] finds "especially curious in view of the fact that theory of belief revision—with the AGM paradigm at its core—has over the past two decades expanded its scope far beyond epistemic logic and philosophy of science to include computer science, artificial intelligence, and economics." Various critiques of BR have appeared [see Aliseda 1997; Boutilier, Friedman, and Halpern 1998, 2008; Nebel 1989: Rott 2000, Darwich and Pearl 1997; Friedman and Halpern 2000; Gilles 2002; Nayak et al. 2003; van Benthem 2004; Jin and Thielscher 2007; Olsson and Enqvist 2011]. In their more recent critique, Friedman and Halpern [2008] identified methodological problems, and argued that careful study of belief change will require explicit ontology or scenario representation of the process.

By the late 1990s, it became clear that BR theory could be related to formal learning theory, as K. Kelly explains in his rationale for bridging between the two theories in "The Learning Power of Belief Revision" [1998, 111]:

> The guiding principle of belief revision theory is to change one's prior beliefs as little as possible in order to maintain consistency with the new information. Learning theory focuses, instead, on learning power: the ability to arrive at true beliefs in a wide range of possible environments. ... learning power depends sharply on details of the methods. Hence, learning power can provide a well-motivated constraint on the design and implementation of concrete belief revision methods.

Perhaps, BR research has been a constructive model of itself? Certainly it has raised core questions that reveal its own theoretical confusion: What is the motivation for revising beliefs (*why* change), beyond maintaining consistency and coherence, and what constitutes new information (*why* is it selected)? Attempts to model theory change provoked the need to explain the *how* of explanatory *coherence*, exposing deeper questions of *why*, what motivates consistency and coherence in representation? While AI research has focused on the *how* questions, the *why* questions have been neglected, and must be addressed in learning theory.

3   Belief as an Instinct

The evolutionary study of human cognition is also an interdisciplinary challenge (involving cognitive and computer sciences, philosophy, economics, and linguistic anthropology). Chomsky's and Pinker's theories of "the language instinct" are now well known [Pinker and Bloom 1999], and can be traced back to W. von Humboldt in the eighteenth century, whose ideas were embraced by nineteenth century anthropologists [Humbolt 1999].

The earliest book-length account of "Evolutionary Psychology" as a discipline is an undergraduate text by evolutionary epistemologist H. Plotkin. His *Evolution in Mind: an Introduction to Evolutionary Psychology* [1998] traces its origins back to Darwin and contemporaries, carried forward by the early pragmatists such as James. Plotkin explains that the early use of *instinct* to account for human behavior was an irresponsible extension of Darwin's theory of the continuity among species.

The hunt for human instincts around the turn of the [18th] century and during its first and second decades marks a low point in human sciences. Without empirical or theoretical justification of any kind, thousands of human instincts were invented, many of them extraordinarily trivial and silly. Worse still, some writers attributed putative characteristics of whole nations to instincts. ... ideology and, in this case chauvinism, intruded into the application of a concept derived from evolutionary theory to human psychology. This early phase of ascribing human action to instincts cannot be called either science or psychology of any description. Although William James himself had come, after a time and with some qualms, to champion the idea of the existence of at least some human instincts, even his great reputation could not save so weak a conceptual edifice. The net effect of the work of the eugenicists and instinct theorists during this sorry episode in the human sciences was not only to discredit the idea of instincts, but by association, seriously to weaken the influence of evolutionary ideas within psychology. [28-29]

Workman and Reader further explain:

... the concept of instinct was dropped from social scientists' terminology in the twentieth century partly because it was considered too imprecise a term to be scientifically meaningful (see Bateson, 2000). Furthermore, many so-called instinctive behaviours are capable of being modified by experience, in which case it is difficult to see where instincts finish and learning begins. A final reason why the concept of instinct fell out of favour is that a new approach to the social sciences denied their existence and saw culture rather than biology as being the principal determiner of human behaviour [11-12].

J. Tooby and L. Cosmides [1992] took issue with the dominant, non-evolutionary model in the social sciences (Standard Social Science Model) for its assumptions, which in turn were a reaction to the preceding biological determinist assumptions [see Workman and Reader 2004, 12]. F. Coolidge and T. Wynn, in *The Rise of Homo Sapiens: The Evolution of Modern Thinking* [2009], explain the motivation for evolutionary psychologists, who over the last two decades "have used reverse engineering to argue for the selective reasons behind a large array of human cognitive abilities, including spatial cognition (Irwin Silverman and Eals, 1992), language (Steve Pinker, 1997), cheater detection (Leda Cosmides, 1989), and even religion (Pascal Boyer, 2001)." All are convinced that "the current structure of human cognition preserves traces of its evolutionary past," features of an "earlier evolutionary adaptedness." A major tenet of evolutionary psychology is that our minds are adapted to a time when humans lived in small hunting and gathering groups, not the modern world, which helps explain many current psychological problems.

Workman and Reader mention that the study of genetics was dominated by "DNA-thinking," but that recently many researchers have conceived non-DNA methods of heritability, in the new field of *epigenetics* ("so new there is still no generally accepted definition of it") [52]. They find support in epigenetics for their proposal: "We now hypothesize that some neural mutation or epigenetic event led to a reorganization of the brain that enabled modern thinking" [55].

Another recent critical introductory text [Swami 2011] evaluates research from "the last decade of dramatic change in our understanding of the way in which the mind operates and the reasons behind a myriad of human behaviours." Evolutionary psychological explanations have supplanted the traditional idea that "nurture trumps nature" in human behavior by positing that shared mental architectures govern our behavior.

Evolutionary psychology (EP) tries to identify human psychological traits that are evolved adaptations ... Applying the same adaptationist thinking about physiological mechanisms common in evolutionary biology, evolutionary psychology argues that the mind has a modular structure similar to the body's. Different modular adaptations serve different functions, so that much of human behavior is the evolutionary result of psychological adaptations to solve recurrent problems in human ancestral environments. As an effort to integrate psychology into the other natural sciences, EP understands psychology as a branch of biology. ... a framework that not only incorporates the evolutionary sciences on a full and equal basis, but that systematically works out all of the revisions in existing belief and research practice that such a synthesis requires. [11]

Most recently, this research trend toward "behavioral genetics" has encouraged some to conclude that beliefs are genetically determined in brain function, not directly but through traits that are, like personality. Psychologist M. Shermer [2011] argues (in *The Believing Brain: how we construct beliefs and reinforce them as truths*) that we may like to think our beliefs come from experience, but instead they come first and then we devise reasons for believing. Our brains are "belief-generating machines," to avoid uncertainty and find patterns to follow. Even scientists operate under paradigms, but science has "built-in self-correcting mechanisms that check belief claims. ... Most guesses are false-positive (low-cost errors), ... even scientists start out with beliefs, which they then try to justify" [278].

Shermer claims that, without science ("the ultimate bias detection machine"), our brains convince us that we are always right. He describes a dozen major tendencies in judgment (biases and effects) identified by researchers.

The Confirmation Bias (*The Mother of All Cognitive Biases*, because it gives birth in one form or another to most of the other heuristics): the tendency to seek and find confirmatory evidence in support of already existing beliefs and ignore or reinterpret disconfirming evidence.

Hindsight Bias (a type of *time-reversal* confirmation bias): the tendency to reconstruct the past to fit with present knowledge. Once an event has occurred, we look back and reconstruct how it happened.

Self-Justification Bias (related to the hindsight bias): the tendency to rationalize decisions after the fact to convince ourselves that what we did was the best thing we could have done. Once we make a decision about something in our lives we carefully screen subsequent data and filter out all contradictory information related to that decision, leaving only evidence in support of the choice we made.

Attribution Bias (several kinds: situational, dispositional, intellectual, and emotional): the tendency to attribute different causes for our own beliefs and actions than that of others (common in political and religious beliefs).

Sunk-Cost Bias: the tendency to believe in something because of the cost sunk into that belief.

Status Quo Bias: the tendency to opt for whatever it is we are used to, that is, the status quo.

Endowment Effect: the tendency to value what we own more than what we do not own.

Framing Effects: the tendency to draw different conclusions based on how data are presented. Framing effects are especially noticeable in financial decisions and economic beliefs.

Anchoring Bias: the tendency to rely too heavily on a past reference or on one piece of information when making decisions, when we have no objective anchor for comparison.

Availability Heuristic: the tendency to assign probabilities of potential outcomes based on examples that are immediately available to us, especially those that are vivid, unusual, or emotionally charged, which are then generalized into conclusions upon which choices are based.

Representative Bias (related to the availability bias): the tendency to judge an event probable to the extent that it represents the essential features of its parent population or generating process.

Inattentional Blindness Bias: the tendency to miss something obvious and general while attending to something special and specific. [Based on Shermer 259-272, and he lists 25 additional biases.]

If our beliefs cause instinctive behavior, evolved under conditions we no longer need to respond to, how can they be "updated," or what is their role in learning?

4   A Peircean Theory of Learning?

No doubt many belief biases and effects were operating among BR researchers, but even our superficial untangling of their confusion points to the pervasive influence of Levi's theory of inquiry (based on Dewey's pragmatism), as R. Hilpinen reminds us:

> Peirce's account of abduction and induction as the main forms of non-demonstrative reasoning has inspired Levi's theory of inquiry and belief revision, articulated in several recent publications [Levi 1997; 1991; 1996; 2000]. In contemporary methodology, abduction is generally recognized as a distinctive form of reasoning, and models of abductive reasoning are being studied in applied logic, cognitive science and artificial intelligence, and in the theory of diagnostic reasoning. (See Josephson and Josephson, 1994; Magnani et al., 1999; Gabbay et al., 2000; Flach and Kakas, 2000.) [2004, 652]

We identify three fundamental ways that Levi's influence prevents BR from effectively modeling human learning as *the improvement of knowledge*, rather than as merely updating a database of biased beliefs. *Learning as inquiry* must challenge assumptions, not "fix" them by maintaining their consistency and coherence as we experience new information. Levi's misconceptions can instruct us how to make better use of Peirce's theory of inquiry.

1. Levi confuses Peirce's *fallibilism* with Popper's *falsificationism* [see Levi 1984, 112], a pervasive problem in philosophy and consequently in AI, especially in efforts to model scientific inquiry. As explained in [Keeler 2008]: "Peirce's inductive fallibility is a metaphysical condition, not to be confused with Popper's falsification, which is strictly a deductive procedure (see Haack, *Evidence and Inquiry*, p. 131)." Misak charges, "Isaac Levi uses C. S. Peirce's fallibilism as a foil for his own 'epistemological infallibilism'" [256]. Levi insists that both the proximal and ultimate purpose of inquiry is to eliminate error (to produce true, maximally consistent belief systems [Levi 1991]). As Misak points out, Peirce's "critical commonsensism" agrees with Levi that we do not doubt what we believe; but she clarifies:

> by "infallibilism" [Peirce] means the position which is opposed to his own fallibilism; the position that our beliefs (or at least some of them) are incorrigible, or not the sort of things that are ever in need of revision. Fallibilism insists that an inquirer must "be at all times ready to dump his whole cartload of beliefs, the moment experience is set against them" (Peirce 1931, 1.55). He cannot have "any such immovable beliefs to which he regards himself as religiously bound to be loyal" (Peirce 1931, 6.3). Such an attitude would block the path of inquiry because our minds would be closed, and hence, we would never be motivated

enough to inquire. One of Peirce's reasons for endorsing fallibilism is the fact that our faculties sometimes fail us, and we cannot be sure when these failures occur. ... Even the greatest mathematicians, he notes, are susceptible of making the simplest mistakes in arithmetic—all it takes is a little lapse of attention. [1987, 259]

2. Levi misconceives Peirce's *abduction* [even dismisses it, see Note 1], which results from his misinterpretation of Peirce's fallibilism and his reliance on Peirce's early essay on inquiry, "The Fixation of Belief" [see Kasser 2011]. Misak describes Peirce's early "doubt-belief" model of inquiry.

The notion of inquiry is central in Peirce's epistemology. He characterizes it as the struggle to rid ourselves of doubt and achieve a state of belief. An agent has a body of settled belief: a set of statements which are not, in fact, doubted. Statements in this body, however, are susceptible to doubt, if it is prompted by some "positive reason," such as a surprising or recalcitrant experience. A body of settled belief is presupposed for the operation of inquiry in that there has to be something settled for surprise to stir up. Doubt is not voluntary, and hence, we cannot simply do it, as Descartes suggests, at will. But when it impinges upon us, it "essentially involves a struggle to escape it" (Peirce 1931, 5.372, n.2) and so, as soon as we are thrown into doubt, inquiry is ignited. It continues until we reach a settled belief—a belief that we regard as "infallible, absolute truth." So Peirce characterizes the path of inquiry as follows: settled belief, doubt, inquiry, settled belief. [259]

However, Peirce's abduction evolved with his theory of inquiry [see Anderson 1986; and for discussions of Peirce's mature theory of abductive reasoning, see Hintikka, 1998, 2007, Hilpinen 2004, and Kapitan 1997]. After careful consideration of BR theory's interpretation of abduction, J. Hintikka concludes that abduction "cannot be thought of as an inference to the best explanation" [2007, 42]. A. Aliseda even advocates finding new terminology for abduction in AI [2010, 9]. Hintikka returns to Peirce's own notion of inference, for clarification: "I call all such inference by the peculiar name, abduction, because its legitimacy depends upon altogether different principles from those of other kinds of inference" [*Collected Papers (CP)* 6.524 (1901]. He points to T. Kapitan's summary of those "different principles."

(1) Inference is a conscious, voluntary act over which the reasoner exercises control (5.109, 2.144).
(2) The aim of inference is to discover (acquire, attain) new knowledge from a consideration of that which is already known (MS 628: 4).
(3) One who infers a conclusion *C* from a premise *P* accepts *C* as a result of both accepting *P* and approving a general method of reasoning according to which if any *P*-like proposition is true, so is the correlated *C*-like proposition (7.536, 2.444, 5.130, 2.773, 4.53–55, 7.459, L232:56).
(4) An inference can be either valid or invalid depending on whether it follows a method of reasoning it professes to and that method is conducive to satisfying the aim of reasoning—namely, the acquisition of truth (2.153, 2.780, 7.444, MS 692: 5). [Kapitan, 479; in Hintikka 2007, 44]

Hintikka explains that Peirce is "going beyond rules of inference that depend on the premise-conclusion relation alone and is considering also rules or principles of inference 'of an altogether different kind.' These rules or principles are justified by the fact that they exemplify a method that is conducive to the acquisition of new knowledge." Furthermore:

the validity of an abductive inference is to be judged by *strategic* principles rather than by *definitory* (move-by-move) rules. This is what makes an abductive inference depend for its legitimacy "upon altogether different principles from those of other kinds of inference." What these "different principles" were in Peirce's mind can be gathered from his various statements. One typical expression of the difference is Peirce's distinction between the validity and the strength of an argument. ... it is only in Deduction that there is no difference between a valid argument and a strong one ("Pragmatism as the Logic of Abduction," p. 17). Thus an argument can be logical but weak. [2007, 44-45]

Hintikka illustrates the "vantage point" of this "interrogative approach":

Peirce's terminology can be claimed merely to follow ordinary usage when he calls an interrogatively interpreted abductive step an inference. The reasoning of the likes of Sherlock Holmes or Nero Wolfe is not deductive, nor does it conform to any known forms of "inductive inference." The "deductions" of great detectives are in fact best thought of as question–answer sequences interspersed with deductive inferences (I have argued). Yet people routinely call them "deductions" or "inferences" accomplished by means of "logic" and "analysis." They now turn out to be right strategically speaking, though not literally (definitorily) speaking. From the strategic vantage point, we can say thus that any seriously asked question involves a tacit conjecture or guess. [2007, 55]

Hilpinen agrees and further explains:

Peirce's distinction between abduction and induction has sometimes been associated with the logical empiricist's distinction between the context of discovery (the discovery or invention of an explanatory hypothesis) and the context of justification (the confirmation or disconfirmation of a hypothesis by empirical evidence) [Reichenbach 1938]. Many logical empiricists regarded only the latter as a proper subject of logical and philosophical investigation, and thought that the study of the discovery of hypotheses belongs to psychology rather than logic. It is clear that Peirce's rules of abduction [**see Note 2**] can be said to "justify" a hypothesis in the way in which inductive reasoning can justify its conclusions: a good abduction justifies a hypothesis as a potential explanation worthy of further empirical testing. In Peirce's words, we can say that *abduction justifies an interrogative attitude towards a hypothesis.* ... an abduction leads to a "conjecture" and can justify only an "interrogative" attitude towards a proposition. According to Peirce, "Induction shows that something actually is operative, Abduction merely suggests that something may be." [652; *CP* 5.171 (1903); emphasis added]

Kapitan's careful analysis concludes that Peirce's abduction as a form of valid inference forces us to broaden the concept of validity [2004, 491] in Peirce's theory of inquiry: "[abduction] is the only logical operation which introduces any new ideas; for induction does nothing but determine a value, and deduction merely evolves the necessary consequences of a pure hypothesis" [*CP* 5.171 (1903); and see "Grounds of Validity of the Laws of Logic ..." *CP* 5.341-357 (1868-93)].

3. Levi's BR theory was originally conceived for and applied to changes in belief of a single individual and in a computerized database. S. Hansson explains five major differences in modeling of scientific knowledge processes that traditional BR theory fails to account for:

The Processes of Change are Collective
The Data/Theory Division
A Partly Accumulative Process

Explanation-Management Rather than Inconsistency-Management
The Irrelevance of Contraction

The transformation of high probabilities to full belief can be described as a process of uncertainty-reduction, or "fixation of belief" (Peirce 1877). It helps us to achieve a cognitively manageable representation of the world, thus increasing our competence and efficiency as decision-makers. This transformation is just as necessary in the collective processes of science as it is in individual cognitive processes. In science as well, our cognitive limitations make it impossible to keep track of an extensive net of interconnected probabilities. We cannot (individually or collectively) deal with a large body of human beliefs such as the scientific corpus in the massively open-ended manner that an ideal Bayesian subject would be capable of. As one example of this, since all measurement practices are theory-laden, no reasonably simple account of measurement would be available to a Bayesian approach (McLaughlin 1970). [In Olsen and Enqvist 2011, *Belief Revision Meets Philosophy of Science*, 48-50.]

However, as Levi and Hintikka agree, "Epistemologists ought to care for the improvement of knowledge rather than its pedigree" [Levi 1980, 1; Hintikka 2007]. Encouraging that direction, T. Deacon's *Incomplete Nature: How Mind Evolved from Matter* gives us a neuroscientist's examination our current "ecology" of cognition.

People tend to be masters of believing incompatible things and acting from mutually exclusive motivations and points of view. Human cognition is fragmented, our concepts are often vague and fuzzy, and our use of logical inference seldom extends beyond the steps necessary to serve an immediate need. This provides an ample mental ecology in which incompatible ideas, emotions, and reasons can long co-exist, each in its own relatively isolated niche. Such a mix of causal paradigms may be invoked in myths and fairy tales, but even here such an extreme discontinuity is seldom tolerated. Science and philosophy compulsively avoid such discontinuities. More precisely, there is an implicit injunction woven into the very fabric of these enterprises to discover and resolve explanatory incompatibilities wherever possible, and otherwise to mark them as unfinished business. Making do with placeholders creates uneasiness, however, and the longer this is necessary, the more urgent theoretical debate or scientific exploration is likely to be. [2011, 63]

Peirce even eliminates belief from the collaborative learning in science.

Full belief is willingness to act upon the proposition in vital crises, opinion is willingness to act upon it in relatively insignificant affairs. But pure science has nothing at all to do with action. The propositions it accepts, it merely writes in the list of premises it proposes to use. Nothing is vital for science; nothing can be. Its accepted propositions, therefore, are but opinions at most; and the whole list is provisional. The scientific man is not in the least wedded to his conclusions. He risks nothing upon them. He stands ready to abandon one or all as soon as experience opposes them. Some of them, I grant, he is in the habit of calling established truths; but that merely means propositions to which no competent man today demurs. It seems probable that any given proposition of that sort will remain for a long time upon the list of propositions to be admitted. Still, it may be refuted tomorrow; and if so, the scientific man will be glad to have got rid of an error. There is thus no proposition at all in science which answers to the conception of belief. [CP 5.635 (1898)]

Extending this view, we argue that Peirce's theory of scientific inquiry represents the logical (not psychological) essence of learning as *collectively engaging in the deliberate, continuous improvement of knowledge.* Hintikka's interpretation of Peirce's abduction clarifies it as a *strategic* procedure for gaining self-critical control of our belief biases, rather than as mere guessing. Furthermore, we must point out, Peirce's abduction also gives us the true "Mother of All Beliefs," the pragmatic aim (or *normative constraint*), the "why" that motivates all learning: the tendency to *hope* that we can *continue to improve* knowledge, by *engaging with the community of inquirers* [see *CP* 5.311 (1878)].

5   Didactic Implications

STEM subjects (science, technology, engineering and maths) are often difficult to teach as demonstrated, for example, by high drop-out and failure rates among first year university students. A significant amount of research has been dedicated to understanding why such subjects are difficult to teach and learn and how to help students in such subjects. As an example, Physics Education Research has come to the conclusion that students have pre-existing "misconceptions" about physics concepts that are counter-intuitive [Hestenes et al., 1992]. Students can usually learn to operate with such concepts in formulas (and thus pass exams) but if their understanding of such concepts is questioned, they fail [Hake, 1998]. Physicists have developed "concept inventories" [Hestenes et al., 1992] which are lists of questions about difficult concepts expressed mostly in everyday language. Using these concept inventories one can measure how much students know at the start and end of a semester. The learning gain in introductory physics courses measured in this manner is often very small if the courses employ standard teaching methods, but apparently students learn much more if "interactive engagement" teaching methods are used [Hake, 1998].

Interactive engagement methods include "peer instruction" (where students explain and discuss concepts with each other [Mazur, 1996], "problem-based" or "inquiry-based" learning, "flipped classroom" (where students read the lectures at home and practice during the lecture time) and "just-in-time teaching" (where lecturers respond to student questions instead of presenting a fixed lecture). But not all of these methods are guaranteed to be successful. For example, Loviscach [2012] reports that his flipped classroom with video-recorded lectures is very popular with students but has not led to significant improvements of students' marks. Noschese and Burk [Noschese, 2011] coin the term "pseudoteaching" for teaching that is on the surface very good, well liked by students and staff and where students think that they are learning a lot, but does still not lead to deep and substantial learning. Thus while not all interactive engagement teaching methods are successful, some are. Therefore the question arises as to what is the reason for the success of some methods.

It should be stressed that this paper is concerned with the learning and understanding of difficult concepts in STEM subjects. Other types of learning (such as learning a skill or learning vocabulary) might require different teaching approaches. But with respect to students overcoming their pre-existing misconceptions, the problem appears to be essentially the problem of "making ideas clear" and "fixing beliefs" as discussed by Peirce. Thus our hypothesis is that teaching methods are successful in changing students' beliefs if they encourage students to conduct inquiry in Peirce's sense. On the one hand, a better understanding of how to practically use Peirce's inquiry could make it easier for educators to predict which teaching methods are likely to be successful. On the other hand, educational methods that help students overcome misconceptions could be studied as examples of successful inquiry. At the moment the main philosophical grounding of interactive engagement methods appears to be constructivism [Ben-Ari, 1998], which is certainly an improvement over Cartesian views of science but could be further improved by a *Peircean* pragmaticist view.

6   Conclusions

According to Peirce, belief relies on assumptions—the attenuation of doubt; while reasoning progresses by suppositions—the perpetuation of doubt. A belief is a cognitive rule (or habit telling us how to think) for guiding action; while reasoning constructs rules of thought to relate facts (explaining why repeatable observations fit together). A belief can be accepted without regard for facts (without asking "why") and, to the extent that such habits are not consciously formed, they can become addictive. Recent evolutionary views of cognition have encouraged psychologists to consider whether our belief capability is instinctive, a form of adaptivity that often limits learning. While we can reason to construct rules to be tested in "learning by experience," belief can lead us to misjudge experience: making us overconfident about what we know, risk-averse to searching for disconfirming evidence, and prone to interpret evidence to preserve established beliefs. Peirce explains that in ordinary everyday life we act from instinct, with just enough reasoning necessary to connect these habits of thought with specific occasions. In learning, however: "when ones purpose lies in the line of novelty, invention, generalization, theory—in a word, improvement of the situation ... —instinct and the rule of thumb manifestly cease to be applicable. The best plan, then, on the whole, is to base our conduct as much as possible on Instinct, but when we do reason to reason with severely scientific logic" [CP 2.176-78 (1902)]. That logic accounts for learning in the conduct of inquiry as the tasks of abduction, induction, and deduction.

We encourage Conceptual Structures researchers to pursue Peirce's theory of inquiry as a logical, rather than psychological, basis for pragmatically (strategically) improving our ability to learn. Hintikka's suggestion of "games of inquiry" [2007:183, 222], and Gärdenfors's "geometry of thought" [see 2000] inspire our hope that Conceptual Structures research can make significant contributions in the future [and see Keeler 2007, 2008, 2010]. Friedman and Halpern's [2008] call for "explicit ontology or scenario representation" in BR presents an invitation to research how a Peircean theory might lead us to more effective human learning.

Note 1. In *The Fixation of Belief and Its Undoing, Changing Beliefs Through Inquiry* [1991], Levi even dismisses abduction: "Peirce devoted substantial effort to characterizing the differences between deduction, abduction, and induction as differences in the formal structure of arguments. At the beginning of the twentieth century he abandoned the project. It would be useful if contemporary writers would take the lessons Peirce learned nearly a century ago to heart." [See p. 172; he gives no references for this claim, but should have been aware at least of Peirce's 1908 essay, "Neglected Argument for the Reality of God," which features his stages of inquiry (CP 6.468-473).]

Note 2. Hilpinen explains that Peirce's "logic of abduction," as rules for good abductions, was an important aspect of his pragmatism. "An abduction is an inference which leads to a conjectured explanation, thus the logic of abduction may be expected to include conditions of adequacy for explanatory hypotheses as well as rules of discovering explanatory hypotheses."

Rule of Abduction 1: The hypothesis (the "conclusion" of an abduction) must be capable of being subjected to empirical testing.

Rule of Abduction 2: The hypothesis must explain the surprising facts. Peirce observes that an explanation may be a deductive explanation which renders the facts "necessary," or it may make the facts "natural chance results, as the kinetic theory of gases does" [*CP* 7.220]. These rules have counterparts in the more recent theories of explanation, for example, in Carl G. Hempel and Paul Oppenheim's account. RA1 and RA2 correspond to Hempel and Oppenheim's "logical conditions of adequacy" for scientific explanations [Hempel and Oppenheim 1948/65, 247-248]. Peirce's logic of abduction also contains a rule which may be called the Principle of Economy:

Rule of Abduction 3: In view of the fact that the hypothesis is one of innumerable possibly false ones, in view, too, of the enormous expensiveness of experimentation in money, time, energy and thought, is the consideration of economy. Now economy, in general, depends upon three kinds of factors: cost, the value of the thing proposed, in itself, and its effect upon other projects. Under the head of cost, if a hypothesis can be put to the test of experiment with very little expense of any kind, that should be regarded as giving it precedence in the inductive procedure. [*CP* 7.220 n. 18); Hilpinen, 651-52]

References

Alchourrón, C., Gärdenfors, P., Makinson, D.: On the Logic of Theory Change: Partial Meet Contradiction and Revision Functions. Journal of Symbolic Logic 50(2), 510–530 (1985)

Aliseda, A.: Abduction Reasoning, Logical Investigations into Discovery and Explanation. Studies in Epistemology, Logic, Methodology, and the Philosophy of Science, Synthese Library (2010),
http://www.scribd.com/doc/8145990/Aliseda-Abductive-Reasoning

Amini, M.: Logical Machines: Peirce on Psychologism. Disputatio: International Journal of Philosophy II, 335–348 (2008)

Anderson, D.: The Evolution of Peirce's Concept of Abduction. Transactions of the Charles S. Peirce Society 22(2), 145–164 (1986)

Ben-Ari, M.: Constructivism in Computer Science Education. SIGCSE Bulletin 30(1), 257–261 (1998)

van Benthem, J.: Dynamic Logic for Belief Revision. Journal of Applied Non-Classical Logics 17(2), 129–155 (2004)

Bacchus, F., Grove, A., Halpern, J., Koller, D.: A Response to Believing on the basis of evidence. Computational Intelligence (1994) (Article first published online April 2, 2007), doi:10.1111/j.1467-8640.1994.tb00141.x

Boutilier, C., Becher, V.: Abduction As Belief Revision: A Model of Preferred Explanations. In: AAAI 1993 Proceedings, pp. 642–648 (1993)

Coolidge, F., Wynn, T.: The Rise of Homo Sapiens: The Evolution of Modern Thinking. Wiley-Blackwell (2009)

Cosmides, L., Tooby, J.: The Adapted Mind: Evolutionary Psychology and the Generation of Culture. Oxford U. Press (1992)

Darwiche, A., Pearl, J.: On the Logic of Iterated Belief Revision. Artificial Intelligence 89(1-2), 1–29 (1997), ftp://cobase.cs.ucla.edu/pub/stat_ser/R202.ps

Deacon, T.: Incomplete Nature: How Mind Emerged from Matter. Norton & Company (2011)

Doyle, J.: Inference and Acceptance. Computational Intelligence 10(1), 46–48 (1994), http://www.csc.ncsu.edu/faculty/doyle/publications/kyburg94.pdf

Friedman, N., Halpern, J.: Belief Revision: A Critique. Journal of Logic and Computation 18(5), 721–738 (2008)

Gärdenfors, P.: Knowledge in Flux: Modeling the Dynamics of Epistemic States. MIT Press (1988)

Gärdenfors, P. (ed.): Belief Revision, Cambridge Tracts in Theoretical Computer Science, vol. 29. Cambridge U. Press (1992)

Gärdenfors, P.: Conceptual Spaces: The Geometry of Thought. MIT Press (2000)

Gillies, A.: Two More Dogmas of Belief Revision: Justification and Justified Belief Change (2002), http://rci.rutgers.edu/~thony/just_and_revision.pdf

Hake, R.: Interactive-engagement Versus Traditional Methods: A six-thousand-student Survey of Mechanics Test Data for Introductory Physics Courses. American Journal of Physics 66(1), 64–74 (1998)

Hansson, B.: Infallibility and Incorrigibility. Philosophical Quarterly 18, 207–215 (2006)

Harper, W.: Rational Belief Change, Popper Functions and Counterfactuals. Synthese 30(1-2), 221–262 (1975)

Hestenes, D., Wells, M., Swackhamer, G.: Force Concept Inventory. Physics Teacher 30, 141–158 (1992)

Hilpinen, R.: Peirce's Logic. In: Handbook of the History of Logic, pp. 611–658. Elsevier BV (2004)

Hintikka, J.: Knowledge and Belief. Cornell U. Press (1962)

Hintikka, J.: Socratic Epistemology: Explorations of Knowledge-Seeking by Questioning. Cambridge U. Press (2007)

Hobbs, J., Stickel, M., Martin, P., Edwards, D.: Interpretation as Abduction. In: ACL 1988 Proceedings, 26th Annual Meeting of Association for Computational Linguistics, pp. 95–103 (1988)

von Humboldt, W.: Humbolt: On Language. Cambridge U. Press (1999)

Kapitan, T.: Peirce and the Structure of Abductive Inference. In: Houser, Roberts, Van Evra (eds.) Studies in the Philosophy of Charles Sanders Peirce, pp. 477–496. Indiana U. Press (1997)

Kasser, J.: Peirce's Supposed Psychologism. Transactions of the Charles S. Peirce Society XXXV, 501–526 (1999)

Kasser, J.: How Settled are Settled Beliefs in 'The Fixation of Belief'? Transactions of the Charles S. Peirce Society 47(2), 226–247 (2011)

Keeler, M.A., Pfeiffer, H.D.: Building a Pragmatic Methodology for KR Tool Research and Development. In: Schärfe, H., Hitzler, P., Øhrstrøm, P. (eds.) ICCS 2006. LNCS (LNAI), vol. 4068, pp. 314–330. Springer, Heidelberg (2006)

Keeler, M.A.: Revelator Game of Inquiry: A Peircean Challenge for Conceptual Structures in Application and Evolution. In: Priss, U., Polovina, S., Hill, R. (eds.) ICCS 2007. LNCS (LNAI), vol. 4604, pp. 443–459. Springer, Heidelberg (2007)

Keeler, M.A., Majumdar, A.: Revelator's Complex Adaptive Reasoning Methodology for Resource Infrastructure Evolution. In: Eklund, P., Haemmerlé, O. (eds.) ICCS 2008. LNCS (LNAI), vol. 5113, pp. 88–103. Springer, Heidelberg (2008)

Keeler, M.A.: Learning to Map the Virtual Evolution of Knowledge. In: Croitoru, M., Ferré, S., Lukose, D. (eds.) ICCS 2010. LNCS (LNAI), vol. 6208, pp. 108–124. Springer, Heidelberg (2010)

Kelly, K.: The Learning Power of Belief Revision. Recommended citation (1998), http://respository.cmu.edu/philosophy/389

Kyburg, H.: Probability and the Logic of Rational Belief. Wesleyan U. Press (1961)

Kyburg, H.: Believing on the Basis of Evidence. Computational Intelligence 10, 3–22 (1994)

Lehmann, D.: Belief Revision, Revised. In: Proceedings of the 14th International Joint Conference on Artificial Intelligence, vol. 2, pp. 1534–1540. Morgan Kaufmann (1995)

Levi, I.: Belief and Action. The Monist 48, 306–316 (1964)

Levi, I.: The Enterprise of Knowledge. The MIT Press, Cambridge (1980)

Levi, I.: Decisions and Revisions. Cambridge University Press (1984)

Levi, I.: Hard Choices: Decision Making Under Unresolved Conflict. Cambridge U. Press (1986)

Levi, I.: The Fixation of Belief and Its Undoing. Cambridge U. Press (1991)

Levi, I.: For the Sake of Argument: Ramsey Test Condtionals, Inductive Inference, and Nonmonotonic Reasoning. Cambridge University Press (1996)

Loviscach, J.: Vorlesungsaufzeichnungen auf YouTube. Herausforderungen, Werkzeuge, Erfahrungen (2012), http://www.youtube.com/watch?v=A34kAqyw8kM

Misak, C.: Peirce, Levi, and the Aims of Inquiry. Philosophy of Science 54(2), 256–265 (1987)

Mazur, E.: Peer Instruction: A User's Manual. Prentice Hall (1996)

Nayak, A., Pagucco, M., Peppas, P.: Dynamic Belief Revision Operators. Artificial Intelligence 146(2), 193–228 (2003)

Nesher, D.: Peirce's Essential Discovery: 'Our Senses as Reasoning Machines' Can Quasi-Prove Our Perceptual Judgments. Transactions of the Charles S. Peirce Society XXXVIII, 175–206 (2002)

Noschese, F.: What is pseudoteaching? (2001), http://fnoschese.wordpress.com/2011/02/21/pt-pseudoteaching-mit-physics/

Olsson, E., Enqvist, S. (eds.): Belief Revision Meets Philosophy of Science. Logic, Epistemology, and the Unity of Science, vol. 21. Springer (2011)

Park, W.: Belief Revision vs. Conceptual Change in Mathematics. In: Magnani, L., Carnielli, W., Pizzi, C. (eds.) Model-Based Reasoning in Science and Technology. SCI, vol. 314, pp. 121–134. Springer, Heidelberg (2010)

Pinker, S., Bloom, P.: Natural language and natural selection. Behavioral and Brain Sciences 13(4), 707 (1990)

Peirce, C.: The Fixation of Belief. Popular Science Monthly (1877)

Peirce, C.: Logical Machines. The American Journal of Psychology I, 165–170 (1887)

Peirce, C.: Collected Papers of Charles Sanders Peirce. In: Hartshorne, C., Weiss, P. (eds.) Elements of Logic, vol. II. Belknap Press of Harvard U. Press, Cambridge (1960)

Plotkin, H.: Evolution in Mind, An Introduction to Evolutionary Psychology. Harvard U. Press (1998)

Rott, H.: Two Dogmas of Belief Revision. The Journal of Philosophy 97(9), 503–522 (2000)

Santos, E.: On the Generation of Alternative Explanations with Implications for Belief Revsion. In: Proceedings of the Seventh Conference on Uncertainty in Artificial Intelligence, pp. 339–347. Morgan Kaufmann (1991)

Shermer, M.: The Believing Brain: How We Construct Beliefs and Reinforce Them as Truths. Henry Holt and Co., LLC (2011)

Subramanian, S., Mooney, R.: Combining Abduction and Theory Revision. Artificial Intelligence 92, 168 (1992)

Swami, V. (ed.): Evolutionary Psychology: A Critical Introduction. British Psychological Society & Blackwell Publishing Ltd. (2011)

Thagard, P.: The Best Explanation: Criteria for Theory Choice. The Journal of Philosophy 75(2), 76–92 (1978)

Thagard, P.: Computational Philosophy of Science. MIT Press (1988)

Thagard, P.: Concepts and Conceptual Change. Synthese 82(2), 255–274 (1990)

Thagard, P.: Conceptual Revolutions. Princeton U. Press (1992)

Van Fraassen, B.: The Pragmatics of Explanation. American Philosophical Quarterly 14(2), 143–150 (1977)

Walliser, B., Zwirn, D., Zwirn, H.: Abductive Logics in a Belief Revision Framework. Language and Information 14(1), 87–117 (2004)

Williams, M.-A., Rott, H.: Frontiers in Belief Revision. Kluwer Academic Publishers (2001)

Workman, L., Reader, W.: Evolutionary Psychology, an Introduction. Cambridge U. Press (2004)

The First-Order Logical Environment

Robert E. Kent

Ontologos

Abstract. This paper describes the first-order logical environment FOLE. Institutions in general (Goguen and Burstall [4]), and logical environments in particular, give equivalent heterogeneous and homogeneous representations for logical systems. As such, they offer a rigorous and principled approach to distributed interoperable information systems via system consequence (Kent [6]). Since FOLE is a particular logical environment, this provides a rigorous and principled approach to distributed interoperable first-order information systems. The FOLE represents the formalism and semantics of first-order logic in a classification form. By using an interpretation form, a companion approach (Kent [7]) defines the formalism and semantics of first-order logical/relational database systems. In a strict sense, the two forms have transformational passages (generalized inverses) between one another. The classification form of first-order logic in the FOLE corresponds to ideas discussed in the Information Flow Framework (IFF [12]). The FOLE representation follows a conceptual structures approach, that is completely compatible with formal concept analysis (Ganter and Wille [2]) and information flow (Barwise and Seligman [1]).

Keywords: schema, specification, structure, logical environment.

1 Introduction

The paper "System Consequence" (Kent [6]) gave a general and abstract solution to the interoperation of information systems via the channel theory of information flow (Barwise and Seligman [1]). These can be expressed either formally, semantically or in a combined form. This general solution closely follows the theories of institutions (Goguen and Burstall [4]),[1] information flow and formal concept analysis (Ganter and Wille [2]). By following the approach of the "System Consequence" paper, this paper offers a solution to the interoperation of distributed systems expressed in terms of the formalism and semantics of first-order logic. It does this be defining FOLE, the first-order logical environment.[2] Since this paper develops a classification form of first order logic as a logical environment, the interaction of information systems expressed in first order logic

---

[1] The technical aspect of this paper is described in the spirit of Goguen's categorical manifesto [3] by using the terminology of mathematical context, passage and bridge in place of category, functor and natural transformation.

[2] A logical environment is a special and more structurally pleasing case of an institution, where the semantics is completely compatible with satisfaction.

H.D. Pfeiffer et al. (Eds.): ICCS 2013, LNAI 7735, pp. 210–230, 2013.

have a firm foundation. Section 2 surveys the architecture of the first-order logical environment FOLE. Section 3 discusses the linguistic/formal and semantic components of FOLE; detailed discussions of the functional base and relational superstructure are given in Appendix A.1 and Appendix A.2, respectively. Section 4 explains how FOLE is a logical environment; a proof of this fact is given in Appendix A.4. Section 5 discusses FOLE information systems. Finally, section 6 summarizes and states future plans for work on these topics.

2   Architecture

Figure 1 is a 3-dimensional visualization of the fibered architecture of the first-order logical environment FOLE. Each node of this figure is a mathematical context, whereas each edge is a passage between two contexts. There is a projection from the 2-D prism below **Struc** representing the relational superstructure (subsec. A.2) to the 2-D prism below **Alg** representing the functional base (subsec. A.1). The front diamond below **Lang** represents the linguistics/formalism, whereas the back diamond below **Struc** represents the semantics. The projective passages from semantics to linguistics/formalism represent the fibration left-to-right and the indexing right-to-left. The vee-shape at the top of each diamond states that the top mathematical context is a product of the side contexts modulo the bottom context. The mathematical contexts on the left side of each diamond form the relational aspect, whereas the mathematical contexts on the right side form the functional aspect that lifts the relational to the (first-order) logical aspect. The 2-D prism below **Log** represents the institutional architecture.

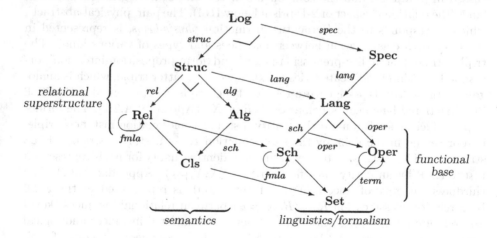

Fig. 1. FOLE Fibered Architecture

3 Components

The architectural components (Fig.1) divide up according to kind and aspect. The outer level describes the kind of component. The indexing kind is a language (type set, relational schema, operator domain, etc.) (front diamond Fig.1), whereas the indexed kind is either a formalism or a semantics (classification, relational structure, algebra, etc.) (back diamond Fig.1). The inner level describes the aspect of component. There are basic, relational, functional and logical aspects (bottom, left, right or top node in either Fig.1 diamond).

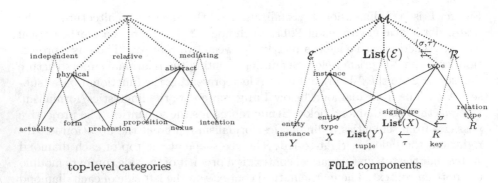

top-level categories   FOLE components

Fig. 2. Analogy

Fig.2 illustrates an analogy between the top-level ontological categories discussed in (Sowa [9]) and the components of the first-order logical environment FOLE (the relational aspect or 2-D prism below **Rel**). The pair 'physical-abstract', which corresponds to the Heraclitus distinction *physis-logos*, is represented in the FOLE by a classification between instances and types of various kinds. The triples (triads) 'actuality-prehension-nexus' and 'form-proposition-intention' correspond to Whitehead's categories of existence. The latter triple, which is analogous to the 'entity type-signature-relation type' triple, is represented in the FOLE by a relational language (schema) $\mathcal{S} = \langle R, \sigma, X \rangle$ (Appendix A.2.1). The former triple, which is analogous to the 'entity instance-tuple-relation instance' triple, is represented in the FOLE by the tuple function $K \xrightarrow{\tau} \mathbf{List}(Y)$ (part of a FOLE structure). The firstness category of 'independent(actuality,form)' is represented in the FOLE by an entity classification $\mathcal{E} = \langle X, Y, \models_{\mathcal{E}} \rangle$ (Appendix A.2.2). The thirdness category of 'mediating(nexus,intention)' is represented in the FOLE by a relation classification $\mathcal{R} = \langle R, K, \models_{\mathcal{R}} \rangle$ between relational instances (keys) and relational types (or a classification between relational instances and logical formula, more generally) (Appendix A.2.2). The secondness category of 'relative(prehension,proposition)' is represented in the FOLE by the list construction of an entity classification $\mathbf{List}(\mathcal{E}) = \langle \mathbf{List}(X), \mathbf{List}(Y), \models_{\mathbf{List}(\mathcal{E})} \rangle$ between tuples and signatures (Appendix A.2.2). Finally, the entire graph of the top-level ontological categories is represented in the FOLE by a (model-theoretic) structure (classification form) $\mathcal{M} = \langle \mathcal{R}, \langle \sigma, \tau \rangle, \mathcal{E} \rangle$, where the relation $\mathcal{R}$ and entity $\mathcal{E}$ classifications are connected by a list designation $\langle \sigma, \tau \rangle : \mathcal{R} \rightrightarrows \mathbf{List}(\mathcal{E})$ (Appendix A.2.2). This is appropriate, since a (model-theoretic) structure represents the knowledge in the local world of a community of discourse.

4   Logical Environment

The FOLE institution (logical system) (Kent [6]) has at its core the mathematical context of first-order logic (FOL) languages **Lang**. For any language $\mathcal{L} = \langle \mathcal{S}, \mathcal{O} \rangle$, there is a set of constraints $\mathbf{fmla}(\mathcal{L})$ representing the formalism at location $\mathcal{L}$, and there is a mathematical context of structures $\mathbf{struc}(\mathcal{L})$ representing the semantics at location $\mathcal{L}$. For any first-order logic (FOL) language morphism $\mathcal{L}_2 = \langle \mathcal{S}_2, \mathcal{O}_2 \rangle \xrightarrow{\langle r,f,\omega \rangle} \langle \mathcal{S}_1, \mathcal{O}_1 \rangle = \mathcal{L}_1$, there is a constraint function $\mathbf{fmla}(\mathcal{L}_2) \xrightarrow{\mathit{fmla}(r,f,\omega)} \mathbf{fmla}(\mathcal{L}_1)$ (Appendix A.2.1) representing flow of formalism in the forward direction, and there is a structure passage $\mathbf{struc}(\mathcal{L}_2) \xleftarrow{\mathit{struc}(r,f,\omega)} \mathbf{struc}(\mathcal{L}_1)$ (Appendix A.2.2) representing flow of semantics in the reverse direction. This structure passage has a relational component $\mathbf{Rel}(\mathcal{S}_2) \xleftarrow{\mathit{rel}\langle r,f \rangle} \mathbf{Rel}(\mathcal{S}_2)$ and a functional (algebraic) component $\mathbf{Alg}(\mathcal{O}_2) \xleftarrow{\mathit{alg}\langle f,\omega \rangle} \mathbf{Alg}(\mathcal{O}_1)$.

FOLE is an institution, since the satisfaction relation is preserved during information flow along any first-order logic (FOL) language morphism $\mathcal{L}_2 = \langle \mathcal{S}_2, \mathcal{O}_2 \rangle \xrightarrow{\langle r,f,\omega \rangle} \langle \mathcal{S}_1, \mathcal{O}_1 \rangle = \mathcal{L}_1$: $\mathbf{struc}(r, f, \omega)(\mathcal{M}_1) \models_{\mathcal{L}_2} (\langle I_2', s_2', \varphi_2' \rangle \xrightarrow{h_2} \langle I_2, s_2, \varphi_2 \rangle)$ iff $\mathcal{M}_1 \models_{\mathcal{L}_1} \mathbf{fmla}(\langle I_2', s_2', \varphi_2' \rangle \xrightarrow{h_2} \langle I_2, s_2, \varphi_2 \rangle)$. In short, "satisfaction is invariant under change of notation". The institution FOLE is a logical environment, since for any language $\mathcal{L} = \langle \mathcal{S}, \mathcal{O} \rangle = \langle R, \sigma, X, \Omega \rangle$, if $\mathcal{M}_2 \xrightarrow{\langle k,g,h \rangle} \mathcal{M}_1$ is a $\mathbf{lang}$-vertical structure morphism over $\mathcal{L}$, then we have the intent order $\mathcal{M}_2 \geq_{\mathcal{L}} \mathcal{M}_1$; that is, $\mathcal{M}_2 \models_{\mathcal{L}} (\varphi \vdash \psi)$ implies $\mathcal{M}_1 \models_{\mathcal{L}} (\varphi \vdash \psi)$ for any $\mathcal{S}$-sequent $(\varphi \vdash \psi)$. In short, "satisfaction respects structure morphisms". (See Appendix A.4 for a proof of this in the relational aspect.)

5   Information Systems

Following the theory of general systems, an information system consists of a collection of interconnected parts called information resources and a collection of part-part relationships between pairs of information resources called constraints. Semantic information systems have logics[3] as their information resources. Just as every logic has an underlying structure, so also every information system has an underlying distributed system. As such, distributed systems have structures for their component parts.

---

[3] A first-order logic $\mathcal{L} = \langle \mathcal{M}, \mathcal{T} \rangle$ in FOLE consists of a first-order structure $\mathcal{M}$ and a first-order specification $\mathcal{T}$ that share a common first-order language $\mathbf{lang}(\mathcal{M}) = \mathbf{lang}(\mathcal{T})$. A logic enriches a first-order structure with a specification. The logic is sound when the structure $\mathcal{M}$ satisfies every constraint in the specification $\mathcal{T}$.

A FOLE distributed system is a passage $\mathcal{M} : \mathbf{I} \to \mathbf{Struc}$ pictured as a diagram of shape $\mathbf{I}$ within the ambient mathematical context of first-order structures. As such, it consists of an indexed family $\{\mathcal{M}_i \mid i \in |\mathbf{I}|\}$ of structures together with an indexed family $\{\mathcal{M}_i \xrightarrow{m_e} \mathcal{M}_j \mid (e : i \to j) \in \mathbf{I}\}$ of structure morphisms. A FOLE (semantic) information system is a diagram $\mathcal{L} : \mathbf{I} \to \mathbf{Log}$ within the mathematical context of first-order logics. This consists of an indexed family of logics $\{\mathcal{L}_i : i \in |\mathbf{I}|\}$ and an indexed family of logic morphisms $\{\mathcal{L}_i \xrightarrow{l_e} \mathcal{L}_j \mid (e : i \to j) \in \mathbf{I}\}$. An information system $\mathcal{L}$ has an underlying distributed system $\mathcal{M} = \mathcal{L} \circ \boldsymbol{struc}$ of the same shape with $\mathcal{M}_i = \boldsymbol{struc}(\mathcal{L}_i)$ for all $i \in |\mathbf{I}|$. An information channel $\langle \gamma : \mathcal{M} \Rightarrow \Delta(\mathcal{C}), \mathcal{C} \rangle$ consists of an indexed family $\{\mathcal{M}_i \xrightarrow{\gamma_i} \mathcal{C} \mid i \in |\mathbf{I}|\}$ of structure morphisms with a common target structure $\mathcal{C}$ called the core of the channel. Information flows along channels. We are mainly interested in channels that cover a distributed system $\mathcal{M} : \mathbf{I} \to \mathbf{Struc}$, where the part-whole relationships respect the system constraints (are consistent with the part-part relationships). In this case, there exist optimal channels. An optimal core is called the sum of the distributed system, and the optimal channel components (structure morphisms) are flow links.

System interoperability is defined by moving formalism over semantics. The fusion (unification) $\coprod \mathcal{L}$ of the information system $\mathcal{L}$ represents the whole system in a centralized fashion. The fusion logic is defined by direct system flow: (i) direct logic flow of the component parts of the information system along the optimal channel over the underlying distributed system to a centralized location (the lattice of logics at the optimal channel core), and (ii) meet product combining the contributions of the parts into a whole. The consequence $\mathcal{L}^{\bullet}$ of the information system $\mathcal{L}$ represents the whole system in a distributed fashion. This is an information system defined by inverse system flow: (i) consequence of the fusion logic, and (ii) inverse logic flow of this consequence back along the same optimal channel, transfering the constraints of the whole system (the fusion logic) to the distributed locations (structures) of the component parts. See Kent [6] for further details.[4]

6   Summary and Future Work

In this paper we have described the first-order logical environment FOLE in classification form. This gives a holistic treatment of first-order logic, by the use of several novel elements: the use of signatures (type lists) for relational arities, in place of ordinal numbers; the use of abstract tuples (relational instances, keys), thus making FOLE compatible with relational databases; the use of classifications for both entities and relations; and the use of relational constraints for the sentences

---

[4] In light of the transformation described in Appendix A.5.2, an information system of sound logics can be regarded as a system of logical/relational databases. The system consequence of such systems represents database interoperabilty. Kent [6] has more details about the information flow of sound logics in an arbitrary logical environment.

of the FOLE institution. FOLE also has an interpretation form (Kent [7]) that represents the formalism and semantics of logical/relational databases, including relational algebra. There are transformational passages between the classification form and a strict version of the interpretation form. Appendix A.5.2 briefly discusses the transformation from sound logics to logical/relational databases.

FOLE has advantages over other approaches to first-order logic: in FOLE the formalism is completely integrated into the semantics; the classification form of FOLE has a natural extension to relational/logical databases, as represented by the interpretation form of FOLE; and FOLE is a logical environment, thus allowing practitioners a rigorously defined approach towards the interoperation of online semantic systems of information resources that include relational databases.

Future work includes: finishing work on the interpretation form of FOLE; further work on defining the transformational passages between the classification and interpretation forms; developing a linearization process from FOLE to sketch-like forms of logic such as Ologs (Spivak and Kent [11]); and linking FOLE with the Common Logic standard.

References

1. Barwise, J., Seligman, J.: Information Flow: The Logic of Distributed Systems. Cambridge University Press, Cambridge (1997)
2. Ganter, B., Wille, R.: Formal Concept Analysis: Mathematical Foundations. Springer, New York (1999)
3. Goguen, J.: A categorical manifesto. Mathematical Structures in Computer Science 1, 49–67 (1991)
4. Goguen, J., Burstall, R.: Institutions: Abstract Model Theory for Specification and Programming. J. Assoc. Comp. Mach. 39, 95–146 (1992)
5. Johnson, M., Rosebrugh, R., Wood, R.: Entity Relationship Attribute Designs and Sketches. Theory and Application of Categories 10(3), 94–112 (2002)
6. Kent, R.E.: System Consequence. In: Rudolph, S., Dau, F., Kuznetsov, S.O. (eds.) ICCS 2009. LNCS, vol. 5662, pp. 201–218. Springer, Heidelberg (2009)
7. Kent, R.E.: Database Semantics (2011), http://arxiv.org/abs/1209.3054
8. Kent, R.E., Spivak, D.I.: Email discussion (2011)
9. Sowa, J.F.: Knowledge Representation: Logical, Philosophical, and Computational Foundations. Brookes/Coles (2000)
10. Sowa, J.F.: ISO Standard for Conceptual Graphs (April 2, 2001), http://users.bestweb.net/~sowa/cg/cgstand.html
11. Spivak, D.I., Kent, R.E.: Ologs: a categorical framework for knowledge representation. PLoS One 7(1), e24274 (2012), http://arxiv.org/abs/1102.1889, doi:10.1371/journal.pone.0024274
12. The Information Flow Framework (IFF), http://suo.ieee.org/IFF/

A   Appendix

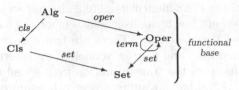

A.1   Functional Base

A.1.1   Linguistics/Formalism

Base Linguistics: **Set.** A set (of entity types) X defines a mathematical context of type lists (signatures) $\mathbf{List}(X) = (\mathbf{Set}{\downarrow}X)$. The FOLE uses type lists for relational arities, instead of ordinal numbers.

The first subcomponent of any linguistic component is a set of entity types (sorts) X. Examples of entity types are 'human' representing the set of all human beings, 'blue' representing the set of all objects of color blue, etc. A type list (signature) $\langle I, s\rangle$ consists of an arity set I and a type map $I \xrightarrow{s} X$ mapping elements of the arity to entity types. This can be denoted by the list notation $(\ldots s_i \ldots)$ or the type declaration notation $(\ldots i{:}s_i \ldots)$ for $i \in I$ and $s_i \in X$. For example, the type list '(make:**String**,model:**String**,year:**Number**,color:**Color**)' is a type list for cars with valence 4, arity set $\{\mathrm{make, model, year, color}\}$, and type map $\{\mathrm{make} \mapsto \mathbf{String}, \cdots\}$. A type list morphism $\langle I_2, s_2\rangle \xrightarrow{h} \langle I_1, s_1\rangle$ is an arity function $I_2 \xrightarrow{h} I_1$ that satisfies the commutative diagram $h \cdot s_1 = s_2$. We say that s_2 is at least as general as s_1.

Given the natural numbers $\aleph = \{0, 1, \cdots\}$, let $\underline{\aleph}$ denote the mathematical context of finite ordinals (number sets) $\underline{n} = \{0, 1, \cdots, n{-}1\}$ and functions between them. This is the skeleton of the mathematical context **Fin** of finite sets and functions. Both represent the single-sorted case where $X = 1$. We have the following inclusion of base language mathematical contexts.[5]

$$\underline{\aleph} \quad \subseteq \quad \mathbf{Fin} \quad \subseteq \quad \mathbf{List}(X)$$
$$\text{skeleton} \qquad \text{single-sorted} \qquad \text{many-sorted}$$

Traditional first-order systems use the natural numbers $\aleph$ for indexing relations. More flexible first-order systems, such as FOLE or relational database systems, use finite sets when single-sorted or type lists when many-sorted.

Algebraic Linguistics: **Oper** $\xrightarrow{set}$ **Set.** A functional language (operator domain) is a pair $\langle X, \Omega\rangle$, where X is a set of entity types (sorts) and Ω is an X-sorted operator domain; that is, $\Omega = \{\Omega_{x,\langle I,s\rangle} \mid x \in X, \langle I, s\rangle \in \overset{*}{\mathbf{List}}(X)\}$ is a collection of sets of function (operator) symbols, where $e \in \Omega_{x,\langle I,s\rangle}$ is a function symbol of entity type (sort) x and finite arity $\langle I, s\rangle$,[6] symbolized by $x \xrightarrow{e} \langle I, s\rangle$. An element $c \in \Omega_{x,\langle \emptyset, 0_X\rangle}$ is called a constant symbol of sort x. Any operator domain $\langle X, \Omega\rangle$ defines a mathematical context of terms $\mathbf{Term}_{\langle X,\Omega\rangle}$, whose objects are X-signatures $\langle I, s\rangle$ and whose morphisms are term vectors $\langle I', s'\rangle \xrightarrow{t} \langle I, s\rangle$, where

---

[5] We use the mathematical context $\overset{*}{\mathbf{List}}(X) = (\mathbf{Fin}{\downarrow}X)$ for type lists of finite arity.

[6] This is a slight misnomer, since $\langle I, s\rangle$ is actually the signature of the function symbol. whereas the arity of e is the indexing set I and the valence of e is the cardinality $|I|$.

$t = \{s'_{i'} \xrightarrow{t_{i'}} \langle I, s\rangle \mid i' \in I'\}$ is an indexed collection (vector) of $\langle I, s\rangle$-ary terms. Terms and term vectors are defined by mutual induction.

A morphism of functional languages is a pair $\langle X_2, \Omega_2 \rangle \xrightarrow{\langle f, \omega\rangle} \langle X_1, \Omega_1 \rangle$, where $X_2 \xrightarrow{f} X_1$ is a function of entity types (sorts) and $\omega : \Omega_2 \to \Omega_1$ is a collection $\{(\Omega_2)_{x_2, \langle I_2, s_2\rangle} \xrightarrow{\omega_{x_2, \langle I_2, s_2\rangle}} (\Omega_1)_{f(x_2), \Sigma_f(I_2, s_2)} \mid x_2 \in X_2, \langle I_2, s_2\rangle \in \mathbf{List}(X_2)\}$ of maps between function symbol sets: ω maps a function symbol $x_2 \xrightarrow{e} \langle I_2, s_2\rangle$ in Ω_2 to a function symbol $f(x_2) \xrightarrow{\omega(e)} \Sigma_f(I_2, s_2) = \langle I_2, s_2 \cdot f\rangle$ in Ω_1. Given any morphism of functional languages $\langle X_2, \Omega_2 \rangle \xrightarrow{\langle f, \omega\rangle} \langle X_1, \Omega_1 \rangle$, there is a term passage $\mathbf{Term}_{\langle X_2, \Omega_2\rangle} \xrightarrow{term_{\langle f, \omega\rangle}} \mathbf{Term}_{\langle X_1, \Omega_1\rangle}$ defined by induction. Let $\mathbf{Oper}$ denote the mathematical context of functional languages (operator domains).

Algebraic Formalism. Let $\mathcal{O} = \langle X, \Omega\rangle$ be an operator domain. An $\mathcal{O}$-equation is a parallel pair of term vectors $\langle I', s'\rangle \xrightarrow{t, t'} \langle I, s\rangle$. We represent an equation using the traditional notation $(t = t')$. An equational presentation $\langle X, \Omega, E\rangle$ consists of an operator domain $\mathcal{O} = \langle X, \Omega\rangle$ and a set of $\mathcal{O}$-equations E. A congruence is any equational presentation closed under left and right term composition. Any equational presentation $\langle X, \Omega, E\rangle$ generates a congruence $\langle X, \Omega, E^{\bullet}\rangle$, which defines a quotient mathematical context of terms $\mathbf{Term}_{\langle X, \Omega, E\rangle}$ with a morphism $\langle I', s'\rangle \xrightarrow{[t]} \langle I, s\rangle$ being an equivalence class of terms. There is a canonical passage $\mathbf{Term}_{\langle X, \Omega\rangle} \xrightarrow{[]} \mathbf{Term}_{\langle X, \Omega, E\rangle}$. A morphism of equational presentations $\langle X_2, \Omega_2, E_2\rangle \xrightarrow{\langle f, \omega\rangle} \langle X_1, \Omega_1, E_1\rangle$ is a morphism of functional languages $\langle X_2, \Omega_2\rangle \xrightarrow{\langle f, \omega\rangle} \langle X_1, \Omega_1\rangle$ that preserves equations: an $\mathcal{O}_2$-equation $\langle I'_2, s'_2\rangle \xrightarrow{t_2, t'_2} \langle I_2, s_2\rangle$ in E_2 is mapped to an $\mathcal{O}_1$-equation $\Sigma_f(I'_2, s'_2) \xrightarrow{\omega^*(t), \omega^*(t')} \Sigma_f(I_2, s_2)$ in the congruence $E_1^{\bullet}$. Hence, there is a term passage $\mathbf{Term}_{\langle X_2, \Omega_2, E_2\rangle} \xrightarrow{term_{\langle f, \omega\rangle}} \mathbf{Term}_{\langle X_1, \Omega_1, E_1\rangle}$ that commutes with canons.

A.1.2   Semantics

Base Semantics: $\mathbf{Cls} \xrightarrow{typ} \mathbf{Set}$. For any entity classification $\mathcal{E} = \langle X, Y, \models_{\mathcal{E}}\rangle$, there is a tuple passage $\mathbf{List}(X)^{\mathrm{op}} \xrightarrow{tup_{\mathcal{E}}} \mathbf{Set}$ defined as the extent of the list classification $\mathbf{List}(\mathcal{E})$. It maps a type list (signature) $\langle I, s\rangle \in \mathbf{List}(X)$ to its extent $tup_{\mathcal{E}}(I, s) = ext_{\mathbf{List}(\mathcal{E})}(I, s) \subseteq \mathbf{List}(Y)$. An entity infomorphism $\langle f, g\rangle :$ $\mathcal{E}_2 \rightleftarrows \mathcal{E}_1$ defines a bridge $tup_{\mathcal{E}_2} \xLeftarrow{\tau_{\langle f, g\rangle}} (\Sigma_f)^{\mathrm{op}} \circ tup_{\mathcal{E}_1}$ between tuple passages. For any source signature $\langle I_2, s_2\rangle \in (\mathbf{Set}\downarrow X_2)$, the tuple function $\tau_{\langle f, g\rangle}(I_2, s_2) = (\text{-}) \cdot g : tup_{\mathcal{E}_1}(\Sigma_f(I_2, s_2)) \to tup_{\mathcal{E}_2}(I_2, s_2)$ is define by composition.

Algebraic Semantics: $\mathbf{Cls} \xleftarrow{cls} \mathbf{Alg} \xrightarrow{oper} \mathbf{Oper}$. A many-sorted algebra $\mathcal{A} = \langle \mathcal{E}, \mathcal{O}, \langle A, \delta\rangle\rangle$ consists of an entity classification $\mathcal{E} = \langle X, Y, \models_{\mathcal{E}}\rangle$, an operator domain $\mathcal{O} = \langle X, \Omega\rangle$, and an $\mathcal{O}$-algebra $\langle A, \delta\rangle$ compatible with $\mathcal{E}$, where

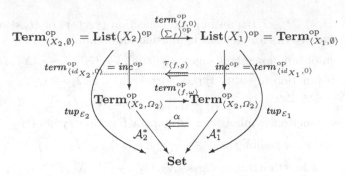

Fig. 3. Functional Base Interpretation

$A = \{A_x \mid x \in X\}$ is an X-sorted set and δ assigns an $\langle I, s\rangle$-ary x-sorted function (operation) $A_x \xleftarrow{\delta_e} A^{\langle I,s\rangle}$ to each function symbol $x \xrightarrow{e} \langle I, s\rangle$ with $A^{\langle I,s\rangle} = \prod_{i\in I} A_{s_i}$ the product set. A many-sorted algebra $\mathcal{A} = \langle \mathcal{E}, \mathcal{O}, \langle A, \delta\rangle\rangle$ defines (by induction) an algebraic interpretation passage $\mathbf{Term}^{\mathrm{op}}_{\langle X, \Omega\rangle} \xrightarrow{\mathcal{A}^*} \mathbf{Set}$, which extends the tuple passage $\boldsymbol{tup}_{\mathcal{E}} = \boldsymbol{inc}^{\mathrm{op}} \circ \mathcal{A}^*$ by compatibility. An algebra $\mathcal{A}$ satisfies an equation $(t = t')$, symbolized by $\mathcal{A} \models (t = t')$, when the interpretation maps the terms to the same function $\mathcal{A}^*(t) = \mathcal{A}^*(t')$. A many-sorted algebraic homomorphism $\mathcal{A}_2 = \langle \mathcal{E}_2, \mathcal{O}_2, \langle A_2, \delta_2\rangle\rangle \xrightarrow{\langle f,g,\omega,h\rangle} \langle \mathcal{E}_1, \mathcal{O}_1, \langle A_1, \delta_1\rangle\rangle = \mathcal{A}_1$ consists of an entity infomorphism $\langle f, g\rangle : \mathcal{E}_2 \rightleftarrows \mathcal{E}_1$, a morphism of many-sorted operator domains $\langle f, \omega\rangle : \mathcal{O}_2 \to \mathcal{O}_1$, and an $\mathcal{O}_2$-algebra morphism $\langle A_2, \delta_2\rangle \xleftarrow{h} \boldsymbol{alg}_{\langle f,\omega\rangle}(A_1, \delta_1)$ compatible with $\langle f, g\rangle$. A many-sorted algebraic homomorphism $\mathcal{A}_2 \xrightarrow{\langle f,g,\omega,h\rangle} \mathcal{A}_1$ defines an algebraic bridge $\mathcal{A}_2^* \xLeftarrow{\alpha} \boldsymbol{term}_{\langle f,\omega\rangle}^{\mathrm{op}} \circ \mathcal{A}_1^*$ between algebraic interpretations, which extends the tuple bridge $\tau_{\langle f,g\rangle} = \boldsymbol{inc}^{\mathrm{op}} \circ \alpha$ by compatibility. Let $\mathbf{Alg}$ denote the mathematical context of many-sorted algebras. (The base semantics embeds into the functional semantics Fig. 3.)

A.2   Relational Superstructure

A.2.1   Linguistics/Formalism

Relational Linguistics: **Sch**.

Schemas. A relational language (schema) $\mathcal{S} = \langle R, \sigma, X\rangle$ has two components: a base and a superstructure built upon the base. The base consists of a set of entity types (sorts) X, which defines the type list mathematical context $\mathbf{List}(X)$. The superstructure consists of a set of relation types (symbols) R and a (discrete) type list passage $R \xrightarrow{\sigma} \mathbf{List}(X)$ mapping a relation symbol $r \in R$ to its type list $\sigma(r) = \langle I, s\rangle$. A relational language (schema) morphism $\mathcal{S}_2 = \langle R_2, \sigma_2, X_2\rangle \xRightarrow{\langle r,f\rangle} \langle R_1, \sigma_1, X_1\rangle = \mathcal{S}_1$ also has two components: a base and a superstructure built upon the base. The base consists of an entity type (sort) function $f : X_2 \to X_1$, which defines the type list passage $\mathbf{List}(X_2) \xrightarrow{\Sigma_f} \mathbf{List}(X_1)$ mapping a type list $(\dots s_{i_2} \dots)$ to the type list $(\dots f(s_{i_2}) \dots)$. The superstructure consists of a relation type function $r : R_2 \to R_1$ which preserves type lists, satisfying the condition $r \cdot \sigma_1 = \sigma_2 \cdot \Sigma_f$. Let $\mathbf{Sch}$ symbolize the mathematical context of relational languages (schemas) with type set projection passage $\mathbf{Sch} \xrightarrow{set} \mathbf{Set}$.

Formulas. For any type list $\langle I, s \rangle$, let $R(I, s) \subseteq R$ denote the set of all relation types with this type list. These are called $\langle I, s \rangle$-ary relation symbols. Formulas form a schema $\boldsymbol{fmla}(\mathcal{S}) = \langle \widehat{R}, \widehat{\sigma}, X \rangle$ that extends $\mathcal{S}$: with inductive definitions, the set of relation types is extended to a set of logical formulas $\widehat{R}$ and the relational type list function is extended to a type list function $\widehat{R} \xrightarrow{\widehat{\sigma}} \mathbf{List}(X)$. For any type list $\langle I, s \rangle$, let $\widehat{R}(I, s) \subseteq \widehat{R}$ denote the set of all formulas with this type list. These are called $\langle I, s \rangle$-ary formulas. Formulas are constructed by using logical connectives within a fiber and logical flow between fibers.

fiber: Let $\langle I, s \rangle$ be any type list. Any $\langle I, s \rangle$-ary relation symbol is an (atomic) $\langle I, s \rangle$-ary formula; that is, $R(I, s) \subseteq \widehat{R}(I, s)$. For any pair of $\langle I, s \rangle$-ary formulas φ and ψ, there are the following $\langle I, s \rangle$-ary formulas: meet $(\varphi \wedge \psi)$, join $(\varphi \vee \psi)$, implication $(\varphi \to \psi)$ and difference $(\varphi \setminus \psi)$. For any $\langle I, s \rangle$-ary formula φ, there is an $\langle I, s \rangle$-ary negation formula $(\neg \varphi)$.

flow: Let $\langle I', s' \rangle \xrightarrow{h} \langle I, s \rangle$ be any type list morphism. For any $\langle I, s \rangle$-ary formula φ, there are $\langle I', s' \rangle$-ary existentially/universally quantified formulas $\Sigma_t(\varphi)$ and $\Pi_t(\varphi)$. For any $\langle I', s' \rangle$-ary formula φ', there is a $\langle I, s \rangle$-ary substitution formula $t^*(\varphi') = \varphi'(t)$.

Formula Fiber Passage. A schema morphism $\mathcal{S}_2 \xrightarrow{\langle r, f \rangle} \mathcal{S}_1$ can be extended to a formula schema morphism $\boldsymbol{fmla}(r, f) = \langle \widehat{r}, f \rangle : \boldsymbol{fmla}(\mathcal{S}_2) = \langle \widehat{R}_2, \widehat{\sigma}_2, X_2 \rangle \Longrightarrow \langle \widehat{R}_1, \widehat{\sigma}_1, X_1 \rangle = \boldsymbol{fmla}(\mathcal{S}_1)$. The formula function $\widehat{r} : \widehat{R}_2 \to \widehat{R}_1$, which satisfies the condition $inc_{\mathcal{S}_2} \cdot \widehat{r} = r \cdot inc_{\mathcal{S}_1}$, is recursively defined in Table 2.

Proposition 1. *There is an idempotent formula passage $\boldsymbol{fmla} : \mathbf{Sch} \to \mathbf{Sch}$ that forms a monad $\langle \mathbf{Sch}, \eta, \boldsymbol{fmla} \rangle$ with embedding.*

Relational Formalism: **Fmla.**

Constraints. Let $\mathcal{S} = \langle R, \sigma, X \rangle$ be a relational schema. A (binary) $\mathcal{S}$-sequent is a pair of formulas $\varphi, \psi \in \widehat{R}$ with the same type list $\widehat{\sigma}(\varphi) = \langle I, s \rangle = \widehat{\sigma}(\psi)$. [7] We represent a sequent using the turnstyle notation $\varphi \vdash \psi$, since we want a sequent to assert logical entailment. A sequent expresses interpretation widening, with the interpretation of φ required to be within the interpretation of ψ. We require entailment to be a preorder, satisfying reflexivity and transitivity (Table 3). Hence, for each type list $\langle I, s \rangle$ there is a fiber preorder $\mathbf{Fmla}_\mathcal{S}(I, s) = \langle \widehat{R}, \vdash \rangle$ consisting of all $\mathcal{S}$-formulas with this type list. In first-order logic, we further

---

[7] We regard the formulas $\widehat{R}$ to be a set of types. Since conjunction and disjunction are used in formulas, we can restrict attention to binary sequents.

Table 1. Lifting Flow

$$
\text{formula flow} \atop \text{logical aspect} \left\{
\begin{array}{lccc}
\text{term vector} & \langle I', s'\rangle & \xrightarrow{\;t\;} & \langle I, s\rangle & \text{in } \mathbf{Term}_{\langle X,\Omega\rangle} \\
\text{operation} & \mathcal{A}^*(I', s') & \xleftarrow{\mathcal{A}^*(t)} & \mathcal{A}^*(I, s) & \\
\text{inverse image} & \mathbf{Rel}_\mathcal{A}(I', s') & \xrightarrow{\;t^*\;} & \mathbf{Rel}_\mathcal{A}(I, s) & \\
\text{quantification} & \mathbf{Rel}_\mathcal{A}(I', s') & \xleftarrow[\forall_t]{\exists_t} & \mathbf{Rel}_\mathcal{A}(I, s) &
\end{array}
\right.
$$

$$\Uparrow \quad \text{functional aspect}$$

$$
\text{formula flow} \atop \text{relational aspect} \left\{
\begin{array}{lccc}
\text{type list morphism} & \langle I', s'\rangle & \xrightarrow{\;h\;} & \langle I, s\rangle & \text{in } \mathbf{List}(X) = \mathbf{Term}_{\langle X,\emptyset\rangle} \\
\text{tuple map} & \mathbf{tup}_\mathcal{E}(I', s') & \xleftarrow{tup_\mathcal{E}(h)} & \mathbf{tup}_\mathcal{E}(I, s) & \\
\text{inverse image} & \mathbf{Rel}_\mathcal{E}(I', s') & \xrightarrow{\;h^*\;} & \mathbf{Rel}_\mathcal{E}(I, s) & \\
\text{quantification} & \mathbf{Rel}_\mathcal{E}(I', s') & \xleftarrow[\forall_h]{\exists_h} & \mathbf{Rel}_\mathcal{E}(I, s) &
\end{array}
\right.
$$

When the relational aspect is lifted along the functional aspect to the first-order aspect (Fig. 1 of Section 2), formula flow is lifted from being along type list morphisms $\langle I', s'\rangle \xrightarrow{h} \langle I, s\rangle$ to being along term vectors $\langle I', s'\rangle \xrightarrow{t} \langle I, s\rangle$. This holds for formula definition (above), formula function definition (Table 2), formula axiomatization (Table 3), formula classification definition (Table 4), satisfaction (Table 5), transformation to databases (Appendix A.5), etc.

Table 2. Formula Function

fiber: type list $\langle I_2, s_2\rangle$

operator			
relation	$\hat{r}(r_2)$	$=$	$r(r_2)$
meet	$\hat{r}(\varphi_2 \wedge_{\langle I_2,s_2\rangle} \psi_2)$	$=$	$(\hat{r}(\varphi_2) \wedge_{\sum_f(I_2,s_2)} \hat{r}(\psi_2))$
join	$\hat{r}(\varphi_2 \vee_{\langle I_2,s_2\rangle} \psi_2)$	$=$	$(\hat{r}(\varphi_2) \vee_{\sum_f(I_2,s_2)} \hat{r}(\psi_2))$
negation	$\hat{r}(\neg_{\langle I_2,s_2\rangle} \varphi)$	$=$	$\neg_{\sum_f(I_2,s_2)} \hat{r}(\varphi)$
implication	$\hat{r}(\varphi \rightarrow_{\langle I_2,s_2\rangle} \psi)$	$=$	$\hat{r}(\varphi) \rightarrow_{\sum_f(I_2,s_2)} \hat{r}(\psi)$
difference	$\hat{r}(\varphi \setminus_{\langle I_2,s_2\rangle} \psi)$	$=$	$\hat{r}(\varphi) \setminus_{\sum_f(I_2,s_2)} \hat{r}(\psi)$

flow: type list morphism $\langle I_2', s_2'\rangle \xrightarrow{h} \langle I_2, s_2\rangle$

operator			
existential	$\hat{r}(\Sigma_h(\varphi_2))$	$=$	$\Sigma_h(\hat{r}(\varphi_2))$
universal	$\hat{r}(\Pi_h(\varphi_2))$	$=$	$\Pi_h(\hat{r}(\varphi_2))$
substitution	$\hat{r}(h^*(\varphi_2'))$	$=$	$h^*(\hat{r}(\varphi_2'))$

require satisfaction of sufficient conditions (Table 3) to described the various logical operations (connectives, quantifiers, etc.) used to build formulas. An indexed $\mathcal{S}$-formula $\langle I, s, \varphi\rangle$ consists of a type list $\langle I, s\rangle$ and a formula φ with signature $\langle I, s\rangle$. An $\mathcal{S}$-constraint $\langle I', s', \varphi'\rangle \xrightarrow{h} \langle I, s, \varphi\rangle$ consists of a type list morphism $\langle I', s'\rangle \xrightarrow{h} \langle I, s\rangle$ and a binary sequent $(\Sigma_h(\varphi) \vdash \varphi')$, or equivalently a binary sequent $(\varphi \vdash h^*(\varphi'))$. The mathematical context $\mathbf{Fmla}(\mathcal{S})$ has indexed $\mathcal{S}$-formula as objects and $\mathcal{S}$-constraints as morphisms. [8] Let $\mathcal{S}_2 \xRightarrow{\langle r,f\rangle} \mathcal{S}_1$ be a schema

---

[8] In some sense, this formula/constraint approach to formalism turns the tuple calculus upside down, with atoms in the tuple calculus becoming constraints here.

morphism. We assume that the function map $\widehat{R}_2 \xrightarrow{\widehat{r}} \widehat{R}_1$ is monotonic (Table 3). Hence, there is a fibered formula passage $\mathbf{Fmla}(\mathcal{S}_2) \xrightarrow{fmla_{\langle r,f \rangle}} \mathbf{Fmla}(\mathcal{S}_1)$ that commutes with the type list projections (Figure 4).

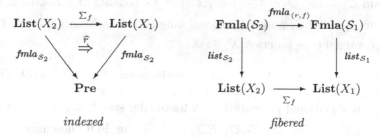

$$\textit{indexed} \qquad\qquad\qquad \textit{fibered}$$

Fig. 4. Indexed-Fibered

Table 3. Axioms

schema: $\mathcal{S}$

fiber: type list $\langle I, s \rangle$

reflexivity :	$\varphi \vdash \varphi$
transitivity :	$\varphi \vdash \varphi'$ and $\varphi' \vdash \varphi''$ implies $\varphi \vdash \varphi''$
meet :	$\psi \vdash (\varphi \wedge \varphi')$ iff $\psi \vdash \varphi$ and $\psi \vdash \varphi'$
	$(\varphi \wedge \varphi') \vdash \varphi$, $(\varphi \wedge \varphi') \vdash \varphi'$
join :	$(\varphi \vee \varphi') \vdash \psi$ iff $\varphi \vdash \psi$ and $\varphi' \vdash \psi$
	$\varphi' \vdash (\varphi \vee \varphi)$, $\varphi' \vdash (\varphi \vee \varphi')$
implication :	$(\varphi \wedge \varphi') \vdash \psi$ iff $\varphi \vdash (\varphi' \rightarrow \psi)$
negation :	$\neg\,(\neg\,(\varphi)) \vdash \varphi$

flow: type list morphism $\langle I', s' \rangle \xrightarrow{h} \langle I, s \rangle$

Σ_h-monotonicity :	$\varphi' \vdash' \psi'$ implies $\Sigma_h(\varphi') \vdash \Sigma_h(\psi')$
h^*-monotonicity :	$\varphi \vdash \psi$ implies $h^*(\varphi) \vdash' h^*(\psi)$
Π_h-monotonicity :	$\varphi' \vdash' \psi'$ implies $\Pi_h(\varphi') \vdash \Pi_h(\psi')$
adjointness :	$\Sigma_h(\varphi') \vdash \psi$ iff $\varphi' \vdash' h^*(\psi)$
	$\varphi' \vdash' h^*(\Sigma_h(\varphi'))$, $\Sigma_h(h^*(\varphi)) \vdash \varphi$

schema morphism: $\mathcal{S}_2 \xRightarrow{\langle r,f \rangle} \mathcal{S}_1$

$\widehat{r}$-monotonicity :	$(\varphi_2 \vdash_2 \psi_2)$ implies $(\widehat{r}(\varphi_2) \vdash_1 \widehat{r}(\psi_2))$

Specifications. A specification $\mathcal{T} = \langle \mathcal{S}, T \rangle$ consists of a schema $\mathcal{S} = \langle R, \sigma, X \rangle$ and a subset $T \subseteq \mathbf{Fmla}(\mathcal{S})$ of $\mathcal{S}$-constraints. As a subgraph, T extends to its consequence $T^{\bullet} \subseteq \mathbf{Fmla}(\mathcal{S})$, a mathematical subcontext, by using paths of constraints. A specification morphism $\mathcal{T}_2 = \langle \mathcal{S}_2, T_2 \rangle \xrightarrow{\langle r,f \rangle} \langle \mathcal{S}_1, T_1 \rangle = \mathcal{T}_1$ is a schema morphism $\mathcal{S}_2 \xRightarrow{\langle r,f \rangle} \mathcal{S}_1$ that preserves constraints: if sequent $\varphi'_2 \vdash h^*(\varphi_2)$ is asserted in T_2, then sequent $\widehat{r}(\varphi'_2) \vdash h^*(\widehat{r}(\varphi_2))$ is asserted in T_1.

First-order Linguistics: **Lang** $\xrightarrow{\text{sch}}$ **Sch**. A first-order logic (FOL) language
$\qquad\qquad\qquad$ **Sch** $\times_{\text{Set}}$ **Oper**
$\mathcal{L} = \langle \mathcal{S}, \mathcal{O} \rangle$ consists of a relational schema $\mathcal{S} = \langle R, \sigma, X \rangle$ and an operator domain
$\mathcal{O} = \langle X, \Omega \rangle$ that share a common type set X. A first-order logic (FOL) language
morphism $\mathcal{L}_2 = \langle \mathcal{S}_2, \mathcal{O}_2 \rangle \xrightarrow{\langle r, f, \omega \rangle} \langle \mathcal{S}_1, \mathcal{O}_1 \rangle = \mathcal{L}_1$ consists of a relational schema
morphism $\mathcal{S}_2 \xrightarrow{\langle r, f \rangle} \mathcal{S}_1$ and a functional language morphism $\mathcal{O}_2 \xrightarrow{\langle f, \omega \rangle} \mathcal{O}_1$ that
share a common type function $X_2 \xrightarrow{f} X_1$.

First-order Formalism. A first-order specification $\mathcal{T} = \langle \mathcal{S}, T, \mathcal{O}, E \rangle$ is an
FOL language $\mathcal{L} = \langle \mathcal{S}, \mathcal{O} \rangle$, where $\langle \mathcal{S}, T \rangle$ is a relational specification and
$\langle \mathcal{O}, E \rangle$ is an equational presentation. A first-order specification morphism $\mathcal{T}_2 = $
$\langle \mathcal{S}_2, T_2, \mathcal{O}_2, E_2 \rangle \xrightarrow{\langle r, f \rangle} \langle \mathcal{S}_1, T_1, \mathcal{O}_1, E_1 \rangle = \mathcal{T}_1$ is an FOL language morphism
$\mathcal{L}_2 = \langle \mathcal{S}_2, \mathcal{O}_2 \rangle \xrightarrow{\langle r, f, \omega \rangle} \langle \mathcal{S}_1, \mathcal{O}_1 \rangle = \mathcal{L}_1$, where $\langle \mathcal{S}_2, T_2 \rangle \xrightarrow{\langle r, f \rangle} \langle \mathcal{S}_1, T_1 \rangle$ is a relational
specification morphism and $\langle \mathcal{O}_2, E_2 \rangle \xrightarrow{\langle f, \omega \rangle} \langle \mathcal{O}_1, E_1 \rangle$ is a morphism of equational
presentations. A first-order specification morphism preserves constraints: if se-
quent $\varphi_2' \vdash [t]^*(\varphi_2)$ is asserted in T_2, then sequent $\widehat{r}(\varphi_2') \vdash [t]^*(\widehat{r}(\varphi_2))$ is asserted
in T_1.

A.2.2   Semantics

Relational Semantics: **Rel** $\xrightarrow{\text{sch}}$ **Sch**.

Structures. A (model-theoretic) relational structure (classification form)
(IFF [12]) $\mathcal{M} = \langle \mathcal{R}, \langle \sigma, \tau \rangle, \mathcal{E} \rangle$ is a hypergraph of classifications — a two dimensional construction consisting of a relation classification $\mathcal{R} = \langle R, K, \models_{\mathcal{R}} \rangle$,
an entity classification $\mathcal{E} = \langle X, Y, \models_{\mathcal{E}} \rangle$ and a list designation $\langle \sigma, \tau \rangle : \mathcal{R} \rightrightarrows$
List$(\mathcal{E})$. [9] Hence, a structure satisfies the following condition: $k \models_{\mathcal{R}} r$ im-
plies $\tau(k) \models_{\textbf{List}(\mathcal{E})} \sigma(r)$. A structure $\mathcal{M}$ has an associated schema $sch(\mathcal{M}) = $
$\langle R, \sigma, X \rangle$.

Formulas. Any structure $\mathcal{M} = \langle \mathcal{R}, \langle \sigma, \tau \rangle, \mathcal{E} \rangle$ has an associated formula struc-
ture $fmla(\mathcal{M}) = \langle \widehat{\mathcal{R}}, \langle \widehat{\sigma}, \tau \rangle, \mathcal{E} \rangle$ with schema $sch(fmla(\mathcal{M})) = \langle \widehat{R}, \widehat{\sigma}, X \rangle$. The
formula classification $\widehat{\mathcal{R}} = \langle \widehat{R}, K, \models_{\widehat{\mathcal{R}}} \rangle$, which extends the relation classification
of $\mathcal{M}$, is directly defined by induction in Table 4.

Satisfaction. Satisfaction is defined in terms of the extent order of the formula
classification. For any $\mathcal{S}$-structure $\mathcal{M} \in \mathbf{Rel}(\mathcal{S})$, two formula $\varphi, \psi \in \widehat{R}$ with
the same type list $\sigma(\varphi) = \sigma(\psi)$ satisfy the specialization-generalization order
$\varphi \leq_{\widehat{\mathcal{R}}} \psi$ when their extents satisfy the containment order $ext_{\widehat{\mathcal{R}}}(\varphi) \subseteq ext_{\widehat{\mathcal{R}}}(\psi)$.
An $\mathcal{S}$-structure $\mathcal{M} \in \mathbf{Rel}(\mathcal{S})$ satisfies an $\mathcal{S}$-sequent $(\varphi \vdash \psi)$ when $\varphi \leq_{\widehat{\mathcal{R}}} \psi$.

---

[9] **List**$(\mathcal{E}) = \langle \mathbf{List}(X), \mathbf{List}(Y), \models_{\mathbf{List}(\mathcal{E})} \rangle$ is the list construction of the entity classification. A tuple $\langle J, t \rangle \in \mathbf{List}(Y)$ is classified by a signature $\langle I, s \rangle \in \mathbf{List}(X)$,
symbolized by $\langle J, t \rangle \models_{\mathbf{List}(\mathcal{E})} \langle I, s \rangle$, when $J = I$ and $t_i \models_{\mathcal{E}} s_i$ for all $i \in I$.

Table 4. Formula Classification

fiber: type list $\langle I, s \rangle$ with interpretation $\boldsymbol{tup}_{\mathcal{E}}(I, s) = \prod_{i \in I} \boldsymbol{ext}_{\mathcal{E}}(s_i)$

operator	definiendum		definiens
relation	$k \models_{\widehat{\mathcal{R}}} r$	when	$k \models_{\mathcal{R}} r$
meet	$k \models_{\widehat{\mathcal{R}}} (\varphi \wedge \psi)$	when	$k \models_{\widehat{\mathcal{R}}} \varphi$ and $k \models_{\widehat{\mathcal{R}}} \psi$
join	$k \models_{\widehat{\mathcal{R}}} (\varphi \vee \psi)$	when	$k \models_{\widehat{\mathcal{R}}} \varphi$ or $k \models_{\widehat{\mathcal{R}}} \psi$
top	$k \models_{\widehat{\mathcal{R}}} \top$		
bottom	$k \not\models_{\widehat{\mathcal{R}}} \bot$		
negation	$k \models_{\widehat{\mathcal{R}}} (\neg \varphi)$	when	$k \not\models_{\widehat{\mathcal{R}}} \varphi$
implication	$k \models_{\widehat{\mathcal{R}}} (\varphi \to \psi)$	when	if $k \models_{\widehat{\mathcal{R}}} \varphi$ then $k \models_{\widehat{\mathcal{R}}} \psi$
difference	$k \models_{\widehat{\mathcal{R}}} (\varphi \setminus \psi)$	when	$k \models_{\widehat{\mathcal{R}}} \varphi$ but not $k \models_{\widehat{\mathcal{R}}} \psi$

flow: type list morphism $\langle I', s' \rangle \xrightarrow{h} \langle I, s \rangle$ with interpretation $\boldsymbol{tup}_{\mathcal{E}}(I', s') \xleftarrow{tup_{\mathcal{E}}(h)} \boldsymbol{tup}_{\mathcal{E}}(I, s)$

(with labels $\widehat{\sigma}(\varphi')$ and $\widehat{\sigma}(\varphi)$ over the morphism)

operator	definiendum		definiens
existential	$k \models_{\widehat{\mathcal{R}}} \Sigma_h(\varphi)$	when	$\tau(k) \in \exists_h(\boldsymbol{R}_{\widehat{\mathcal{M}}}(\varphi))$
universal	$k \models_{\widehat{\mathcal{R}}} \Pi_h(\varphi)$	when	$\tau(k) \in \forall_h(\boldsymbol{R}_{\widehat{\mathcal{M}}}(\varphi))$
substitution	$k \models_{\widehat{\mathcal{R}}} h^*(\varphi')$	when	$\tau(k) \in h^{-1}(\boldsymbol{R}_{\widehat{\mathcal{M}}}(\varphi'))$
			where $\boldsymbol{R}_{\widehat{\mathcal{M}}}(\varphi) = \wp\tau(\boldsymbol{ext}_{\widehat{\mathcal{R}}}(\varphi))$

An $\mathcal{S}$-structure $\mathcal{M} \in \mathbf{Rel}(\mathcal{S})$ satisfies an $\mathcal{S}$-constraint $\varphi' \xrightarrow{h} \varphi$, symbolized by $\mathcal{M} \models_{\mathcal{S}} (\varphi' \xrightarrow{h} \varphi)$, when $\mathcal{M}$ satisfies the sequent $(\Sigma_h(\varphi) \vdash \varphi')$; that is, when $\Sigma_h(\varphi) \leq_{\widehat{\mathcal{R}}} \varphi'$; equivalently, when $\varphi \leq_{\widehat{\mathcal{R}}} h^*(\varphi')$. This can be expressed in terms of implication as $(\Sigma_h(\varphi) \to \varphi') \equiv \top$; equivalently, $(\varphi \to h^*(\varphi')) \equiv \top$. When converting structures to databases, the satisfaction relationship $\mathcal{M} \models_{\mathcal{S}} (\varphi \xrightarrow{h} \varphi')$ determines the morphism of $\mathcal{E}$-relations $\boldsymbol{R}_{\widehat{\mathcal{M}}}(\varphi) \xleftarrow{h} \boldsymbol{R}_{\widehat{\mathcal{M}}}(\varphi')$ in $\mathbf{Rel}(\mathcal{E})$ and a morphism of $\mathcal{E}$-tables $\boldsymbol{T}_{\widehat{\mathcal{M}}}(\varphi) \xleftarrow{\langle h, k \rangle} \boldsymbol{T}_{\widehat{\mathcal{M}}}(\varphi')$ in $\mathbf{Tbl}(\mathcal{E})$. (The operators $\boldsymbol{R}_{\widehat{\mathcal{M}}}$ and $\boldsymbol{T}_{\widehat{\mathcal{M}}}$ are defined in Appendix A.5.1. Satisfaction is summarized in Table 5.)

Structure Morphisms. A (model-theoretic) structure morphism (IFF [12])

$$\langle r, k, f, g \rangle : \mathcal{M}_2 = \langle \mathcal{R}_2, \langle \sigma_2, \tau_2 \rangle, \mathcal{E}_2 \rangle \rightleftarrows \langle \mathcal{R}_1, \langle \sigma_1, \tau_1 \rangle, \mathcal{E}_1 \rangle = \mathcal{M}_1$$

is a two dimensional construction consisting of a relation infomorphism $\langle r, k \rangle :$ $\mathcal{R}_2 = \langle R_2, K_2, \models_{\mathcal{R}_2} \rangle \rightleftarrows \langle R_1, K_1, \models_{\mathcal{R}_1} \rangle = \mathcal{R}_1$, an entity infomorphism $\langle f, g \rangle :$ $\mathcal{E}_2 = \langle X_2, Y_2, \models_{\mathcal{E}_2} \rangle \rightleftarrows \langle X_1, Y_1, \models_{\mathcal{E}_1} \rangle = \mathcal{E}_1$, and a list classification square

$$\langle \langle r, k \rangle, \mathbf{List}_{\langle f, g \rangle} \rangle : \langle \mathcal{R}_2 \overset{\langle \sigma_2, \tau_2 \rangle}{\rightrightarrows} \mathbf{List}(\mathcal{E}_2) \rangle \rightleftarrows \langle \mathcal{R}_1 \overset{\langle \sigma_1, \tau_1 \rangle}{\rightrightarrows} \mathbf{List}(\mathcal{E}_1) \rangle,$$

where the list infomorphism of the entity infomorphism is the vertical target of the list square. Hence, a structure morphism satisfies the following conditions.

infomorphisms

$$k_1 \models_{\mathcal{R}_1} r(r_2) \quad \underline{\text{iff}} \quad k(k_1) \models_{\mathcal{R}_2} r_2$$
$$y_1 \models_{\mathcal{E}_1} f(x_2) \quad \underline{\text{iff}} \quad g(y_1) \models_{\mathcal{E}_2} x_2$$
$$t_1 \cdot g = \Sigma_g(J, t_1) \models_{\mathbf{List}(\mathcal{E}_2)} \langle I, s_2 \rangle = s_2 \quad \underline{\text{iff}} \quad t_1 = \langle J, t_1 \rangle \models_{\mathbf{List}(\mathcal{E}_1)} \Sigma_f(I, s_2) = s_2 \cdot f$$

list preservation

$$r \cdot \sigma_1 = \sigma_2 \cdot \Sigma_f$$
$$k \cdot \tau_2 = \tau_1 \cdot \Sigma_g$$

Table 5. Satisfaction

$$
\mathcal{M} \models_{\mathcal{S}} (\varphi' \xrightarrow{h} \varphi)
$$
$$
\text{when } \Sigma_h(\varphi) \leq_{\widehat{\mathcal{R}}} \varphi'
$$
$$
\text{iff } \forall_{k \in K} \left(k \models_{\widehat{\mathcal{R}}} (\Sigma_h(\varphi) \rightarrow \varphi') \right)
$$
$$
\text{iff } \forall_{k \in K} \left(k \models_{\widehat{\mathcal{R}}} \Sigma_h(\varphi) \text{ implies } k \models_{\widehat{\mathcal{R}}} \varphi' \right)
$$
$$
\text{implies } \exists_h (\boldsymbol{R}_{\widehat{\mathcal{M}}}(\varphi)) \leq \boldsymbol{R}_{\widehat{\mathcal{M}}}(\varphi')^a
$$
$$
\text{implies } \exists_k \left(\Sigma_h(\boldsymbol{T}_{\widehat{\mathcal{M}}}(\varphi)) \xrightarrow{k} \boldsymbol{T}_{\widehat{\mathcal{M}}}(\varphi') \right)
$$

[a] For relational structure $\mathcal{M} = \langle \mathcal{R}, \langle \sigma, \tau \rangle, \mathcal{E} \rangle$, the fibered mathematical context $\mathbf{Rel}(\mathcal{E})^{\mathrm{op}} \xrightarrow{list} \mathbf{List}(X)$ of $\mathcal{E}$-relations is determined by the indexed preorder $\mathbf{List}(X)^{\mathrm{op}} \xrightarrow{rel} \mathbf{Pre}$, which maps a type list $\langle I, s \rangle$ to the fiber relational order $\mathbf{Rel}_{\mathcal{E}}(I, s) = \langle \wp tup_{\mathcal{E}}(I, s), \subseteq \rangle$ and maps a type list morphism $\langle I', s' \rangle \xrightarrow{h} \langle I, s \rangle$ to the fiber monotonic function $\exists_h = \exists_{tup_{\mathcal{E}}(h)} : \mathbf{Rel}_{\mathcal{E}}(I', s') \leftarrow \mathbf{Rel}_{\mathcal{E}}(I, s)$. Similarly, for the fibered context $\mathbf{Tbl}(\mathcal{E})^{\mathrm{op}} \xrightarrow{pr} \mathbf{Term}(X)$ of $\mathcal{E}$-tables.

Structure morphisms compose component-wise. Let $\mathbf{Rel}$ denote the mathematical context of relational structures and structure morphisms. A structure morphism $\langle r, k, f, g \rangle : \mathcal{M}_2 \rightleftarrows \mathcal{M}_1$ has an associated schema morphism $sch(r, k, f, g) = \langle r, f \rangle : sch(\mathcal{M}_2) = \langle R_2, \sigma_2, X_2 \rangle \Longrightarrow \langle R_1, \sigma_1, X_1 \rangle = sch(\mathcal{M}_1)$. Hence, there is a schema passage $sch : \mathbf{Rel} \to \mathbf{Sch}$.

Formula. Any structure morphism $\langle r, k, f, g \rangle$: $\langle \mathcal{R}_2, \langle \sigma_2, \tau_2 \rangle, \mathcal{E}_2 \rangle$ $\rightleftarrows$ $\langle \mathcal{R}_1, \langle \sigma_1, \tau_1 \rangle, \mathcal{E}_1 \rangle$ has an associated formula structure morphism

$$
fmla(r, k, f, g) = \langle \widehat{r}, k, f, g \rangle : fmla(\mathcal{M}_2) = \langle \widehat{\mathcal{R}}_2, \langle \sigma_2, \tau_2 \rangle, \mathcal{E}_2 \rangle \rightleftarrows \langle \widehat{\mathcal{R}}_1, \langle \sigma_1, \tau_1 \rangle, \mathcal{E}_1 \rangle = fmla(\mathcal{M}_1)
$$

with schema morphism $sch(fmla(r, k, f, g)) = \langle \widehat{r}, f \rangle : \langle \widehat{R}_2, \widehat{\sigma}_2, X_2 \rangle \Rightarrow \langle \widehat{R}_1, \widehat{\sigma}_1, X_1 \rangle$.

Hence, there is a formula passage $fmla : \mathbf{Rel} \to \mathbf{Rel}$. [10] Between any structure and its formula extension is an embedding structure morphism $\eta_{\mathcal{M}} = \langle inc_{\mathcal{M}}, 1_K, 1_{\mathcal{E}} \rangle : \mathcal{M} \Longrightarrow fmla(\mathcal{M})$. The formula operator commutes with embedding: $\eta_{\mathcal{M}_2} \circ fmla(r, k, f, g) = \langle r, k, f, g \rangle \circ \eta_{\mathcal{M}_1}$. There is an embedding bridge $\eta : id_{\mathbf{Rel}} \Rightarrow fmla$.

Proposition 2. *There is an idempotent formula passage $fmla : \mathbf{Rel} \to \mathbf{Rel}$ that forms a monad $\langle \mathbf{Rel}, \eta, fmla \rangle$ with embedding.*

Structure Fiber Passage. Let $\mathcal{S}_2 = \langle R_2, \sigma_2, X_2 \rangle \xrightarrow{\langle r, f \rangle} \langle R_1, \sigma_1, X_1 \rangle = \mathcal{S}_1$ be a schema morphism. There is a structure passage $\mathbf{Rel}(\mathcal{S}_2) \xleftarrow{rel_{\langle r, f \rangle}} \mathbf{Rel}(\mathcal{S}_2)$ defined as follows. Let $\mathcal{M}_1 = \langle \mathcal{R}_1, \langle \sigma_1, \tau_1 \rangle, \mathcal{E}_1 \rangle \in \mathbf{Rel}(\mathcal{S}_1)$ be an $\mathcal{S}_1$-structure

---

[10] The schema and formula passages commute: $fmla \circ sch = sch \circ fmla$ (Fig. 1).

with a relation classification $\mathcal{R}_1 = \langle R_1, K_1, \models_{\mathcal{R}_1} \rangle$, an entity classification $\mathcal{E}_1 = \langle X_1, Y_1, \models_{\mathcal{E}_1} \rangle$ and a list designation $\langle \sigma_1, \tau_1 \rangle : \mathcal{R}_1 \rightrightarrows \mathbf{List}(\mathcal{E}_1)$. Define the inverse image $\mathcal{S}_2$-structure $\mathbf{rel}_{\langle r, f \rangle}(\mathcal{M}_1) = \langle r^{-1}(\mathcal{R}_1), \langle \sigma_2, \tau_1 \rangle, f^{-1}(\mathcal{E}_1) \rangle \in \mathbf{Rel}(\mathcal{S}_2)$ with $r^{-1}(\mathcal{R}_1) = \langle R_2, K_1, \models_r \rangle$, $f^{-1}(\mathcal{E}_1) = \langle X_2, Y_1, \models_f \rangle$ and a list designation $\langle \sigma_2, \tau_1 \rangle : r^{-1}(\mathcal{R}_1) \rightrightarrows f^{-1}(\mathcal{E}_1)$. From the definitions of inverse image classifications, we have the two logical equivalences (1) $k_1 \models_r r_2$ iff $k_1 \models_{\mathcal{E}_1} r(r_2)$ and (2) $\langle J_1, t_1 \rangle \models_{\Sigma_f} \langle I_2, s_2 \rangle$ iff $\langle J_1, t_1 \rangle \models_{\mathbf{List}(\mathcal{E}_1)} \Sigma_f(I_2, s_2)$. Hence, $k_1 \models_r r_2$ implies $\tau_1(k_1) \models_{\Sigma_f} \sigma_2(r_2)$. There is a bridging structure morphism

$$\mathbf{rel}_{\langle r, f \rangle}(\mathcal{M}_1) = \langle r^{-1}(\mathcal{R}_1), \langle \sigma_2, \tau_1 \rangle, f^{-1}(\mathcal{E}_1) \rangle \overset{\langle r, 1_K, f, 1_Y \rangle}{\rightleftarrows} \langle \mathcal{R}_1, \langle \sigma_1, \tau_1 \rangle, \mathcal{E}_1 \rangle = \mathcal{M}_1$$

with relation and entity infomorphisms $r^{-1}(\mathcal{R}_1) \overset{\langle r, 1_K \rangle}{\rightleftarrows} \mathcal{R}_1$ and $f^{-1}(\mathcal{E}_1) \overset{\langle f, 1_Y \rangle}{\rightleftarrows} \mathcal{E}_1$.

First-order Semantics: $\mathbf{Rel} \overset{rel}{\longleftarrow} \mathbf{Struc} \overset{lang}{\longrightarrow} \mathbf{Lang}$. The mathematical context of first-order structures $\mathbf{Struc}$ is the product of the context $\mathbf{Rel}$ of relational structures and the context $\mathbf{Alg}$ of algebras modulo the context $\mathbf{Cls}$ of classifications. A first-order logic (FOL) structure is a "pair" $\mathcal{M} = \langle \mathcal{R}, \langle \sigma, \tau \rangle, \mathcal{E}, \langle \Omega, A, \delta \rangle \rangle$ consisting of a relational structure $\langle \mathcal{R}, \langle \sigma, \tau \rangle, \mathcal{E} \rangle$ and an algebra $\langle \mathcal{E}, \langle \Omega, A, \delta \rangle \rangle$ that share a common entity classification $\mathcal{E}$. The algebra is the semantic base and the relational structure is the superstructure. Given a FOL language $\mathcal{L} = \langle \mathcal{S}, \mathcal{O} \rangle$ and an $\mathcal{L}$-structure $\mathcal{M}$ with relational $\mathcal{S}$-structure $\mathbf{rel}(\mathcal{M})$ and $\mathcal{O}$-algebra $\mathbf{alg}(\mathcal{M})$, $\mathcal{M}$ satisfies an $\mathcal{L}$-equation $\langle I', s' \rangle \overset{(t=t')}{\longrightarrow} \langle I, s \rangle$, symbolized by $\mathcal{M} \models_{\mathcal{L}} (t = t')$, when $\mathbf{alg}(\mathcal{M}) \models_{\mathcal{L}} (t = t')$; and $\mathcal{M}$ satisfies an $\mathcal{L}$-constraint $\varphi' \overset{[t]}{\longrightarrow} \varphi$, symbolized by $\mathcal{M} \models_{\mathcal{L}} (\varphi' \overset{[t]}{\rightarrow} \varphi)$, when $\mathbf{rel}(\mathcal{M}) \models_{\mathcal{S}} (\varphi' \overset{t}{\rightarrow} \varphi)$ for any representative term vector $\hat{\sigma}(\varphi') = \langle I', s' \rangle \overset{t}{\rightarrow} \langle I, s \rangle = \hat{\sigma}(\varphi)$. A first-order logic (FOL) structure morphism $\langle \mathcal{R}_2, \langle \sigma_2, \tau_2 \rangle, \mathcal{E}_2, \langle \Omega_2, A_2, \delta_2 \rangle \rangle \overset{\langle \langle r, k \rangle, \langle f, g \rangle, \langle \omega, h \rangle \rangle}{\longrightarrow} \langle \mathcal{R}_1, \langle \sigma_1, \tau_1 \rangle, \mathcal{E}_1, \langle \Omega_1, A_1, \delta_1 \rangle \rangle$ consists a relational structure morphism $\langle \mathcal{R}_2, \langle \sigma_2, \tau_2 \rangle, \mathcal{E}_2 \rangle \overset{\langle \langle r, k \rangle, \langle f, g \rangle \rangle}{\longrightarrow} \langle \mathcal{R}_1, \langle \sigma_1, \tau_1 \rangle, \mathcal{E}_1 \rangle$ and an many-sorted algebraic homomorphism $\langle \mathcal{E}_2, \mathcal{O}_2, \langle A_2, \delta_2 \rangle \rangle \overset{\langle f, g, \omega, h \rangle}{\longrightarrow} \langle \mathcal{E}_1, \mathcal{O}_1, \langle A_1, \delta_1 \rangle \rangle$ that share a common entity infomorphism $\langle f, g \rangle : \mathcal{E}_2 \rightleftarrows \mathcal{E}_1$.

A.3   Examples

Conceptual Graphs: Consider the English sentence "John is going to Boston by bus" [9]. We describe its representation in a FOLE logic language $\mathcal{L} = \langle R, \sigma, X, \Omega \rangle$. By representing the verb as a ternary relation, a graphical representation is

$$[Person : John] \overset{agnt}{\longleftarrow} (Go) \overset{dest}{\longrightarrow} [City : Boston]$$
$$\downarrow inst$$
$$[Bus]$$

Formally, we have the following elements: three entity types $\texttt{Person}, \texttt{City},$ $\texttt{Bus} \in X$; a relation type $\texttt{Go} \in R$ with signature $\sigma(\texttt{Go}) = \langle I, s \rangle$ having valence 3, arity $I = \{\texttt{agnt}, \texttt{dest}, \texttt{inst}\}$ and signature function $I \xrightarrow{s} X$ mapping $\texttt{agnt} \mapsto \texttt{Person}, \texttt{dest} \mapsto \texttt{City}, \texttt{inst} \mapsto \texttt{Bus}$; a constant symbol $\texttt{John} \in \Omega_{\texttt{Person}, \langle \emptyset, 0_X \rangle}$ of sort $\texttt{Person}$ and a constant symbol $\texttt{Boston} \in \Omega_{\texttt{City}, \langle \emptyset, 0_X \rangle}$ of sort $\texttt{City}$.[11] In a conceptual graph representation, the logic language $\mathcal{L} = \langle R, \sigma, X, \Omega \rangle$ corresponds to a CG module $\langle X, R, C \rangle$ with type hierarchy X, relation hierarchy R and catalog of individuals $C \subseteq \Omega$. A CG representation is

```
[Go]-
    (agnt)->[Person: John]
    (dest)->[City: Boston]
    (inst)->[Bus].
```

Formally (compare this linear form to 11), we have the following elements: four entity types $\texttt{Go}, \texttt{Person}, \texttt{City}, \texttt{Bus} \in X$; three relation types $\texttt{agnt}, \texttt{dest}, \texttt{inst} \in R$ with signatures $\sigma(\texttt{agnt}) = \langle \mathbf{2}, s_{\texttt{agnt}} \rangle$, $\sigma(\texttt{dest}) = \langle \mathbf{2}, s_{\texttt{dest}} \rangle$, $\sigma(\texttt{inst}) = \langle \mathbf{2}, s_{\texttt{inst}} \rangle$ having valence 2, arity $\mathbf{2} = \{0, 1\}$ and signatures $s_{\texttt{agnt}}, s_{\texttt{dest}}, s_{\texttt{inst}} : \mathbf{2} \to X$, where $s_{\texttt{agnt}}(0) = s_{\texttt{dest}}(0) = s_{\texttt{inst}}(0) = \texttt{Go}$, $s_{\texttt{agnt}}(1) = \texttt{Person}, s_{\texttt{dest}}(1) = \texttt{City}$, and $s_{\texttt{inst}}(1) = \texttt{Bus}$; and two constants as above.

Quantification: The universal quantification '$\forall_{x \in X} P(x{:}X, y{:}Y, z{:}Z)$' is traditionally viewed as formula flow along the type list inclusion $\{y, z\} \subseteq \{x, y, z\}$. FOLE handles existential/universal quantification and substitution in terms of formula flow (Table 1) along type list morphisms in the relational aspect or along term vectors in the logical aspect. Given a morphism of type lists $\langle I', s' \rangle \xrightarrow{h} \langle I, s \rangle$, for any table $\langle K, t \rangle \in \mathbf{Tbl}_{\mathcal{E}}(I, s)$, you can get two tables $\Sigma_h(K, t), \Pi_h(K, t) \in \mathbf{Tbl}_{\mathcal{E}}(I', s')$ as follows. Given any possible row (or better, tuple) $t' \in \boldsymbol{tup}_{\mathcal{E}}(I', s')$, you can ask either an existential or a universal question about it: for example, "Does there *exist* a key $k \in K$ in T with image t'?" ($\boldsymbol{tup}_h(t_k) = t'$) or "Is it the case that *all* possible tuples $t \in \boldsymbol{tup}_{\mathcal{E}}(I, s)$ with image t' are present in T?" ([8])

Relation/Database Joins: The joins of $\mathcal{E}$-relations (or $\mathcal{E}$-tables) are represented in FOLE in terms of fibered products — products modulo some reference. If an $\mathcal{S}$-span of constraints $\langle I_1, s_1, \varphi_1 \rangle \xleftarrow{h_1} \langle I, s, \varphi \rangle \xrightarrow{h_2} \langle I_2, s_2, \varphi \rangle$ holds in a relational structure $\mathcal{M} = \langle \mathcal{R}, \langle \sigma, \tau \rangle, \mathcal{E} \rangle$, it is interpreted as an opspan of

---

[11] According to (Sowa [9]), every participant of a process is an entity that plays some role in that process. There is a "linearization" procedure that converts a binary/relational logical representation (FOLE, conceptual graphs) to a unary/functional logical representation (Sketches [5], Ologs [11]). In this example, linearization would define *functional* roles, changing the ternary relation type (process) to an entity type $\texttt{Go} \in X$ and converting its arity elements (participant roles) to function types $\texttt{agnt} \in \Omega_{\texttt{Person}, \langle 1, \texttt{Go} \rangle}$, $\texttt{dest} \in \Omega_{\texttt{City}, \langle 1, \texttt{Go} \rangle}$ and $\texttt{inst} \in \Omega_{\texttt{Bus}, \langle 1, \texttt{Go} \rangle}$.

$\mathcal{E}$-relations (or $\mathcal{E}$-tables). Then the join of $\mathcal{E}$-relations (or $\mathcal{E}$-tables) is represent by the formula $\iota_1{}^*(\varphi_1) \wedge_{\langle \widehat{I}, \widehat{s} \rangle} \iota_2{}^*(\varphi_2)$, where $\langle I_1, s_1 \rangle \xrightarrow{\iota_1} \langle \widehat{I}, \widehat{s} \rangle \xleftarrow{\iota_2} \langle I_2, s_2 \rangle$ is the fibered sum of type lists. In general, the join of an arbitrary diagram of $\mathcal{E}$-relations (or $\mathcal{E}$-tables) is obtained by substitution followed by conjunction.

A.4   Logical Environment

Let $\mathcal{S}_2 = \langle R_2, \sigma_2, X_2 \rangle \xRightarrow{\langle r, f \rangle} \langle R_1, \sigma_1, X_1 \rangle = \mathcal{S}_1$ be a schema morphism, with structure fiber passage $\mathbf{Struc}(\mathcal{S}_2) \xleftarrow{\mathbf{struc}_{\langle r, f \rangle}} \mathbf{Struc}(\mathcal{S}_2)$ and bridging structure morphism

$$\mathbf{struc}_{\langle r, f \rangle}(\mathcal{M}_1) = \langle r^{-1}(\mathcal{R}_1), \langle \sigma_2, \tau_1 \rangle, f^{-1}(\mathcal{E}_1) \rangle \overset{\langle r, 1_K, f, 1_Y \rangle}{\rightleftarrows} \langle \mathcal{R}_1, \langle \sigma_1, \tau_1 \rangle, \mathcal{E}_1 \rangle = \mathcal{M}_1$$

with relation and entity infomorphisms $r^{-1}(\mathcal{R}_1) \overset{\langle r, 1_K \rangle}{\rightleftarrows} \mathcal{R}_1$ and $f^{-1}(\mathcal{E}_1) \overset{\langle f, 1_Y \rangle}{\rightleftarrows} \mathcal{E}_1$.

Proposition 3. *The (formula) interpretation of the inverse image structure is the inverse image of the (formula) interpretation.*

Fact 1. *The formula classification of the inverse image relation classfication is the inverse image classfication of the formula relation classification:*

$$r^{-1}\widehat{(\mathcal{R}_1)} = \langle R_2, \widehat{K_1}, \models_r \rangle = \langle \widehat{R}_2, K_1, \models_{\widehat{r}} \rangle = \widehat{r}^{-1}(\widehat{\mathcal{R}}_1).$$

Proof. The proof is by induction on formulas $\varphi_2 \in \widehat{R}_2$.

Fact 2. *The formula structure morphism of the bridging structure morphism is:*

$$\langle \widehat{r}, 1_K, f, 1_Y \rangle : \langle r^{-1}\widehat{(\mathcal{R}_1)}, \langle \sigma_2, \tau_1 \rangle, f^{-1}(\mathcal{E}_1) \rangle \rightleftarrows \langle \widehat{\mathcal{R}}_1, \langle \sigma_1, \tau_1 \rangle, \mathcal{E}_1 \rangle.$$

*Its (**inst**-vertical) relation infomorphism*

$$\langle \widehat{r}, 1_K \rangle : r^{-1}\widehat{(\mathcal{R}_1)} = \langle R_2, \widehat{K_1}, \models_r \rangle = \langle \widehat{R}_2, K_1, \models_{\widehat{r}} \rangle \rightleftarrows \langle \widehat{R}_1, K_1, \models_{\widehat{\mathcal{R}}_1} \rangle = \widehat{\mathcal{R}}_1$$

is the bridging infomorphism of the formula relation classification, with the infomorphism condition $k_1 \models_{r^{-1}\widehat{(\mathcal{R}_1)}} \varphi_2$ *iff* $k_1 \models_{\widehat{\mathcal{R}}_1} \widehat{r}(\varphi_2)$. *The extent monotonic function* $\widehat{r} : \mathbf{ext}(r^{-1}\widehat{(\mathcal{R}_1)}) \to \mathbf{ext}(\widehat{\mathcal{R}}_1)$ *is an isometry:* $\varphi \leq_{r^{-1}\widehat{(\mathcal{R}_1)}} \psi$ *iff* $\widehat{r}(\varphi) \leq_{\widehat{\mathcal{R}}_1} \widehat{r}(\psi)$.

Proposition 4. *Satisfaction is invariant under change of notation; that is, for any schema morphism* $\mathcal{S}_2 = \langle R_2, \sigma_2, X_2 \rangle \xRightarrow{\langle r, f \rangle} \langle R_1, \sigma_1, X_1 \rangle = \mathcal{S}_1$ *the following satisfaction condition holds:*

$$\mathbf{struc}_{\langle r, f \rangle}(\mathcal{M}_1) \models_{\mathcal{S}_2} (\varphi_2 \xrightarrow{h} \varphi_2') \quad \textit{iff} \quad \mathcal{M}_1 \models_{\mathcal{S}_1} (\widehat{r}(\varphi_2) \xrightarrow{h} \widehat{r}(\varphi_2')) = \mathbf{fmla}_{\langle r, f \rangle}(\varphi_2 \vdash \varphi_2').$$

Proof. But this holds, since $r^{-1}\widehat{(\mathcal{R}_1)} = \widehat{r}^{-1}(\widehat{\mathcal{R}}_1)$. In more detail,

$\mathbf{struc}_{\langle r, f \rangle}(\mathcal{M}_1) \models_{\mathcal{S}_2} (\varphi_2 \xrightarrow{h} \varphi_2') \text{ iff } \Sigma_h(\varphi_2') \leq_{r^{-1}\widehat{(\mathcal{R}_1)}} \varphi_2$

iff $\widehat{r}(\Sigma_h(\varphi_2')) \leq_{\widehat{\mathcal{R}}_1} \widehat{r}(\varphi_2)$ iff $\Sigma_h(\widehat{r}(\varphi_2')) \leq_{\widehat{\mathcal{R}}_1} \widehat{r}(\varphi_2)$

iff $\mathcal{M}_1 \models_{\mathcal{S}_1} (\widehat{r}(\varphi_2) \xrightarrow{h} \widehat{r}(\varphi_2')) = \mathbf{fmla}_{\langle r, f \rangle}(\varphi_2 \vdash \varphi_2')$.

Proposition 5. *The institution* $\langle \mathbf{Sch}, \mathbf{fmla}, \mathbf{struc} \rangle$ *is a logical environment, since it satisfies the bimodular principle "satisfaction respects structure morphisms": given any schema* $\mathcal{S} = \langle R, \sigma, X \rangle$, *if* $\langle 1_R, k, 1_X, g \rangle : \mathcal{M}_2 \rightleftarrows \mathcal{M}_1$ *is a $\mathbf{sch}$-vertical structure morphism over* $\mathcal{S}$, *then we have the intent order* $\mathcal{M}_2 \geq_{\mathcal{S}} \mathcal{M}_1$; *that is,* $\mathcal{M}_2 \models_{\mathcal{S}} (\varphi \vdash \psi)$ *implies* $\mathcal{M}_1 \models_{\mathcal{S}} (\varphi \vdash \psi)$ *for any $\mathcal{S}$-sequent* $(\varphi \vdash \psi)$. [12]

Proof. The $\mathbf{typ}$-vertical formula morphism $\langle 1_{\widehat{R}}, k, 1_X, g \rangle : \widehat{\mathcal{M}}_2 \rightleftarrows \widehat{\mathcal{M}}_1$ over $\widehat{\mathcal{S}}$ has the $\mathbf{typ}$-vertical relation infomorphism $\langle 1_{\widehat{R}}, k \rangle : \widehat{\mathcal{R}}_2 \rightleftarrows \widehat{\mathcal{R}}_1$ over $\widehat{R}$.
$\mathcal{M}_2 \models_{\mathcal{S}} (\varphi \vdash \psi)$ $\underline{\text{iff}}$ $\varphi \leq_{\widehat{\mathcal{R}}_2} \psi$ $\underline{\text{implies}}$ $\varphi \leq_{\widehat{\mathcal{R}}_1} \psi$ $\underline{\text{iff}}$ $\mathcal{M}_1 \models_{\mathcal{S}} (\varphi \vdash \psi)$
for any $\mathcal{S}$-sequent $(\varphi \vdash \psi)$.

A.5   Transformation to Databases

A.5.1   Relational Interpretation.

Let $\mathcal{M} = \langle \mathcal{R}, \langle \sigma, \tau \rangle, \mathcal{E} \rangle$ be a (model-theoretic) relational structure. The relation classification $\mathcal{R}$ is equivalent to the extent function $\mathbf{ext}_{\mathcal{R}} : R \to \wp K$, which maps a relational symbol $r \in R$ to its $\mathcal{R}$-extent $\mathbf{ext}_{\mathcal{R}}(r) \subseteq K$. The list classification $\mathbf{List}(\mathcal{E})$ is equivalent to the extent function $\mathbf{ext}_{\mathbf{List}(\mathcal{E})} : \mathbf{List}(X) \to \wp \mathbf{List}(Y)$, a restriction of the tuple passage $\mathbf{tup}_{\mathcal{E}} : \mathbf{List}(X)^{\mathrm{op}} \to \mathbf{Set}$, which maps a type list $\langle I, s \rangle \in \mathbf{List}(X)$ to its $\mathbf{List}(\mathcal{E})$-extent $\mathbf{tup}_{\mathcal{E}}(I, s) \subseteq \mathbf{List}(Y)$. The list designation satisfies the condition $k \models_{\mathcal{R}} r$ implies $\tau(k) \models_{\mathbf{List}(\mathcal{E})} \sigma(r)$ for all $k \in K$ and $r \in R$; so that $k \in \mathbf{ext}_{\mathcal{R}}(r)$ implies $\tau(k) \in \mathbf{ext}_{\mathbf{List}(\mathcal{E})}(\sigma(r)) = \mathbf{tup}_{\mathcal{E}}(\sigma(r))$. Hence, $\wp \tau(\mathbf{ext}_{\mathcal{R}}(r)) \subseteq \mathbf{tup}_{\mathcal{E}}(\sigma(r))$ for all $r \in R$. Thus, we have the function order $\mathbf{ext}_{\mathcal{R}} \cdot \wp \tau \subseteq \sigma \cdot \mathbf{ext}_{\mathbf{List}(\mathcal{E})}$.

The relational interpretation function $\mathbf{R}_{\mathcal{M}} : R \to |\mathbf{Rel}(\mathcal{E})|$ maps a relational symbol $r \in R$ with type list $\sigma(r) = \langle I, s \rangle$ to the set of tuples $\mathbf{R}_{\mathcal{M}}(r) = \wp \tau(\mathbf{ext}_{\mathcal{R}}(r)) \in \wp \mathbf{tup}_{\mathcal{E}}(I, s) = \mathbf{Rel}_{\mathcal{E}}(I, s)$. The tabular interpretation function $\mathbf{T}_{\mathcal{M}} : R \to |\mathbf{Tbl}(\mathcal{E})| = |(\mathbf{Set} \downarrow \mathbf{tup}_{\mathcal{E}})|$ maps a relational symbol $r \in R$ with type list $\sigma(r) = \langle I, s \rangle$ to the pair $\mathbf{T}_{\mathcal{M}}(r) = \langle K(r), t_r \rangle$ consisting of the key set $K(r) = \mathbf{ext}_{\mathcal{R}}(r) \subseteq K$ and the tuple function $K(r) \xrightarrow{t_r} \mathbf{tup}_{\mathcal{E}}(I, s)$, a restriction of the tuple function $\tau : K \to \mathbf{List}(Y)$, which maps a key $k \in K_r$ to the tuple $t_r(k) = \tau(k) \in \mathbf{tup}_{\mathcal{E}}(I, s)$. Applying the image passage $\mathbf{im}_{\mathcal{E}}(I, s) : \mathbf{Tbl}_{\mathcal{E}}(I, s) \to \mathbf{Rel}_{\mathcal{E}}(I, s)$, the image of the table interpretation is the relation interpretation $\mathbf{im}_{\mathcal{E}}(I, s)(\mathbf{T}_{\mathcal{M}}(r)) = \mathbf{R}_{\mathcal{M}}(r)$ for any relation symbol $r \in R$. Using the combined image passage $\mathbf{im}_{\mathcal{E}} : \mathbf{Tbl}(\mathcal{E}) \to \mathbf{Rel}(\mathcal{E})$, we get the composition $\mathbf{R}_{\mathcal{M}} = R \xrightarrow{T_{\mathcal{M}}} |\mathbf{Tbl}(\mathcal{E})| \xrightarrow{|im_{\mathcal{E}}|} |\mathbf{Rel}(\mathcal{E})|$. Note that $t_r : K_r \to \mathbf{R}_{\mathcal{M}}(r) \to \mathbf{tup}_{\mathcal{E}}(I, s)$, is a surjection-injection factorization of the tuple function. [13]

---

[12] For any classification $\mathcal{A} = \langle X, Y, \models_{\mathcal{A}} \rangle$, the intent order $int(\mathcal{A}) = \langle Y, \leq_{\mathcal{A}} \rangle$ is defined as follows: for two instances $y, y' \in Y$, $y \leq_{\mathcal{A}} y'$ when $int_{\mathcal{A}}(y) \supseteq int_{\mathcal{A}}(y')$; that is, when $y' \models_{\mathcal{A}} x$ implies $y \models_{\mathcal{A}} x$ for each $x \in X$.

[13] Two tables are informationally equivalent when they contain the same information; that is, when their image relations are equivalent in $\mathbf{Rel}_{\mathcal{E}}(I, s) = \wp \mathbf{tup}_{\mathcal{E}}(I, s)$. In particular, the table $\mathbf{T}_{\mathcal{M}}(r)$ and relation $\mathbf{R}_{\mathcal{M}}(r)$ of a relational symbol are informationally equivalent.

A.5.2   Relational Logics. A relational logic $\mathcal{L} = \langle \mathcal{S}, \mathcal{M}, T \rangle$ consists of a relational structure $\mathcal{M} = \langle \mathcal{R}, \langle \sigma, \tau \rangle, \mathcal{E} \rangle$ and a relational specification $T = \langle \mathcal{S}, T \rangle$ that share a common relational schema $\mathbf{sch}(\mathcal{M}) = \mathcal{S}$. The logic is sound when the structure $\mathcal{M}$ satisfies every constraint in the specification T. A sound relational logic enriches a relational structure with a specification. For any sound logic $\mathcal{L} = \langle \mathcal{S}, \mathcal{M}, T \rangle$, there is an interpretation passage $\widehat{\mathbf{R}}^{\mathrm{op}} \xrightarrow{T_{\mathcal{L}}} \mathbf{Tbl}(\mathcal{E}) = (\mathbf{Set}{\downarrow}\mathbf{tup}_{\mathcal{E}})$, where $\widehat{\mathbf{R}} \subseteq \mathbf{Fmla}(\mathcal{S})$ is the consequence of T. Sound logics are important in the transformation of structures to databases (below). A relational logic morphism $\mathcal{L}_2 = \langle \mathcal{S}_2, \mathcal{M}_2, T_2 \rangle \xrightarrow{\langle \langle r,k \rangle, \langle f,g \rangle \rangle} \langle \mathcal{S}_2, \mathcal{M}_2, T_2 \rangle = \mathcal{L}_2$ consists of a relational structure morphism $\mathcal{M}_2 \xrightarrow{\langle \langle r,k \rangle, \langle f,g \rangle \rangle} \mathcal{M}_1$ and a relational specification morphism $T_2 = \langle \mathcal{S}_2, T_2 \rangle \xrightarrow{\langle r,f \rangle} \langle \mathcal{S}_1, T_1 \rangle = T_1$ that share a common relational schema morphism $\mathbf{sch}(\langle r,k \rangle, \langle f,g \rangle) = \mathcal{S}_2 \xrightarrow{\langle r,f \rangle} \mathcal{S}_1$.

Any sound relational logic $\mathcal{L} = \langle \mathcal{S}, \mathcal{M}, T \rangle$ with structure $\mathcal{M} = \langle \mathcal{R}, \langle \sigma, \tau \rangle, \mathcal{E} \rangle$ and specification $T = \langle \mathcal{S}, T \rangle$ has an associated logical/relational database $\boldsymbol{db}(\mathcal{L}) = \langle \mathcal{S}, \mathcal{E}, \boldsymbol{K}, \tau \rangle$ with category of formulas $\widehat{\mathbf{R}} \subseteq \mathbf{Fmla}(\mathcal{S})$ (the consequence of T), signature passage $\boldsymbol{S} : \widehat{\mathbf{R}} \to \mathbf{List}(X)$, entity classification $\mathcal{E}$, key passage $\boldsymbol{K} : \widehat{\mathbf{R}}^{\mathrm{op}} \to \mathbf{Set}$, tuple bridge $\tau : \boldsymbol{K} \Rightarrow \boldsymbol{S}^{\mathrm{op}} \circ \boldsymbol{tup}_{\mathcal{E}}$, and table interpretation passage $\widehat{\mathbf{R}}^{\mathrm{op}} \xrightarrow{T} \mathbf{Tbl}(\mathcal{E}) = (\mathbf{Set}{\downarrow}\boldsymbol{tup}_{\mathcal{E}})$, where $\tau = \boldsymbol{T}\tau_{\mathcal{E}}$. Any sound

$$\mathcal{L}_2 = \langle \mathcal{S}_2, \mathcal{M}_2, T_2 \rangle \xrightarrow{\langle \langle r,k \rangle, \langle f,g \rangle \rangle} \langle \mathcal{S}_2, \mathcal{M}_2, T_2 \rangle = \mathcal{L}_2$$

$$\Downarrow \boldsymbol{db}$$

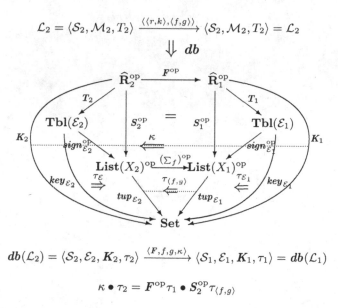

$$\boldsymbol{db}(\mathcal{L}_2) = \langle \mathcal{S}_2, \mathcal{E}_2, \boldsymbol{K}_2, \tau_2 \rangle \xrightarrow{\langle F, f, g, \kappa \rangle} \langle \mathcal{S}_1, \mathcal{E}_1, \boldsymbol{K}_1, \tau_1 \rangle = \boldsymbol{db}(\mathcal{L}_1)$$

$$\kappa \bullet \tau_2 = \boldsymbol{F}^{\mathrm{op}} \tau_1 \bullet \boldsymbol{S}_2^{\mathrm{op}} \tau_{\langle f,g \rangle}$$

Fig. 5. From Sound Logics to Logical/Relational Databases

relational logic morphism $\mathcal{L}_2 = \langle \mathcal{S}_2, \mathcal{M}_2, T_2 \rangle \xrightarrow{\langle \langle r,k \rangle, \langle f,g \rangle \rangle} \langle \mathcal{S}_2, \mathcal{M}_2, T_2 \rangle = \mathcal{L}_2$ with structure morphism $\mathcal{M}_2 \xrightarrow{\langle \langle r,k \rangle, \langle f,g \rangle \rangle} \mathcal{M}_1$ and specification morphism $T_2 = \langle \mathcal{S}_2, T_2 \rangle \xrightarrow{\langle r,f \rangle} \langle \mathcal{S}_1, T_1 \rangle = T_1$ has an associated (strict) logical/relational database morphism $db(\langle r, k \rangle, \langle f, g \rangle) = \langle \boldsymbol{F}, f, g, \kappa \rangle : db(\mathcal{L}_2) = \langle \mathcal{S}_2, \mathcal{E}_2, \boldsymbol{K}_2, \tau_2 \rangle \rightarrow \langle \mathcal{S}_1, \mathcal{E}_1, \boldsymbol{K}_1, \tau_1 \rangle = db(\mathcal{L}_1)$ with (strict) database schema morphism $\langle \boldsymbol{F}, f \rangle :$ $\mathcal{S}_2 \rightarrow \mathcal{S}_1$, entity infomorphism $\langle f, g \rangle : \mathcal{E}_2 \rightleftarrows \mathcal{E}_1$, and key natural transformation $\kappa : \boldsymbol{F}^{\mathrm{op}} \circ \boldsymbol{K}_1 \Rightarrow \boldsymbol{K}_2$, which satisfy the condition $\kappa \bullet \tau_2 = \boldsymbol{F}^{\mathrm{op}} \tau_1 \bullet \boldsymbol{S}_2^{\mathrm{op}} \tau_{\langle f, g \rangle}$. The passage $\widehat{\mathbf{R}}_2 \xrightarrow{F} \widehat{\mathbf{R}}_1$ from formula subcontext $\widehat{\mathbf{R}}_2 \subseteq \mathbf{Fmla}(\mathcal{S}_2)$ to formula subcontext $\widehat{\mathbf{R}}_1 \subseteq \mathbf{Fmla}(\mathcal{S}_1)$ is a restriction of the fibered formula passage $\mathbf{Fmla}(\mathcal{S}_2) \xrightarrow{fmla_{\langle r, f \rangle}} \mathbf{Fmla}(\mathcal{S}_1)$. (Kent [7] has more details on relational database semantics.)

Designing Learning to Research the Formal Concept Analysis of Transactional Data

Martin Watmough, Simon Polovina, and Simon Andrews

Conceptual Structures Research Group
Communication and Computing Research Centre
Faculty of Arts, Computing, Engineering and Sciences
Sheffield Hallam University, Sheffield, UK
{M.Watmough,S.Polovina,S.Andrews}@shu.ac.uk

Abstract. Transactional systems are core to much business activity; however leveraging any advantage from the data in these enterprise systems remains a challenging task for businesses. To research and discover the hidden semantics in transactional data, Sheffield Hallam University has incorporated Formal Concept Analysis (FCA) into two of its degree courses. We present a learning, teaching and assessment (LTA) method that integrates with this research. To make it reflect industrial practice and to further the state of the art of the research, this method includes the use of ERPsim. This large scale, real-world business simulation software is based on the Enterprise Resource Planning (ERP) enterprise system by SAP A.G., a global business software vendor. Together with a mix of individual and group work approaches, FCA tools (namely FCA BedRock, In-Close and Concept Explorer) and comparisons with alternative approaches, it is emerging that FCA can fulfil an important role in transactional systems and enhance its role in Business Intelligence (BI).

Keywords: FCA (Formal Concept Analysis), LTA (Learning, Teaching and Assessment), ERPsim (ERP Simulation Software), BI (Business Intelligence), Transactional Data.

1 Introduction

Transactional systems provide a core function and support considerable business activity within organisations around the world; however, they are intrinsically complex and significant effort is required to understand and manage them effectively. Based on the current outlook, system landscapes are evolving and becoming more flexible and agile. Therefore, analysis techniques must follow suit.

Our research interests arise from how to discover the hidden semantics within transactional systems i.e. how useful information or knowledge can be identified from mainstream database systems, by applying and combining analysis techniques. To assist, at Sheffield Hallam University we have incorporated this research into two Computing degree course modules. As such we aim for the research to be informed by the student's experiences, whilst enriching the student's

H.D. Pfeiffer et al. (Eds.): ICCS 2013, LNAI 7735, pp. 231–238, 2013.

knowledge in this topical area. Accordingly this paper focuses on the incremental design of the learning environment in the pursuit of discovering hidden semantics and the results achieved.

By applying and developing the approach to teaching transactional systems and analysis, two benefits are envisaged. Firstly, an insight into how learning these methods benefits the modules and students. Secondly, to engender a creative arena that encourages open answers from the students. Formal Concept Analysis (FCA) is a technique for analysing data in order to discover information and knowledge. FCA is particularly attractive in that offers an automated means of eliciting these concepts from the data [1] [2]. Therefore, FCA was selected as our underlining technique for designing learning in order to research the hitherto hidden semantics in transactional data.

2   Background

Enterprise Resource Planning (ERP) systems are typically transactional systems that support the core functions within an organisation. SAP A.G. are one of the leading providers [3]. We therefore base the analysis on ERPsim [4], an SAP A.G. ERP based simulation game that features competitive behaviour and increasing levels of complexity in a highly immersive and demanding atmosphere that reflects industrial practice.

ERPsim has a strong pedagogic foundation that has been adopted and applied during the development of the degree modules. ERPsim is designed for active learning in that it achieves long-term retention. ERPsim takes advantage of Situation Cognitive Theory and Problem-based Learning [5].

Situation theory states that activities, tasks, and understanding do not exist in isolation, but rather are part of broader relation systems and that situated cognition is associated with a higher level of engagement and motivation in learners, thus generally leads to a better understanding and transfer of knowledge [6]. Problem-based learning is a widely applied technique that has its origins dating back to 1966 in medical education [7]. It is a teaching strategy to promote self-directed learning and critical thinking through problem solving in which active participation and challenging problems in a relevant context are key [8].

Furthermore, the learning environment created by ERPsim has been carried into the analysis of its output by comparative techniques. These techniques, described later, are used in order to evaluate the comparative value of FCA for transactional data.

To assess the effectiveness of our learning, teaching and assessment (LTA) we examined the marks achieved and learning objectives; the findings and feedback from the students have also been considered. It should be clarified that the students on these modules' did not have a significant mathematical background; rather the modules focussed on the business application of FCA. For this reason and to preserve consistency we ensured that the FCA tools were explained and applied according to their understanding. The fact that the raw data structure was constant aided the process.

3   Method

Our method was based on previous work that applied Biggs' Constructive Alignment and Yin's Case Study Method [9], [10], [11]. We modified this method so that it could better support the learning outcomes and theory relevant to the analysis of business transactional data using FCA tools.

Yin's method was applied to capture and learn from a number of case studies, where each case study represents the relevant modules' assignments. There were four case studies, one from each module for the academic years 2011-11 and 2011-12. An overview is contained in Table 1. Two aspects have been used for evaluation, firstly, the assignment marks per section have been compared with the teaching and learning techniques applied. Secondly, the student's evaluations and conclusions have been used for qualitative analysis. Biggs' Constructive Alignment has two basic concepts; learners construct meaning from what they do to learn and that the teacher makes an alignment between learning activities and learning outcomes [10]. The combination of Biggs' constructive alignment and Yin's Case Study Method provides an overall method for aligning the learning activities and learning outcomes for the benefit of future students [9]. It was also envisaged that an insight into the introduction of FCA into an organisation's Business Intelligence capability would be gained.

The FCA tools used for our study were FcaBedrock [12], In-Close [13] and Concept Explorer [14]. An overview of the steps is described shortly and a comprehensive description is provided elsewhere [15].

We now describe how we applied our method. To begin with, ERPsim stores the raw data from the game in a relational database. This raw data was extracted directly into a Microsoft (MS) Access database that mirrors the tables and relationships of the ERPsim system. MS Access queries were then used to extract data into a comma seperated values (CSV) text format that contained the key attributes and meta data. From this file, FcaBedrock was used to created a Formal Context. In-Close was then used to provide minimum support by reduce the number of formal concepts, before graphically presenting its results using Concept Explorer as a concept lattice. Without the application of In-Close the output from FcaBedrock can be too complex for meaningful visualisations; in effect the less dominant relationships are removed.

The techniques summaried in Table 1 were employed in order to develop the teaching and assessment methods. These have included learning in conjunction with ERPsim, a mix of individual and group work approaches and comparisons with alternative approaches.

The graph in Figure 1 indicates how the assignment marks deviated from the average mark for each module. Taking case study 4 as an example, the students achieved higher than the average percentage for the introduction and lower for the FCA sections. The perfect line would run through zero with each student achieving the same percentage for each section of the assignment; as this is based on the average the performance does not differentiate between high and low achieving students.

Table 1. Chronology of Teaching Methods and Results

Case Study	1	2	3	4
Module	SA 2010-11	ES 2010-11	SA 2011-12	ES 2011-12
Average Mark	56.6	58.4	66.8	58.6
Standard Deviation	15.3	21	3.8	11.5
Data Preparation Demonstrated	X	X		X
End to End Data Prepared			X	X
Graphical presentation	X			
Document		X	X	X
Excel and FCA	X	X	X	
BI, Excel and FCA				X
Group discussion	X	X		
Group work		X		
Jigsaw based approach			X	X
Horizontal and Vertical Group work				X
Re-use (multi company)	X	X		
BPM Integration			X	X

4   Case Study Review

Beginning with Case Study 1, generic processing steps were intended to be reused over different subsets of the data incorporated. This however appeared to have only generated repetition and not an improvement in marks or learning. For comparison, the students were required to target the same data with FCA and Excel. The results achieved did not differ to a noticeable extent. Comments by the students suggested that significantly more time was required in order to apply and understand FCA, although its graphical nature did lend itself effectively to creating content for inclusion in the assignment.

Reducing or simplifying the quantity of data preparation was suggested by the students, but it was not clear if the challenge of performing this task was beneficial to the learning process. Instead of eliminating the data preparation task it was decided to reduce the individual workload through group work. Therefore, the group would still retain any learning from the experience while sub-dividing the manual effort required.

Case Study 2 generated clear comments and queries during the module indicating that a method for reducing the amount of time for preparing data is required, even with a group-based approach as collaboratively preparing data proved difficult to achieve. The majority of students managed the task but probably at the expense of actually performing and evaluating the analysis. An approach to improving group work and networking was also identified as the use of communication tools, predominantly the discussion boards and blogs were limited.

Learning the principles of generic design and reuse was successful but repeating the analysis for multiple scenarios did not add significant value. There were enough opportunities to repeat and tailor the analysis in a single section of the

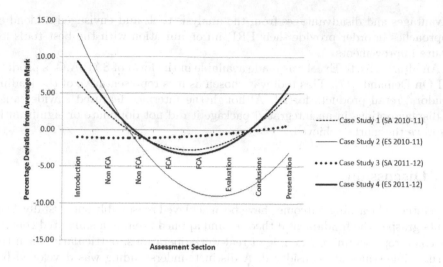

Fig. 1. Deviation per Section from Student Average Mark

assignment to support the learning outcomes. Reintroducing group based assignments would be an interesting choice to return to as the FCA tools develop and focus can be shifted to collaborative or even social topics; however, in the current context it did not have a positive effect upon the results.

Case Study 3 applied a technique for cooperative learning called Jigsaw [16] that encourages participation and emphasises the value of every student's contribution towards the outcome. Jigsaw was intended to develop group problem-solving while maintain the individual's contribution to the task. An emergent outcome was identified in that this group cooperation has parallels with working in current or future workplaces that feature more diverse skills requirements, physical distributed teams and all manners of collaboration and communication mechanisms.

A complete set of data was prepared for each group through all stages, including instructions about how to modify, refine and enhance the analysis. Preparing the initial data for each individual team resulted in less creativity and fewer variations across the assignments. However, the average mark for FCA did improve.

Rule definition was introduced as a mechanism for expressing the findings as a formula or in a logic form. Deriving rules from the analysis was challenging for the students, however it appeared to complete the cycle back to source transactional data. The students demonstrated an understanding of the relationships discovered and how they could be applied to ERPsim processes.

Case Study 4 delivered the most comprehensive learning judging by feedback from the students. The Jigsaw [16] structure employed resulted in the most frequent use of the discussion boards and collaboration. The groups were organised in two directions, vertically to promote interaction and team work within the group and horizontally to create cross group knowledge sharing, almost like expert communities between groups. A number of students found advantages and disadvantages from the analysis tools and envisaged a blend of approaches in order provide their ERPsim organisation with the best tools for future improvements.

An alternative to Excel was made available in the form of SAP A.G.'s product 'BI On Demand' [17]. This tool was chosen as it is representative of the leading vendors' retail products for BI. Although the interface featured a wide range of display options in an integrated package it did not dominate or significantly improve the marks achieved.

5   Discussion

The intended learning outcomes have been achieved reasonably successfully. Students grasped the fundamental theories and applied them in a simulated context that is a representative example of real-work operations, particularly when the actual time scales are considered. A distinct understanding was developed between the simplicity of models and the challenge of identifying useful data and outcomes from a large data set.

The context and energy developed during ERPsim is inherently valuable in achieving the learning outcomes; it promotes the group dynamics, rapid learning and knowledge retention. The complete cycle, including a range of contemporary through to research level analysis methods was key to achieving the learning outcomes. Data preparation was a highly cited problem; however, it is anticipated that this will simplified into a data selection task with future generations of the software. The difference between industry produced solutions such as 'BI On demand' and the FCA tools was quickly highlighted by the students.

A number of unintended but valued learning outcomes were also highlighted in line with Biggs Constructive Alignment. There emerged an inherent value in the analysts (students) being involved in the data preparation, despite their raising this as an issue. Rather than just implying that it was unduly time consuming they appreciated the value in understanding the context, source and calculations that help discern towards extracting the transactional data. In passing there was little to differentiate the results from the analysis for the case study 1 where students complete the whole preparation task with case study 4 where students modified a generic preparation routine, thus enabling the students to focus on FCA. A further unintended but valued learning outcome was how effective the tools would be when used in conjunction with each other instead of the separation of the tools as originally directed by the assignments.

Certain students found it difficult to grasp an 'incomplete' picture. The idea of determining rules helped somewhat but the data only provided a fraction of the complete set of rules. Partial cognitive models will probably be more common than a comprehensive understanding as the rate of change and volume of information increases making this potentially a topic for further research.

The capability of FCA for discovering the concepts and relationships in transactional data was repeatedly identified as a key reason for applying it. Confidence was also cited by the students, particularly in the context of understanding what the analysis actually indicated. It frequently required that a number of repetitions were needed to clarify and subsequently accept the result. There were also some interesting remarks that expressed unexpected negatives about more familiar tools (Excel) when considering large data sets or potential 'big data' problems. This demonstrated that the key messages of the modules had been learnt and applied usefully in comparison with FCA.

6   Conclusion

The method and tools applied in Case Study 3 represents the most successful teaching methods to date for FCA in Sheffield Hallam's modules in terms of marks. There was an evident improvement in the marks of the FCA sections across both modules as Figure 1 illustrates, but there are still opportunities to develop and improve the application of FCA based tools. The learning environment largely succeeded in providing students with those meaningful experiences that business analysts need. In particular it equipped them with a well rounded experience, which is a significant factor. This reinforces Gartner's findings that analysis will be controlled by business units and not technical experts [18].

The propositions of Presthus in describing why teaching Business Intelligence is challenging from the perspective of students and lecturers also emerged [19]. It is interesting (and comforting) to note that the approaches taken in our study happened to address to an extent these propositions. The propositions included providing a mechanism for reducing the level of abstraction when teaching and demonstrating the business value of BI. This lead to generating interest, effective learning based on suitable data sets, and the value of case studies.

Our research indicates how FCA could be integrated into BI; a novel research outcome arising from our LTA (Learning, Teaching, and Assessment) approach. As such, it has been shown how FCA could fulfil an important role within BI solutions for transactional data. The need for integrated tools that support knowledge discovery in a collaborative model from a rich source of data is clear; the success of these tools will be based on far more than their discrete technical capabilities.

The challenges faced through teaching and also those experienced by the students has clear parallels with the implementation and adoption of such tools in the workplace. Comparing and contrasting the techniques that have proven to be successful in the classroom to the business world would be an interesting research topic, as would addressing the problem of managing incomplete information and models. From our education experiences in the interim, we can envisage that FCA has an important role to play.

References

1. Wille, R.: Conceptual Graphs and Formal Concept Analysis, Technische Hochschule Darmstadt and Fachbereich Mathematik. Springer (1997)
2. Wolff, K.R.: A First Course in Formal Concept Analysis. In: Faulbaum, F. (ed.) SoftStat 1993. Advances in Statistical Software, vol. 4, pp. 429–438. Gustac Fisher Verlag, Stuttgart (1994)

3. SAP, About SAP AG (2012), http://www.sap.com/about-sap/about-sap.epx (cited October 8, 2012)
4. Leger, P.-M., Robert, J., Babin, G., Pellerin, R., Wagner, B.: ERPsim, ERPsim Lab, HEC Montréal, Montréal, Qc (2007)
5. Feldstein, H.: Interview with Harvey Feldstein ERPsim (2011), http://erpsim.hec.ca/learning/#/curriculum/153 (viewed August 20, 2012)
6. Leger, P.-M., Robert, J., Babin, G., Pellerin, R., Lyle, D., et al.: ERP Simulation Game with SAP ERP: Logistics Game (Platinum Version), ERPsim Lab, HEC Montréal, 44 p. (2011) ISBN: 978-0-9866653-2-5
7. Hillen, H., Scherpbier, A., Wijnen, W.: History of problem-based learning in medical education. Oxford University Press (2010)
8. Ginty, A.: Problem Based Learning. Higher Education Academy, Escalate (2007)
9. Andrews, S.: Aligning the Teaching of FCA with Existing Module Learning Outcomes. In: Andrews, S., Polovina, S., Hill, R., Akhgar, B. (eds.) ICCS 2011. LNCS (LNAI), vol. 6828, pp. 394–401. Springer, Heidelberg (2011)
10. Biggs, J.: Teaching for Quality Learning at University. SRHE and Open University Press, Buckingham (1999)
11. Yin, R.: Case Study Research: Design and Methods, 4th edn. Applied Social Research Methods Series, vol. 5. SAGE, Thousand Oaks (2009)
12. Andrews, S., Orphanides, C.: FcaBedrock, a Formal Context Creator. In: Croitoru, M., Ferré, S., Lukose, D. (eds.) ICCS 2010. LNCS, vol. 6208, pp. 181–184. Springer, Heidelberg (2010)
13. Andrews, S.: In-Close2, a High Performance Formal Concept Miner. In: Andrews, S., Polovina, S., Hill, R., Akhgar, B. (eds.) ICCS 2011. LNCS (LNAI), vol. 6828, pp. 50–62. Springer, Heidelberg (2011)
14. ConExp, Concept Explorer (2006), http://sourceforge.net/projects/conexp/ (cited April 1, 2012)
15. Andrews, S., Orphanides, C., Polovina, S.: Visualising Computational Intelligence through Converting Data into Formal Concepts. In: Xhafa, F., Barolli, L., Nishino, H., Aleksy, M. (eds.) Proceedings of the 5th International Conference on P2P, Parallel, Grid, Cloud and Internet Computing (3GPCIC), pp. 302–307. IEEE Computer Society (2010) ISBN 978-0-7695-4237-9/10
16. Aronson, E.: Jigsaw Classroom: Overview of the Technique Social Psychology Network, (2012), http://www.jigsaw.org/overview.html (cited August 15, 2012)
17. SAP, BI on Demand powered by Hana SAP (2012), http://www.biondemand.com/businessintelligence (SAP viewed August 15, 2012)
18. Gartner, Gartner reveals five business intelligence predictions for 2009 and beyond (2009), http://www.gartner.com/it/page.jsp?id=856714 (cited August 15, 2012)
19. Presthus, W.: Never giving up: Challenges and solutions when teaching Business Intelligence, The Norwegian School of IT (2012)

Cross-Domain Inference Using Conceptual Graphs in Context of Laws of Science

Shreya Inamdar

Birla Institute of Technology and Science,
Pilani, Rajasthan, India
shreyainamdar3141@gmail.com

Abstract. Knowledge bases, as conceptual graphs, are considered to be brittle as they are highly domain specific. This paper attempts to get some flexibility by predicting the possible nodes, using the other existing graphs. Graph theory principles of maximum common sub-graph and minimum common super-graph for labelled graphs, allow extension of a given conceptual graph. This paper attempts to solve this problem for laws of science. Given a few fundamental equations of two different domains, but similar mathematical structure, equations can be converted to a common set of dummy variables. These transformed equations will be the labels for further set operations. Extending the two graphs using the minimum common super-graph and maximum common super-graph, we then convert these transformed equations back to their original variables. Then, apply constraints to check the feasibility and finalize this extension. Thus we have inferred some part of the knowledge base from other domains.

Keywords: Labeled graph, Maximum Common sub-graph, Minimum Common super-graph, Cross-Domain Inference.

1 Introduction

In Science, we come across several instances where two phenomena are explained on the same mathematical arguments. For example, in physics, the resistor-capacitor-inductor system is analogous to spring-mass-dashpod system. They are governed by the same second order differential equations. Automated capturing of this analogy is the task at hand. Conceptual Graphs are chosen for this because of their universal representation of information as triples and the ease with which all graph operations can be applied to them. This opens it as a field of interest to computer scientists as well as mathematicians.

In general, Knowledge bases are considered to be brittle due to their domain specific nature[1]. In the given example, we refer to electrical and mechanical systems as different domains. This paper proposes a novel algorithm for cross-domain inference. For a graph with given concepts (nodes) and relations among these concepts (edges), this paper proposes an algorithm to find the possible nodes and relations which are not given. This is the inference part. The method

H.D. Pfeiffer et al. (Eds.): ICCS 2013, LNAI 7735, pp. 239–244, 2013.

uses graph theory extensively. We can compare the two graphs, by bringing them to common set of labels. This can be done by pattern matching or by using measures that give structural similarity between graphs, graph isomorphism. We then generate a dictionary that maps the common labels to the specific labels for each graph to get back to original variables from the inferred graph. Our comparison is driven by the degree of similarity of the two labeled graphs with common set of labels. This degree of similarity is estimated by their maximum common subgraph. If there is a significant match between the graphs,The inference is done by changing each graph to the minimum common super graph of the two graphs. But each of these graphs now has common labels. The common labels are converted back to original variables by using the dictionary generated previously. Given some restrictions on variable values, some of the wrongly inferred nodes and edges can then be discarded.

This paper is divided into 6 sections. The second section covers introduction to conceptual graphs and the definitions of graph theory that are used in this paper. Sections third through fifth cover the steps of the algorithm in order. Sixth section covers the work done so far and its future scope.

2   Concepts and Definitions

2.1   Conceptual Graphs

For Graph theoretic treatment of Conceptual Graphs, they can be visualized as labeled graphs with nodes as concepts and edges as relations. They are directed for non-symmetric relations and undirected for symmetric relations. In this paper, we will use directed graphs for generating the common label set and un-directed graphs for the generating the maximum common subset and minimum common super set. For the common label generation, the direction of flow of variables in the graph is important. So we use directed graphs. But our graphs that we wish to extend have no direction. They are just related nodes. So, we use un-directed graphs here.

In general, conceptual graphs are in wide use because they present an easy way to store data in a more language independent format so that it can be processed without ambiguity. The node-type and class of a node are treated as a don't care here because we have a mathematical basis of comparison when we are dealing with equations. The class of the node and it's type become important when we are dealing with generic graphs which does not have a clearly defined basis for generating the common label set.

2.2   Graph Theory Concepts

There are several measures of similarity between two graphs. Graph Isomorphism, Graph edit distance, Maximum Sub-graph etc. We choose maximum sub-graph measure because of the following reasons:

- This has a complementary minimum common super-graph for further operations of inference. Also, Minimum Common Super-graph requires computing the

maximum common subgraph. So, anyways we needed to compute this measure.
- Graph isomorphism doesn't use our information of labels and depends only on
structure. On the other hand, maximum common sub-graph uses this information. So, it is expected to give better results.
- As the complexity of our graph decreases, the complexity of the algorithm also
decreases. For instance, for computing the maximum common sub-tree, linear
running time algorithms have been proposed.

Finding the maximum common sub-graph is a NP hard problem, but there
are algorithms to solve it in linear time[2] if the graph is given to be planar. So,
we make the assumption that the graph is planar.

Let L denote the finite set of labels for nodes and edges. Each node and edge
has a node label and an edge label. So, a labeled graph $A = (V, E, L, G)$.

V: finite set of vertexes $E \subseteq V \times V$: set of edges L : set of labels G : reduced
set of labels

Label sets of each edge (vertex) are stored as attributes of that edge (vertex).

Maximum Common Sub-graph. Let $g_1(V_1, E_1, L_1, G_1)$ and $g_2(V_2, E_2,$
$L_2, G_2)$ be graphs. A common subgraph of g_1 and g_2 , $cs(g_1, g2)$, is a graph
$g(V, E, L, G)$ such that there exist subgraph isomorphisms from g to g_1 and
from g to g_2. We call g a maximum common subgraph of g_1 and g_2, $mcs(g_1, g_2)$,
if there exists no other common subgraph of g_1 and g_2 that has more nodes
than g [3].

Minimum Common Super-graph. Let $g_1(V_1, E_1, L_1, G_1)$ and $g_2(V_2, E_2,$
$L_2, G_2)$ be graphs. A common subgraph of g_1 and g_2 , $cs(g_1, g2)$, is a graph
$g(V, E, L, G)$ such that there exist subgraph isomorphisms from g_1 to g and from
g_2 to g. We call g a minimum common supergraph of g_1 and g_2 , $MCS(g_1, g_2)$,
if there exists no other common supergraph of g_1 and g_2 that has fewer nodes
and, for a given set of nodes, fewer edges than g [3].

Embedding of One Graph in Other. Let $g_1(V_1, E_1, L_1, G_1)$ and $g_2(V_2, E_2,$
$L_2, G_2)$ be graphs with $g_1 \subseteq g_2$. The embedding of g_1 in g_2, $emb(g_1, g_2)$, is the
set of edges that connect g_1 and g_2 - g_1, i.e.,

$$emb(g_1, g_2) = (V_1 \times (V_2 - V_1)) \cup ((V_2 - V_1) \times V_1), \qquad (1)$$

where the edge labels are same for any edge in $emb(g_1, g_2)$[3]. Henceforth, we
refer to $emb(g_1, g_2)$ as e_2.

2.3   Procedure

The procedure is described as an algorithm. The following sections will take-up
each step in detail.

Algorithm 1. crossDomainInference(graph g1, graph g2)

$(g1', g2') \leftarrow convertToCommonLabels(g1, g2)$
$g \leftarrow maximumCommonSubgraph(g1', g2')$
$G \leftarrow minimumSuperGraph(g1', g2')$
$G1' \leftarrow extend(G, g1')$
$G2' \leftarrow extend(G, g2')$
$G1 \leftarrow reLabel(G1')$
$G2 \leftarrow reLabel(G2')$
$G1 \leftarrow validate(G1)$
$G2 \leftarrow validate(G2)$

3   Convert to Common Labels

In general, this is a pattern matching problem. But this can again be solved using graphs. We have two types of nodes. One is the variables and the other are unary operators, eg. D - the differential operator. We have Inv for multiplicative inverse as an operator. For additive inverse, we have aInv. The edge labels are binary operations, eg. $+$, $*$. We have not taken subtraction and division as binary operations. Instead, we made them unary operators. This is because they are not commutative. Also, in this step we are not concerned with the meaning of the variables. If there are two contesting variables, they might as well be interchanged, because we are not computing values from the equations. The operator nodes need to match at all locations. These will serve as check points to generate the dictionary. Also, there is no interlinking between different equations, as we can see in the figure. We have tree/linear structures. The tree has directed edges. Tree matching is done on two equations. If the two equations, one from g_1 and other from g_2 match, a dictionary is generated. This maps the variables into reduced or common variables used for further set operations. This dictionary generation is not discussed in detail here.

4   Generating the Maximum Common Subgraph

The maximum common sub-graph can be interpreted as the overlapping region of the two graphs. In simpler words, the common information that is given in both the domains. The author of [4] analyses two famous algorithms, McGregor and Koch algorithm for computing the maximum common sub-graph for labeled graphs. Going by its results, we choose Koch algorithm for Computing the maximum common sub-graph. The algorithm by Koch transforms the maximal common subgraph problem to the maximal clique problem and searches for the maximal clique using a branch-and-bound algorithm.

5   Generating the Minimum Common Super-Graph

[3] Proves that the minimum common super-graph(mcs) computation can be solved by means of the maximum common sub-graph(MCS) computation as follows

$$MCS(g_1, g_2) = mcs(g_1, g_2) \underset{e_1}{\bigcup}(g_1 - mcs(g_1, g_2)) \underset{e_2}{\bigcup}(g_2 - mcs(g_1, g_2)) \quad (2)$$

where $e_1 = emb(mcs(g_1, g_2), g_1)$ and $e_2 = emb(mcs(g1, g2), g2)$.

The minimum common super graph can be interpreted as the inferred graph. This is still in terms of the common variables.

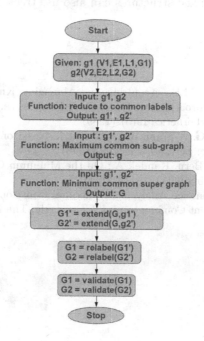

Fig. 1. Flowchart for the algorithm showing inputs and outputs

6   Relabeling and Validation

The maximum common super graph can be relabeled to original variables using the dictionary generated during reduction. The formed graph's nodes, which contain equations, be checked for constraint satisfaction and possible erroneous extrapolations. Thus the final graph can be inferred using the other similar graphs in the knowledge base.

7   Work So Far and Future Scope

This is a novel idea which is in its initial stages of development. Currently, the work on generating the common dictionary is progress. The approach used is to generate a common set of rules for standardized representation of the equation and then use the operators as milestones to get the correct match. We assume that the set of variables and operators is a given in both the graphs. This is a valid assumption as all nodes and edges are stored before entering them for representation in the software. Currently we are sorting out certain issues with Gephi software.

This concept of cross-domain inference can be extended to the general knowledge bases using classes instead of mathematical expressions for dictionary generation. The sparsely linked structures can also use trees, instead of graphs, for simpler computations.

References

1. Berg-Cross, G., Price, M.E.: Acquiring and Managing Knowledge Using a Conceptual Structures Approach: Introduction and Framework. IEEE Transactions on Systems, Man, and Cybernetics 19(3) (1989)
2. Bunke, H., Shearer, K.: A graph distance metric based on the maximal common subgraph. Pattern Rec. Lett. 19, 255–259 (1998)
3. Bunke, H., Jiang, X., Bern, Kandel, A.: On the Minimum Common Supergraph of Two Graphs. Computing 65(1), 13–15 (2000)
4. Welling, R.: A Performance Analysis on Maximal Common Subgraph Algorithms. In: 15th Twente Student Conference on IT, Enschede, The Netherlands. University of Twente (2011)

Summarizing Conceptual Graphs
for Automatic Summarization Task

Sabino Miranda-Jiménez, Alexander Gelbukh, and Grigori Sidorov

Natural Language and Text Processing Laboratory,
Center for Computing Research, National Polytechnic Institute
Av. Juan de Dios Bátiz, s/n, esq. Mendizábal,
Col. Nueva Industrial Vallejo, 07738, Mexico City, Mexico
sabino@sagitario.cic.ipn.mx, www.gelbukh.com,
www.cic.ipn.mx/~sidorov

Abstract. We propose a conceptual graph-based framework for abstractive text summarization. While syntactic or partial semantic representations of texts have been used in literature, complete semantic representations have not been explored for this purpose. We use a complete semantic representation, namely, conceptual graph structures, composed of concepts and conceptual relations. To summarize a conceptual graph, we remove the nodes that represent less important content, and apply certain operations on the resulting smaller conceptual graphs. We measure the importance of nodes on weighted conceptual graphs by the HITS algorithm, augmented with some heuristics based on VerbNet semantic patterns. Our experimental results are promising.

Keywords: Automatic summarization, conceptual graphs, graph-based ranking algorithms, HITS algorithm.

1   Introduction

With the overwhelming amount of information available today on the Internet and elsewhere, summarization technologies are essential to improve the access to this information. High-quality automatic text summarization is a challenging task that involves text analysis, text understanding, the use of domain information, and natural language generation.

Summarization approaches can be categorized as extractive and abstractive. The limitations of extractive approach are well known: in the first place, low quality of the generated summaries. On the other hand, abstractive summaries have not been sufficiently explored because of the need in a deeper text analysis required for understanding the texts, and complexity associated with it. Such a deep analysis is indispensible to improve the quality of summaries [1].

We propose a method for single-document abstractive summarization, based on conceptual graphs as the underlying text representation [8]. This kind of representation has not been used for automatic summarization so far. We focus on ranking nodes and applying a kind of pruning operation, namely, selecting the most important

H.D. Pfeiffer et al. (Eds.): ICCS 2013, LNAI 7735, pp. 245–253, 2013.
© Springer-Verlag Berlin Heidelberg 2013

nodes according to HITS algorithm [5] over weighted conceptual graphs and using other heuristics based on semantic patterns of VerbNet [13]. The summary at semantic level is the resulting structure of selected nodes. Automatic generation of conceptual graphs from text is beyond the scope of this paper.

This paper is organized as follows. Section 3 describes our approach. Section 4 presents the experimental results. Finally, Section 5 gives the conclusions and future work.

2   Related Work

In recent years, there has been an increase in the interest to graph-based methods in Natural Language Processing. Graph-based approaches such as LexRank [2] and TextRank [3] have been used for keyword extraction for extractive summarization. In these approaches, graphs are usually considered undirected and unweighted; their nodes are either sentences, words, or other kind of units, and edges are defined by overlaps of the content between units. In these approaches, well-known iterative algorithms are used such as HITS or PageRank to rank the nodes in order to select salient ones. The selected nodes represent the summary; non-salient nodes are removed from the graph.

Other approaches use word order to create the graphs [6]. The graphs are directed. Nodes are words, and the edges represent the precedence of the word in the sentence, that is, the word in the word order is important. The resulting graph is ranked similar to TextRank approach.

In [3] the notion of weighting edges was introduced in HITS algorithms. Overlap of sentences was used as a kind of weight, but because of an unnatural way of using weights, the study was mainly on undirected and unweighted graphs. In contrast, a conceptual graph can be considered as a weighted graph having sense because conceptual relations between concepts provide a semantic flow through the graph, namely, the semantic flow over agent relations, object relations, attribute relations, etc. Another feature in our model is the preference of the node in order to select concepts (nodes) which the users are interested (see Section 3.3 and Section 3.4.).

There have been attempts to use the semantics of the document, such as in Semantic Graphs approach [4, 7]. This approach uses triplets (subject—predicate—object). Each triple is characterized by a rich set of linguistic, statistical, and graph attributes. A Support Vector Machine classifier is used to identify important triples to generate the summary. Nevertheless, a real and complete, fine-grained semantic representation is not used.

3   Approach Using Conceptual Graph

3.1   Conceptual Graphs Formalism

Conceptual Graphs (CGs) [8] are structures for knowledge representation based on first-order logic. They are natural, simple, and fine-grained semantic representations to depict texts. A conceptual graph is a finite, connected and bipartite graph. It has two kinds of nodes: *conceptual relations* (ovals) and *concepts* (rectangles) (Fig. 1) [8]. A concept is connected to a related concept by conceptual relation. Each conceptual relation must be linked to some other concept.

In our approach, by concepts, we consider content words (that is, all except for stop words); by conceptual relations we consider semantic roles [11]: agent, causer, instrument, experiencer, patient, location, time, object, source, and goal, as well as some other relations, such as attribute, quantity, measure, etc.— approximately 30 relations used in [8].

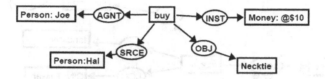

Fig. 1. Conceptual graph for sentence: *Joe buys a necktie from Hal for $10*

Other element of CGs is concept types. Concept types represent classes of entities (*Person, Money,* Fig. 1), attribute, state and event. It is also called concept type hierarchy that represents an AKO (is-a-kind-of) hierarchy, and it is used to map concepts into the hierarchy for inference purposes [8, 15]. For example, in Fig. 1, *Person:Joe* denotes the concept type *Person*, and its referent *Joe* is an instance of *Person*.

CG Framework allows graph-based operations for reasoning. A number of operations such as: restriction, simplification, unification (join), graph matching (projection), and indexing can be performed to create, manipulate and retrieve large sets of conceptual graphs [8, 15].

3.2   Construction of Conceptual Graphs

The construction of a conceptual graph from a text is not direct. It requires an additional process to discover relationships among text units. Approaches have been proposed for automatically generating conceptual graphs such as in [10], but tools are not available. Thus, we manually created the collection of conceptual graphs based on news of DUC-2003 competition in order to prove our ideas.

We use simple conceptual graphs (without negations, situations, or contexts) to simplified our task. For instance, the conceptual graphs for the following news are shown in Figure 2: *"Typhoon Babs weakened into a severe tropical storm Sunday night after it triggered massive flooding and landslides in Taiwan and slammed Hong Kong with strong winds. The storm earlier killed at least 156 people in the Philippines and left hundreds of thousands homeless."*

In Figure 2, we use a notation '(number)' for a concept that would be referred, and '#' for a co-reference to the concept marked with the specific number; for instance, #3 refers to the concept *Typhoon-Babs* (3). In addition, we use the hierarchy of WordNet [12] to map a referent to its concept type. For instance, *Hong-Kong, Taiwan* is mapped to *City*.

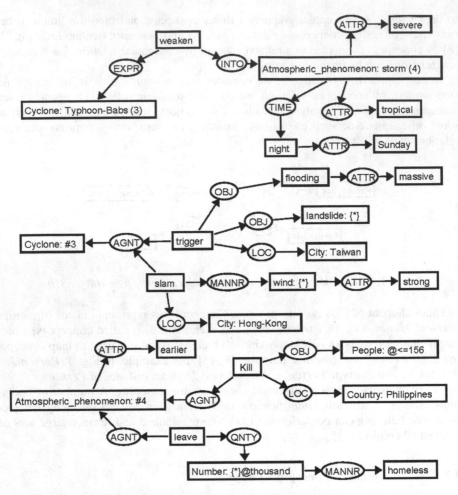

Fig. 2. Example of news as conceptual graph

3.3 Weighted Conceptual Graphs

We introduce a weighted conceptual graph (Fig. 3). The idea behind these kinds of conceptual graphs is the interest in the semantic flow of graphs. In our approach, edges and nodes have weights. The edge weights are assigned according to the semantic flow in the graph—flow through conceptual relations—, and node weight measures the degree of interest of the topics to the user.

Thus, if the interest is on some semantic flows such as agents, locations, attributes, or other thematic roles, the edge weight that pass through them should be increased in order to reward the flow that pass through them such as in Fig 3. Similar to node preference, a value greater than 1 rewards the topic preference; a value less than 1 penalizes the preference; a value equal to 1 for no reward.

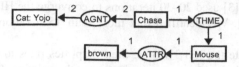

Fig. 3. A weighted conceptual graph for sentence: *The cat Yojo is chasing a brown mouse*

For example, if we are interested in the flows that pass through agent relations (AGNT) the incoming and outgoing edges for these conceptual relations are set to value of 2 (see Fig 3).

3.4   Ranking Algorithm

HITS [5] is an iterative algorithm that takes into account both in-degree and out-degree of nodes for ranking. The algorithm makes a distinction between authorities (nodes with a large number of incoming links) and hubs (nodes with a large number of outgoing links). For each node, HITS produces two sets of scores: **AUTH**ority and **HUB**. We use the authority score (means that the node is good as information source) in order to choose the nodes that will take part in the summary. We used a modified version of HITS algorithm similar to the proposed in [3].

The equations (1) and (2) are used to compute authorities and hubs scores. Where I is the set of incoming links for node V_i; O is the set of outgoing links for node V_i; W_{ki} is the weight of semantic flow of edge; and PREF is the node preference.

$$AUTH(V_i) = \sum_{V_k \in I(V_i)} W_{ki} \cdot HUB(V_k) \cdot PREF(V_k) \tag{1}$$

$$HUB(V_i) = \sum_{V_k \in O(V_i)} W_{ik} \cdot AUTH(V_k) \cdot PREF(V_k) \tag{2}$$

3.5   Ranking Algorithm of Conceptual Graphs

In order to select the important nodes in CGs, we carry out the following steps:

1. Set hub and authority scores associated to each node a value of 1.
2. Apply the operation Authority, equation (1).
3. Apply the operation Hub, equation (2).
4. Normalize the Authority and Hub values by Euclidian norm.
5. Repeat from 2–4 up to convergence or N iterations.
6. Sort nodes by authority values in descending order.
7. Expand the connected concepts for each selected conceptual relation.
8. Expand the associated nodes for each selected concept (verb concept) according to its semantic pattern.
9. Select the top concepts according to a threshold in order to prune the graph.

Mihalcea and Tarau [3] used 20–30 iterations to converge the HITS algorithm; others use one iteration [6]. We identified that more than 15 iterations are enough in our collections of graphs.

Steps 1–6 calculate the HITS scores. Step 7 applies rules to expand the concepts that a conceptual relation connects; for instance, the relation *OBJ(trigger,flooding)* (see Table 1) is expanded into two concepts *flooding* and *trigger*. Step 8 applies the verb pattern rules in order to keep coherent structures.

The semantic patterns of verb concepts were extracted from VerbNet [13]. For example, the pattern for the *chase* concept (Fig. 3) is identified in the VerbNet class ID *chase-51.6*. The pattern is *NP V NP* (Noun Phrase, Verb, Noun Phrase), and the verb is *Basic Transitive*. The role for the first NP is *agent,* and the second NP is *Theme.* Both of them are required for the concept *chase* because it is defined as transitive verb. Thus, the *agent* and the *theme* must be included in a summary.

After applying steps 1–8, Step 9 applies the pruning operation by means of a threshold set by user. It selects nodes without duplicates according to the threshold. The selected nodes represent the summary at the semantic level (see Table 2).

4   Experimental Results

We carried out our experiments on the collection of news articles provided by the DUC 2003 [9]. We selected news with length from 40 to 60 words. For each article, there are 3 summaries on average made by humans.

We created three groups of documents from DUC: 2-senteces, 3-sentences, and 4-sentences length such as news in Fig 2. Each group consists of 4 documents represented as conceptual graphs. We set the threshold for pruning operation to 20% of concepts of the original document. As a baseline, we selected the first concepts beginning at the first paragraphs up to the established threshold (except stop words). We set the semantic flow value for agent relations to value of 2. Standard metrics (precision and recall) are used to evaluate our method. **Recall** is the fraction of concepts chosen by the human that were also correctly identified by the method. **Precision** is the fraction of concepts chosen by the method that were correct. **F-measure** is the harmonic mean of precision and recall.

Table 1 shows the selected nodes by ranking method including conceptual relations. Also, expansions of conceptual relations are shown such as object relations (OBJ). Table 2 shows the selected concepts by the method that are part of the summary considering their interrelationships between them; (**req**) indicates that the concept was added because the verb pattern requires it, i.e., *kill* pattern requires its Object *(People:@lt=156)*.Table 3 shows the average of the evaluation of the approach for the three collections of graphs.

Our method slightly outperforms the baseline. It is because text documents are very short and the baseline covers the concepts in a good way. Although other approaches have demonstrated that the first and last sentences in the paragraphs are good indicators to find relevant information [14], our method uses all the net and outperforms the baseline. It demonstrates that the method in huge graphs could operate equally as in small graphs. Finally, the selected concepts in Table 2 represents the summary; according to the CG representation in Fig 2. It could be read: *"Typhoon-Babs triggered flooding and landslides in Taiwan. The storm killed at least 156 people. Typhoon-Babs slammed in Hong Kong."*

Table 1. Selected concepts and conceptual relations by ranking method with expansion of conceptual relations

NODE	RELATION EXPANSION	AUTH	HUB
Cyclone:Typhoon-Babs	-	0.729	0.3E-16
Atmospheric_phenomenon:storm	-	0.680	0.70E-03
AGNT(trigger-Cyclone:Typhoon-Babs)	trigger/:Typhoon-Babs	0.054	0.147
OBJ(trigger-flooding)	trigger/ flooding	0.027	0.10E-04
OBJ(trigger-landslide)	trigger/landslide	0.027	0.67E-05
LOC(trigger-City:Taiwan)	trigger/:Taiwan	0.027	0.137
AGNT(kill- Atmospheric_ phenomenon:storm)	kill/:storm/ People:@lt=156 (req)	0.022	0.147
AGNT(slam-Cyclone:Typhoon-Babs)	slam/:Typhoon-Babs/ City:Hong Kong (req)	0.022	0.67E-05
LOC(kill-Country:Philippines)	kill/:Philippines	0.011	0.38E-16

Table 2. Final selected concept by the ranking method

NODE	AUTH	HUB
Cyclone:Typhoon-Babs	0.729	0.3E-16
Atmospheric_phenomenon:storm	0.680	0.70E-03
trigger	0.054	0.147
flooding	0.027	0.10E-04
landslide	0.027	0.67E-05
City:Taiwan	0.027	0.137
kill	0.022	0.147
slam	0.022	0.67E-05
City:Hong Kong	0.022	0.147
People:@lt=156	0.022	0.70E-03
Country:Philippines	0.011	0.38E-16

Table 3. Evaluation of the system

	Precision		Recall		F-Measure	
	Baseline	System	Baseline	System	Baseline	System
Group I (2-sentences)	0.38	0.50	0.38	0.50	0.38	0.50
Group II (3-sentences)	0.11	0.25	0.13	0.22	0.12	0.23
Group III (4-sentences)	0.43	0.50	0.45	0.50	0.43	0.50

5   Conclusions

We have proposed a novel graph-based approach for single-document summarization. Our approach is based on the Hub-Authority framework and conceptual graphs as underlying semantic text representation. It combines the text content with semantic roles into graph-based ranking algorithms. The method uses semantic patterns from VerbNet to keep coherent structures when a threshold is applied in order to prune the nodes. Furthermore we introduced a weighted conceptual graph to provide a flexible schema to focus on certain semantic flows or topics by means of weights and preferences. We evaluate our method on DUC-2003 data. The results show that our approach is promising.

In future work, we plan to apply operations such as generalization and join on resulting conceptual graphs in order to improve the quality of the generated summaries. Also, we expect to improve the results on larger conceptual graphs, 500–1000 words per document.

Acknowledgments. This work was done under partial support of the Mexican Government (SNI, COFAA-IPN, PIFI-IPN, SIP-IPN 20121823 and 20120418, CONACYT 50206-H and 83270), CONACYT-DST India (122030, "Answer Validation through Textual Entailment"), Mexico City Government (ICYT PICCO10-120), and European project WIQ-EI 269180.

References

1. Spärck Jones, K.: Automatic summarising: The state of the art. Information Processing & Management 43(6), 1449–1481 (2007)
2. Erkan, G., Radev, D.: LexRank: Graph-based Lexical Centrality as Salience in Text Summarization. Journal of Artificial Intelligence Research 22(1), 457–479 (2004)
3. Mihalcea, R., Tarau, P.: TextRank: Bringing Order into Texts. In: Proceedings of the Conference on Empirical Methods in Natural Language Processing (EMNLP 2004), Barcelona, Spain, pp. 404–411 (2004)
4. Leskovec, J., Grobelnik, M., Milic-Frayling, N.: Learning Semantic Graph Mapping for Document Summarization. In: Proceedings of ECML/PKDD 2004, Workshop on Knowledge Discovery and Ontologies, Pisa, Italy, pp. 1–6 (2004)
5. Kleinberg, J.: Authoritative Sources in a Hyperlinked Environment. Journal of the ACM 46(5), 604–632 (1999)
6. Litvak, M., Last, M.: Graph-based keyword extraction for single-document summarization. In: Proceedings of the Workshop on Multi-source Multilingual Information Extraction and Summarization, Manchester, United Kingdom, pp. 17–24 (2008)
7. Tsatsaronis, G., Varlamis, I., Nørvåg, K.: SemanticRank: ranking keywords and sentences using semantic graphs. In: Proceedings of the 23rd International Conference on Computational Linguistics, Beijing, China, pp. 1074–1082 (2010)
8. Sowa, J.F.: Conceptual Structures: Information Processing in Mind and Machine. Addison-Wesley, Reading (1984)
9. DUC. Document Understanding Conference (2003),
   http://duc.nist.gov/pubs.html#2003

10. Hensman, S., Dunnion, J.: Automatically Building Conceptual Graphs Using VerbNet and WordNet. In: Proceedings of the 3rd International Symposium on Information and Communication Technologies, Las Vegas, USA, pp. 115–120 (2004)
11. Jackendoff, R.: Semantic Interpretation in Generative Grammar. MIT Press, Cambridge (1972)
12. Fellbaum, C.: WordNet: An Electronic Lexical Database. MIT Press, Cambridge (1998)
13. Kipper, K., Trang Dang, H., Palmer, M.: Class-Based Construction of a Verb Lexicon. In: Proceedings of Seventeenth National Conference on Artificial Intelligence (AAAI 2000), Austin, TX, pp. 691–696 (2000)
14. Hovy, E., Chin-Yew, L.: Automating Text Summarization in SUMMARIST. In: Mani, I., Maybury, M.T. (eds.) Advances in Automatic Text Summarization, pp. 81–94. MIT Press, Cambridge (1999)
15. Chein, M., Mugnier, M.-L.: Graph-based Knowledge Representation: Computational Foundations of Conceptual Graphs. Springer, London (2009)

Logical Form *vs.* Logical Form: How Does the Difference Matter for Semantic Computationality?

Prakash Mondal

Indian Institute of Technology Delhi, Hauz Khas
New Delhi 110016, India
mndlprksh@yahoo.co.in

Abstract. This paper aims at pointing out a range of differences between logical form as used in logic and logical form (LF) as used in the minimalist architecture of language. The differences will be shown from different angles based on the ways in which they differ in form and represent some natural language phenomena. The implications as following on from such differences will be then linked to the issue of whether semantic realization in mind/brain is computational. It will be shown that the differences between logical form as used in logic and logical form (LF) as used in the minimalist architecture of language will help us latch on to the realization that there is no determinate way in which semantics can be computational or computationally realized.

Keywords: Logical form, logic, minimalist architecture, semantic realization.

1   Introduction

Is semantics really computationally realized? How much of meaning can be computationally realized? And how much cannot? The path toward an answer to such questions can be tremendously difficult to follow given the fact that such questions are still faintly understood or grasped given a huge dearth in understanding what meaning really is. Here in this paper such a path will be traced through tracking the differences between logical form as used in logic and logical form in the minimalist architecture of language which will be extrapolated to approach the question of how such differences can throw light on whether meaning can be computationally realized. Both logical form as used in logic and logical form in the minimalist architecture of language *represent* semantics of natural language. If they can really represent meaning in natural language, their computability can have ramifications over how and to what extent semantics is computational. The question is whether semantics or meaning in language can be computational or computationally realized in mind/brain at all on the basis of concrete facts that the differences between logical form as used in logic and logical form in the minimalist architecture of language will provide us with. An important proviso has to be added right at this juncture. The question is not to scout out and magnify differences between logical form in logic and logical form as used in generative grammar. Such differences are quite well-known. The focus is

H.D. Pfeiffer et al. (Eds.): ICCS 2013, LNAI 7735, pp. 254–265, 2013.

rather on how such differences matter for and unlock aspects of semantic computationality given that both meta-languages *represent* natural language semantics. Why do differences in formal representation actually matter given that logical form as used in logic and logical form in the minimalist architecture of language formally *represent* semantics of natural language in different ways? One can write numbers in binary or decimal and countless other ways. That does not certainly change the fact that arithmetical operations are algorithmic; the details of the algorithm just vary appropriately based on the representation used. But the case in point is here semantics, *not* mathematical facts. Semantics is different from mathematical facts both in form and nature. The latter may well lie in the Platonic realm, but the former cannot perhaps be such given the fact that the very metalanguages that encode or represent semantics are not uniform in their representational faithfulness. This is *not* true of mathematical facts or objects as one can really write numbers in binary or decimal formats without any differences in the faithfulness with which decimal format or binary format can represent numbers. This is what will be shown below in the section 2 and these differences are crucial as far as the matter about the question of whether semantics is computational or not is concerned.

1.1   Logical Form in Logic and Logical Form (LF) in Generative Grammar

Logical Form in Logic. In brief, logical form of sentences of natural language is what determines their logical properties and logical relations. Logical form of natural language sentences is always constructed relative to a theory of logical form in the language of a theory of logic (say, first order logic) [1]. To schematize what we have in mind when we talk about logical form in logic, we can have

$$T = \{T_1 \dots T_n\}, L = \{L_1 \dots L_m\}, \mathcal{L} = \{\mathcal{L}_1 \dots \mathcal{L}_k\}$$
$$A = \{A_1 \dots A_j\}, B = \{B_1 \dots B_i\} \tag{1}$$

Here T is the set of theories of logical form; L is the set of all possible logical forms and $\mathcal{L}$ is the set of theories of logic. $L\mathcal{L}_k^{Tm}$ is the set of logical forms relative to a theory of logical form T_m and in the formulas of a theory of logic $\mathcal{L}_k$. Then:

$$L\mathcal{L}_k^{Tm} \subseteq L \quad \& \quad \Psi : L\mathcal{L}_k^{Tm} \to A \times B \tag{2}$$

Here A is the set of logical properties and B is the set of logical relations. Logical properties consist in truth values as fixed by the terms assigned to predicates and logical relations are relations between sentences which are linked by chains of different types of inferences; so entailment, implication, equivalence etc. are different kinds of logical relations which are defined with respect to a set of sentences which must not be a singleton set. The function Ψ ensures a proper mapping from a set of logical forms to ordered pairs containing logical properties and logical relations. The mapping Ψ might be a little idealized given certain approximations that might exist at the interface between what we call logical properties and what we brand as logical relations. In sum, logical forms are a way of making out what logical properties and logical relations are. The following example can perhaps make it clearer.

(1) Peter danced and Clare sang. [D(p) ∧ S(c)]
(2) Peter danced. [D(p)]

Here we have two sentences with their logical forms alongside. The logical form of each determines whether each is true if the given circumstances hold true of them. This is what logical property is. And then the sentence in (1) entails the sentence in (2). Again it can be determined by looking at the logical forms of both (1) and (2). This falls under logical relations. It now becomes clear that the logical forms of the sentences in (1-2) lead us to the logical properties and logical relations in question.

Logical Form (LF) in Generative Grammar. Logical Form (LF) as used in generative grammar has a little different, if not too distant, sense embedded in it. From now onwards 'LF' will be used to denote logical form as used in generative grammar and the full form will be reserved for the identical term in logic, just to avoid any confusion. So to come to the point of discussion, Logical form (LF) in an architecture of grammar is a level of syntactic representation which is interpreted by semantics. LF represents properties of syntactic form relevant to semantic interpretations or aspects of semantic structure that are expressed syntactically [2]. As May [2] argues, it all starts with Russell's and Freges' concerns with the relation of logical form to the syntax of natural language in that the logical form representing the semantic structure is not akin to the syntactic form of natural language (in fact, it dates back to the Greek thinkers including Aristotle who bothered about this mismatch, and then it lasted well into the twentieth century pervading all thinking about language and logic). Logical form is masked by the syntactic structure of natural language. An example can be given to show this:

(3) Coffee grows in Africa.

Here one might want to say the grammatical subject is 'coffee' and the rest is the predicate. But logically, 'be in Africa' characterizes the property-so it is the logical predicate and 'the growth of coffee' is the logical subject [3]. It can be written as:

(4) + **P** (Be in Africa) ([growth of coffee])

In this sense LF has a similarity of purpose with logical form in logic. More on this will be drawn up later. So they are two different strata of representational structures. Since LF is a syntactic level of representation, the question of representations at this level and meanings assigned to structures at this level are of paramount significance. In reality such a level gains its theoretical justification through the existence of a number of independent descriptive levels each of which has its own well-formedness conditions and formal representations as maintained throughout the main streams of thought in generative grammar. Overall, LF attempts to characterize the extent to which a class of semantic interpretations that can be assigned to syntactic structures at this level are a function of their grammatical properties; but it does not mean that LF has any commitment to all possible semantic interpretations that can be assigned to syntactic structures which are derived out of their grammatical properties. LF was actually motivated on facts about quantificational NP movement which fulfills the purpose of showing the difference between surface structure and covert structure in natural languages. An epitome of such a case is the following:

(5) Every woman loves a man.

(6) i. $[_S$ a man$_2$ $[_S$ every woman$_1$ $[_S$ e$_1$ loves e$_2$ $]]]$

ii. $[_S$ every woman$_1$ $[_S$ a man$_2$ $[_S$ e$_1$ loves e$_2$ $]]]$

(5) can have two different LF representations based on the two different scopal interpretations shown in (6i-ii). So much for LF. Let's now look at the parallels between logical form and LF. This will give us a handle to an exploration of the ways they differ from each other in their fundamental nature and form too.

1.2   Some Parallels between Logical Form and LF

Given that we have got a rough outline of what logical form and LF are, here are some parallels between them that can be brought out. Throughout this article we will be using first-order logic for any discussion on logical form; it is not due to any bias toward it but because of its more widespread use. However, these parallels can be highlighted on three grounds: (i) They are both aimed at uncovering the semantic/logical properties masked by grammatical forms; (ii) They are both translations of natural language sentences in a kind of meta-form. (iii) They are both 'paraphrases of natural language sentences', to use Quine's words [4]. A simple example can exhibit the parallels most succinctly.

(7) Every boy likes a game.

Logical Form (Logic): (i) $\forall x \exists y [L(x,y)]$, (ii) $\exists y \forall x [L(x,y)]$

Logical Form (LF): (i) $[_S$ a game$_2$ $[_S$ every boy$_1$ $[_S$ e$_1$ likes e$_2$ $]]]$
(ii)$[_S$ every boy$_1$ $[_S$ a game$_2$ $[_S$ e$_1$ likes e$_2$ $]]]$

Even if such parallels between logical form and LF might seem to be apparently evident, they mask the fundamental differences between them. Let's now turn to those differences.

1.3   Of the Differences between Logical Form and LF

The differences between logical form and LF can now be put forward. They will be traced out from a number of angles in terms of how they behave. Extrapolating Quine's [4] postulated difference between logical form and deep structure, let's say that logical form and LF are used for quite different purposes. Logical form of natural language sentences is used in logic for logical calculations and inferential implications. Whereas LF is a level of syntactic representation generated through a sequence of derivational operations used for further semantic interpretations. This leads us to a much better characterization of the differences. Here's is how. Logical form as used in logic is externally motivated, but LF is internally motivated as it is a part of internalist architecture of grammar in that LF is a part or component of an internalist architecture of language regarded as the faculty of language which is itself a part or component of mind/brain. LF is a level of syntactic operations/computations which feeds semantic interpretations at Conceptual-Intentional (C-I) interface in the minimalist architecture of the language faculty. Logical form is not anchored to any such system; so logical form cannot be characterized in that way. The differences between them can now be zoomed in on from a number of focal perspectives.

Differences in Ontology. Logical form and LF have differences in ontology too. Logical forms are constructed in the language of theory of logic which consists of two quantifiers (existential and universal). LFs in natural language represent a whole range of quantifiers apart from the two, like 'most', 'many', 'two', 'few', 'likely', 'seem' etc. etc [5]. Such differences in ontology pave the way for fundamental differences between logical form and LF come into a clearer view.

Differences in Formal Representations. Logical forms and LF have remarkable differences in formal representations which unmask the distinction in terms of their fundamental forms. The examples below show this clearly enough.

(8) Sam killed every tiger.
    LF: $[_S$ every tiger$_1$ $[_S$ Sam killed e$_1$ $]]$
    Logical Form: $\forall x$ [Tiger(x) $\rightarrow$ Killed (s, x)]

(9) Most linguists sleep.
    LF: $[_S$ most linguists$_1$ $[_S$ e$_1$ sleep]]
    Logical Form: (most x: x is a linguist) [sleep(x)]

(10) Few philosophers like cats.
    LF: $[_S$ few philosophers$_1$ $[_S$ e$_1$ sleep]]
    Logical Form: (few x: x is a philosopher) $\exists y$ [Cat(y) $\wedge$ Like(x, y)]

But the point to be noted is that even if the formulas in (9-10) use restricted quantification, it is done through an extension of natural language quantifiers like 'few', 'most' into logic. Hence it cannot be said that logical forms do *not* lack all quantifiers found in natural languages.

Differences in Restriction. It is quite well known that that in logical forms quantifiers range over a universe of individuals, as in (11) below.

(11) Every linguist drinks.
$\forall x [\text{linguist}(x) \rightarrow \text{drinks}(x)]$

But in LF the range is restricted by the head noun, as in 'few good girls' by 'good girls'.

Differences as Seen from the Phenomenon of Crossed Binding. The phenomenon of crossed binding is interesting because it opens a window onto the crucial differences between logical form and LF. Crossed binding is a problem for LF as has been seen in Bach-Peters sentence. Let's see how in the example in (12) taken from May[2].

(12) Every pilot who shot at it hit some MIG that chased him.

(13) i. [[Every pilot who shot at it]$_1$ [[some MIG that chased him]$_2$ [e$_1$ hit e$_2$]]]
    ii. [[some MIG that chased him]$_2$ [[Every pilot who shot at it]$_1$ [e$_1$ hit e$_2$]]]

This sentence in (12) can have two LF representations in (13).As can be seen above, in (13i) the pronoun 'him' is bound by the hierarchically higher antecedent 'every pilot...'; but in (13ii) only the pronoun 'it' is bound by the antecedent 'some MIG...'. Both these two bindings are not represented in any single LF representation. To alleviate this situation, May has proposed 'absorption':

$$... [NP_i ... [NP_j ... \rightarrow ... [NP_i \; NP_j]_{i,j} ... \tag{3}$$

Such a representation turns (n-tuples of) unary quantifiers into binary (n-ary) quantifiers. Crossed binding is not a problem for logical forms. Need for absorption does not arise either. Let's see how:

(14) i. $\forall x \exists y [[\text{pilot}(x) \wedge [\text{MIG}(y) \wedge \text{Shot at}(x, y)]] \rightarrow [\text{Hit}(x, y) \wedge \text{Chased}(y, x)]]$
    ii. $\exists y \forall x [[\text{pilot}(x) \wedge [\text{MIG}(y) \wedge \text{Shot at}(x, y)]] \rightarrow [\text{Hit}(x, y) \wedge \text{Chased}(y, x)]]$

Interesting to note is the fact that the LF representations in (13) can be mapped in a partial manner to the ones in (14). Thus (13i) can be mapped to (14i) and (13ii) to (14ii), but crossed binding is reflected in the either of logical forms in (14i-ii), but not in any of (13i-ii). The LF representation with the mechanism of 'absorption' applied can be mapped to both in (14i-ii). Hence again, it will be a case of partial homology if we try to map LF structures to logical forms. Meaning representation is blocked due to a bottleneck in the mapping process itself. This will have its repercussions across other cases to be discussed below.

Differences as Seen from the Phenomenon of Crossover. Other differences between logical form and LF can be telescoped through the phenomenon of crossover. Let' see look at the sentences below.

(15) *His$_i$ cat loves every boy$_i$.
(16) *Her$_i$ friend loves some spinster$_i$.

The indexes indicate co-reference between the NPs. As has been argued and shown throughout the generative literature, this is due to the covert movement of the QNPs (quantificational noun phrases) 'every boy' and 'some spinster'. So the LF representations will look like:

(17) [$_S$ every boy$_1$ [$_S$ his$_i$ cat loves e$_1$]]]
(18) [$_S$ some spinster$_1$ [$_S$ her$_i$ friend loves e$_1$]]]

Logical forms do not reflect such problems so much so that we can have perfectly fine logical forms for (15) and (16), contrary to facts in natural language as shown below:

(19) $\forall x$ [Boy(x) $\rightarrow$ $\exists y$[x's cat(y) $\wedge$ Loves(y, x)]]
(20) $\exists x$ [Spinster(x) $\wedge$ $\exists y$[x's friend(y) $\wedge$ Loves(y, x)]]

Again this reveals the fact that logical forms can sometimes overgenerate or overrepresent natural language sentences, LF do not. One could, of course, argue that some further syntactic rules may be added to formal logic to capture some constraints that will bar the constructions in (15-16); but this begs the question as the lack of existence of these to-be-postulated syntactic constraints or rules is the reason why we find (19) and (20) to be problematic as far as logical form is concerned. Again this reveals the fact that logical forms can sometimes overgenerate or overrepresent natural language sentences, LF do not. Of course, in the case of (19), one may argue that it is a representation for the sentence "Every boy is loved by his cat". But a fact that is basic and obvious but not of trivial significance can be driven home from this. It is that logical form does *not* distinguish between the two. That is where the problem lies.

Differences as Seen from the Phenomenon of Binding. Further evidence can be accumulated regarding the nature of differences between logical form and LF. This can come from further facts about binding. The examples taken from Miyagawa [6] below exhibit this clearly.

(21) Some students from his$_i$ class appear to every professor$_i$ to be idiots.
(22) Jack$_i$'s mother seems to him$_i$ to be wonderful.

(23) $\exists x$ [Student(x) $\wedge$ $\forall y$ [Professor(y) $\rightarrow$ $\exists z$ [y's class(z) $\wedge$ From (x, z) $\wedge$
Appear-to-be-idiot(x, y)]]]

(24) $\exists x$ [j's mother(x) $\wedge$ Seem-to-be-wonderful(x, j)]

The logical forms of (21-22) are (23-24). (23) does not represent the fact that the surface form and LF do not coincide in (21) in that the QNP 'some students…' has moved from the position below 'every professor'; and (24) does not reflect the fact that in (22) the surface form and LF correspond with each other as had it not been the case the sentence would have created a violation of binding principle C when the referring expression 'Jack' if lowered is c-commanded by the pronoun 'him'. This has significant implications for the differences between LF and logical form. LFs are thus sequence-dependent and sensitive to levels of representations in an architecture of grammar; logical forms are not sequence dependent in this way and are self-contained.

2   What Does It All Reveal?

It is now the time to wrap up the differences between logical form and LF into a space of important generalizations and implications on the differences as shown above. Let's now flesh them out. LF is a stage in a derivational sequence of computations. Let's call it $<D_1 … D_n>$ where each D_i is computed from the output generated by D_{i-1}. Let's assume that D_n is the stage where LF is computed. Since $<D_1 … D_n>$ is driven by computational considerations of locality, economy and other syntactic constraints (global or local), LF is also sensitive to such constraints. LFs are constructed on the basis of the computational operations as required by the derivations. But logical forms are constructed without any reference to any prior or posterior stages in a sequence of operations. Hence the differences in representational forms too! Logical forms cannot be specified this way. Hence the problems above that LF faces do not get reflected in logical forms. LFs are also sensitive to the requirement of generating licit sentences, while logical forms are not. Moreover *logical conservatism* which goes in for an economy in extensions in a logical theory and *ontological conservatism* which favors fewer ontological commitments constrain the form of logical forms [1]; LF, on the other hand, is constrained by *computational parsimony* which favors fewer computations in Merge operations. *Inclusive Condition* which bans entities not present in the Numeration (selected items from the lexicon) as defined in the minimalist architecture of grammar can at best carry the tenets of ontological conservatism but it is more global if we want to draw some parallels between the constraints governing logical form and those governing LF.

2.1   Fodor's Isomorphy, Logical Form and LF

To see how the differences between logical form and LF play out at the level of semantic representations in mind/brain, it is necessary to look into the roles they each play in Fodor's [7] postulated supervenience of logical forms of propositional attitudes on the syntactic properties of mental representations. This supervenience is also a sort of isomorphy as whatever the nature of logical form of a propositional attitude like belief is, the corresponding mental representation will have the same syntactic property. Such an isomorphy runs into fiendish problems in assignment of meanings. Let's see how.

(25) a. M1~ John walks. $\longrightarrow$ F(j)
     b. M2 ~ Max walks. $\longrightarrow$ F(m)
(26) a. M1~ Crystal is bright. $\longrightarrow$ G(c)/ $\exists x\ [C(x) \wedge G(x)]$
     b. M2 ~ John is bright. $\longrightarrow$ G(j)
(27) a. M1~ Crystal is bright. $\longrightarrow$ G(c) $\Big|\ \exists x\ [C(x) \wedge G(x)]$
     b. M2~ Summer is bright. $\longrightarrow$ G(s)

Here *M* refers to mental representation. The logical forms of the sentences are placed alongside the sentences as indicated by arrows. In (25), the logical forms are different based on a difference in terms in that in (25a) the logical form contains 'John' and in (25b) it is 'Max' the predicate is about. So the corresponding mental representations M1 and M2 will also have different syntactic properties aligned with the respective logical forms. What happens in (26) is pretty interesting. In (26a), the sentence will have two possible logical forms based on whether we interpret 'Crystal' as a common noun or a proper name. But in (26b) this problem does not arise. What is of significance is that the indeterminacy present in (26a) cannot be resolved from within the sentence in question; it needs context which is not a syntactic property. Overall, on one hand, logical form does not supervene on the syntactic property of the mental presentation as we see in (26a); on the other logical form supervenes on the syntactic property of the mental representation as in (26b). The case in (27) leads to inconsistency in that both the sentences, on one hand, have two different logical forms and on the other possess the same logical form too. The inconsistency is again due to the unavailability of context which is not a syntactic property.

This suggests that the postulated isomorphy between logical forms of propositional attitudes and the syntactic properties of mental presentations is misleading and based on a shaky ground. Interestingly LFs may not run into this problem as it is anchored to C-I (Conceptual-Intentional) system. In generative grammar semantic structures-whatever their form is- are determined by syntactic computations. Hence the relation between syntax and semantics is much more restricted and constrained than is supposed to be. Much of semantics has been pushed into the mapping between C-I (Conceptual-Intentional) interface and the domain of concepts, meanings. C-I system might resolve the indeterminacy when the pairs are interpreted at C-I system after being shipped to it. If this is the case and the fact that only logical form-syntax isomorphy runs into problems as shown above but LF does not, then it follows that logical form and LF are different in kind and phenomenology, a fortiori.

3   What Does It Mean for Semantics to Be Computational?

The question of what it means for semantics to be computational needs to be keyed onto how computationality is involved in semantics. This needs a little more elaboration given a faint understanding and absence of a full grasp of what meaning is. The same can be said about the notion of computation. What is it that is meant when a question on whether semantics is computational or not is asked? Computation is one of the most confounded and unclear notions employed in cognitive science [8], [9]. So when a question on whether semantics is computational or not is asked, much hinges on the fact that the right concept of computation is applied to the phenomenon that is under the scan of the evaluation criteria of computation [8]. So here the notion of semantic computationality will be used in the classical sense of computation where inputs are mapped to outputs according to some well-defined rules by means of symbolic manipulation of digital vehicles in the form of linguistic strings. This notion of computation is the narrowest in the hierarchy of notions of digital computation [8]. The reason behind the employment of this notion of computation is that this is the very notion of computation that has been keyed to much of generative linguistics. The question of whether semantics is computational in the analog sense of computation or in the *generic sense* [8] that encompasses both digital and analog computation will not be touched upon here in that the differences between logical form as used in logic and logical form in the minimalist architecture of language are targeted as the pedestal on which the issue of semantics being computational will be teased apart. And these are the two metalanguages that represent apparently intangible semantics, which is what has been capitalized on for the sake of an investigation into whether semantics is computationally realized or not. But of course, the question of whether semantics is computational in the analog sense of computation or in the *generic sense* can be sharpened to a larger degree from the following discussion as we will see below.

4   Semantic Computationality, Logical Form and LF

Now let's turn to the issue of semantic computationality as we gather the implications derived from what we have shown above so far. The discussion above indicates that LFs are semantically more accessible and transparent, while logical forms are not, given the problems pointed out above. Apart from that, the correspondence between logical forms and syntactic properties might be a case of partial homology, but not a full correlation. If this is so, logical forms weaken the case for semantic computationality as they are in a patchy correspondence with syntactic properties which are actually computational properties. LFs do strengthen the case for semantic computationality as LF is anchored to the C-I system thereby being more semantically accessible and LF representations for cases in (25-27) will be identical in parts which contain the subjects that will be treated uniformly as the same in being all amenable to interpretation only at the later stage at C-I system. In addition, the fact that at LF aspects of semantic structure supervene on syntactic properties, qua Fodor [10] further regiments aspects of meaning being computational by dint of being represented at LF. But if we co-opt Fodor's analysis and view of what is computational, we may run into severe problems as we have already encountered problems of other kinds derived from his isomorphy of logical form with syntactic properties. It is because he has allowed for the possibility that C-I system intentions, beliefs and other inferential (global) processes supervene on syntactic properties (internal or internal), qua Fodor [7]. So by that means, such processes are also computational. Therefore everything at and beyond LF becomes computational and it is computational all around! Such a conclusion seems to be unwarranted and uncalled for given that it forces the case for globality in semantic computationality all the way throughout the entire gamut of cognition. It also leads to the absurd conclusion that semantic properties or aspects of semantics are actually syntactic! Nothing then prevents intentions, beliefs and other inferential (global) processes from flowing into (narrow) syntax. In fact Fodor's notion of computation is based on his classical notion of computation; and hence if it is Turing or Von Neumann style of computation, then semantics at LF are computational, but those from the C-I system are not. But as Langendoen and Postal [11] show in their NL (Natural Language) Non-Constructivity Theorem, this is also flawed under a closer analysis in that there is no Turing machine-style-constructive-procedure for generating either syntactic rules or semantic rules at LF if NL (Natural Language) Non-Constructivity Theorem is to be believed to be true.

But there are a range of views of computation-causal, functional and semantic [12]; and in addition, there are a lot of problems inherent in the notion of computation itself that blocks the path that differentiates computing systems from non-computing systems [13]. This borders on Searle's [14] conclusion that any physical system computes an algorithm leading to an emptiness in the notion of computation. Semantic computationality becomes a non-issue over which we are all perhaps cudgeling our brains as everything computed by the brain is computational. The matter becomes more complicated as we try to home in on the issue of semantic realization being computational. LF, on one hand, trivializes the notion of semantic computationality by overextending, *over*specifying its domain; logical form, on the other, *under*specifies it. LF cannot act as a standard against which we can assess semantic computationality chiefly because there are other parallel logical systems including logical form which do a similar job of representing meaning and these parallel logical systems project a different picture of semantic computationality. We thus fall into the trap of a relativistic notion of semantic computationality. Worse than that is the fact that LF and logical form are not *inter*convertible and *inter*translatable without any change in meaning as shown in the sections above as they cannot be mapped onto one another without much readjustment in meanings. Therefore, there cannot even appear any sense in which we can assess or examine semantic computationality as this issue cannot be checked against any standard, nor can it be put on the pedestal of test as both the notational/representational languages (LF and logical form) project varying pictures of semantic computationality even though Steedman and Stone [15] have defended a realist interpretation of semantics within which semantics can be conceived of in computational terms by keeping semantics from aspects of processing. For semantics to be representable, we need some representational (meta)language which can be checked for how much space it allows for so that semantics can be seen to be computationally realizable. But the discussion above shows that no such representational (meta)language is consistent and uniform with respect to the way meaning can be shown to be computationally realized. Hence, this fact extends to conceptual graphs as well which also represent meaning. The problem of logical form equivalence in cases of semantic distinctions [16] makes it more unlikely that logical representational (meta)languages can actually represent meaning fully, let alone reveal the extent of semantic computationality. There is then no determinate way to determine whether semantics is computationally realized in mind/brain.

5   Conclusion

To conclude, semantic computationality is very much an issue to be determined through a thorough consideration of the representational devices that represent semantics. But these very devices or systems of representation do not provide a constant testing ground on which semantic computationality can be scrutinized. Rather what we find is that either there is an exaggeration of semantic computationality or there cannot be any case for semantic computationality at all. We lose out in both ways. Further thinking and research can clarify which path is the better one.

References

1. Menzel, C.: Logical Form. In: Craig, E. (ed.) Routledge Encyclopedia of Philosophy, vol. 5. Routledge, London (1998)
2. May, R.: Logical Form: Its Structure and Derivation. MIT Press, Cambridge (1985)
3. Seuren, P.A.: Language in Cognition, vol. 1. Oxford University Press, New York (2009)
4. Qunie, W.V.O.: Methodological Reflections on Current Linguistic Theory. Synthese 21(3-4), 386–398 (1970)
5. Harman, G.: Deep Structure as Logical Form. Synthese 21, 275–297 (1970)
6. Miyagawa, S.: Why Agree? Why Move? MIT Press, Cambridge (2010)
7. Fodor, J.: The Mind does not Work that Way. MIT Press, Cambridge (2000)
8. Piccinini, G., Scarantino, A.: Information Processing, Computation and Cognition. Journal of Biological Physics 37, 1–38 (2011)
9. Fresco, N.: Concrete Digital Computation: What does it Take for a Physical System to Compute? Journal of Logic, Language and Information 20, 513–537 (2011)
10. Fodor, J.: LOT2. Oxford University Press, New York (2008)
11. Langendoen, D.T., Postal, P.: The Vastness of Natural Languages. Basil Blackwell, Oxford (1984)
12. Fresco, N.: Explaining Computation without Semantics: Keeping it Simple. Minds and Machines 20, 165–181 (2010)
13. Shagrir, O.: Why we View the Brain as a Computer. Synthese 153, 393–416 (2006)
14. Searle, J.: The Rediscovery of the Mind. MIT Press, Cambridge (1992)
15. Steedman, M., Stone, M.: Is Semantics Computational? Theoretical Linguistics 32(1), 73–89 (2006)
16. Shieber, S.M.: The Problem of Logical Form Equivalence. Computational Linguistics 19, 179–190 (1993)

Model for Knowledge Representation
of Multidimensional Measurements Processing Results
in the Environment of Intelligent GIS

Alexander Vitol[1], Nataly Zhukova[2], and Andrey Pankin[2]

[1] Saint-Petersburg State Technical University, Information Systems Dept, Russia
[2] Saint-Petersburg Institute for Informatics and Automation of the Russian Academy
of Sciences, Research Laboratory of Object-Oriented Geo-Information Systems, Russia
abg-778a@yandex.ru, {gna,pankin}@oogis.ru

Abstract. The paper describes models for knowledge representation, i.e. extracted at different steps of multidimensional measurements processing procedure, in the context of JDL data fusion model. Models are developed taking into account the requirements of geo information systems environment. As a case study system of conditions lighting which implements the models for oceanographic data processing is described.

Keywords: knowledge representation, multidimensional measurements, data processing, data fusion, intelligent geo information systems.

1 Introduction

Over the past few years in the field of information systems development much attention is paid to the development of geographic information systems (GIS), in particular intelligent geographic information systems (IGIS). Intellectual geoinformation system refers to GIS, which includes means of artificial intelligence (AI) as a component. Attention to IGIS increased due to the requirements of end users to informtiveness and ease of use for information provided by the system on the researched phenomena, entities, environments, and the results of their processing. AI techniques can also be used to justify variants of complex problems solutions that require consideration of different factors, their interconnections and mutual influence. Due to efficiency of use of IGIS most part of multi-dimensional measurements processing and analysis systems are based on intelligent geographic information systems technologies. As measurements various environmental and technical parameters measurements are considered. Recent active development of the area of measurements processing is associated with increasing number of measuring instruments, the appearance of new types of measuring instruments that provide data for geographic information systems. It results in a substantial increase in the volume of data that must be processed while maintaining the requirements of efficiency and accuracy. Measurements that are to be processed usually have rather bad quality, aren't coordinated in time and space and are implemented as non-stationary time

H.D. Pfeiffer et al. (Eds.): ICCS 2013, LNAI 7735, pp. 266–276, 2013.

series. To solve the problem of measurements processing and analyses in automation mode, a unified adaptive approach is proposed. It is based on the use of ontologies, providing automated construction of the processing chains and analysis of multi-dimensional structures, based on the structure of processed data, requirements to the formed decisions, and available processing algorithms [1]. Application of a unified adaptive approach involves formalizing the processed data description and the results of its processing, the formalization of the applicable processing algorithms and methods. The formalized description of algorithms and processing methods described in [2, 3] and can be efficiently used for the organization of multidimensional measurements processing. For modeling geospatial data on the base of Semantic Web technologies many approaches are developed[4], including approaches based on Semantic Web technologies [5]. For the description of the data and the results of its processing and analysis the JDL data fusion model is widely used [6, 7]. Unfortunately, measurements processing results formed at different steps of measurements processing and analyses cannot be completely described using this model. For example, representation of time series in the form of a sequence of stationary time series blocks, clusters and patterns of measurements that are formed during measurements processing need specialized models for presentation. In the paper we propose a formal definition and implementation of JDL model for multidimensional data for intelligent GIS systems. In section 2 of the paper we propose a formal model for initial measurements and measurements results processing representation. The proposed model is to be considered as detailing of JDL data fusion model. In section 3 geospatial data representation model that is used to represent data about sources of measurements is described. JDL model specification for measurements processing provides possibility to use a uniform approach to presentation of different types of data and knowledge in GIS.

2   Measurements and Processing Results Representation Models

According to the general concept of multi-dimensional measurement processing and analysis three main stages of processing are being proposed: measurements harmonization, measurements integration and measurements fusion. Harmonization of measurements involves defining the basic concepts and their relationships (ontology) on the relevant domain area and/or areas of responsibility. Measurements integration provides merging of information from different sources and access to information resources for applications. Measurements fusion is defined as the process of merging data from different sources, which can provide a new quality of information and discard unnecessary data.

To formally describe the measurement data and knowledge resulting from its processing according to the defined stages four types of data and knowledge representation levels have been developed: representation baseline measurements (Table 1, baseline measurements representation models), representation of measurements harmonization (Table 1, models for structural representation of measurements), representation of measurements integration results (semantic measurements representation models and multidimensional measurements representation models (Table 1), models describing qualitative and quantitative data (Table 2)), presentation of data fusion results (model for combined representation of different types of data (Table 3)). Data represented with the use of described models is used as input data for level one of JDL data fusion model. Such input data representation allows effectively estimate objects of subject domain as not only measurements but also knowledge about them is provided.

Table 1. Time series presentation models

N	Level	Description	Basic algorithms	Presentation methods
Baseline measurements representation models				
1	$L^l 0-0$	Structured binary data stream	-	XML/XML Schema
2	$L^l 0-1$	Initial set of measurements	-	XML/XML Schema
3	$L^l 0-2$	Measurements after preprocessing	Data preprocessing algorithms	XML/XML Schema
Models for structural representation of measurements				
4	$L^l 1-0$	Measurement as a set of segments	Segmentation algorithms	RDF/ RDF Schema
5	$L^l 1-1$	Measurements as a set of object states	Clustering algorithms	RDF/ RDF Schema
Semantic measurements representation models				
6	$L^l 2-0$	Alphabetical representation of measurements	Algorithms of alphabet constructing	RDF/ RDF Schema
7	$L^l 2-1$	Measurement templates	Statistical processing algorithms	RDF/ RDF Schema
8	$L^l 2-2$	Measurement images	Image constructing algorithms	RDF/ RDF Schema
Multidimensional measurements representation models				
9	$L^l 3-0$	Measurements as a state transition graph	Sequential analysis algorithms	RDF/ RDF Schema
10	$L^l 3-1$	Measurements as a set of behavior dependencies	Association algorithms	OWL

1. Baseline measurements are transmitted in the form of structured binary stream containing metadata about the measurements, and the measurements themselves. Original binary stream will be denoted as G, original measurements contained in a binary stream as C. For data processing the original binary stream is converted into a set of separate time series represented as plain text. For converting the binary stream to the set of separate measurements, the binary stream structure description Ω, is required. Then structured binary stream of baseline measurements can be represented as: $G = < C\ \Omega\ F >$, where F - function, providing binary-to-text conversion.

2. Baseline measurements are set of time series $C = < C_1,...,C_M >$, C_i - i-th time series. $C_i = \{c_1(t_j),...,c_k(t_j),...,c_N(t_j)\}_{j=1}^{N}$, M - the total number of time series contained in a binary stream, N - number of measurements contained in the time series, $c_k(t_j) = c_k(x(t_j), y(t_j), z(t_j))$, $x(t_j)$, $y(t_j)$, $z(t_j)$ - latitude, longitude, and depth at which the measurements were made at time t_j. Set of measurements is a set ordered by time: $\min(I) \le t_1 \le t_2 \le ... \le t_N \le \max(I)$, where I is a measurement interval, intervals between measurements are equal: $\Delta t_i = t_i - t_{i-1}; \Delta t_2 = ... = \Delta t_N$.

3. The format of measurements representation after pre-processing, involving removal of outliers and noise, filling missing values, is represented in the format of baseline measurements.

4. Presentation of measurements as a set of segments requires the description of time series C_i as a sequence of blocks. Measurements within one block represent stationary time series according to defined criteria set. Each block can be defined as: $B(l,r) = \{c_l, c_{l+1},...,c_r\}$, where l and r are indices of the initial and final measurements in the sequence. Partition of an interval I is a set of disjoint blocks: $P(I) = \{N_{blocks}, \{B_k\}_{k=1}^{N_{blocks}}\}$, where N_{blocks} is a number of blocks. Set of allocated segments can be divided into three subsets: $B_k \in \{B_T, B_P, B_S\}$, where B_T is typical segments, B_P is template segments, B_S is special segments. Typical segments are segments that can be allocated using a single criterion such as change in the sign of the first and second derivative, change of frequency characteristics, etc. Template segments are defined a priori and reflect one of the states of the data source. Special segments are segments allocated during the segmentation which belong neither to typical segments, nor template segments.

5. For cases where the observed behavior of time series in a single block can uniquely determine the state of the data source, blocks are labeled, each label corresponds to one of the possible states of the source. Set of states, if necessary, can be broken down into classes. Presentation of the measurements as a set of states requires a formed presentation of measurements as a sequence of segments and can be

described   as: $((B_1, m_i), ..., (B_{N_{blocks}}, m_j))$, $\forall (B_l, m_i), (B_k, m_j)$, $0 \leq l < k < N_{blocks}$: $m_i \neq m_j$, where $m_i, m_j \in M$ - set of all distinct states of the data source.

6. Alphabetic representation allows determining the set of the characteristics to be calculated for each of the possible states of the data source. Alphabetical representation is defined as $A = \{< L_k >\}_{k=1}^{N_{blocks}}$, $L_k = < B_k, m_i, D_k >$ with a set of characteristics $D_k = (\alpha_k, \beta_k, ..., \omega_k)$ defined for each block, where k - number of block, α, β, ω - calculated characteristics. Types T can be defined for time series. Each type $t \in T$ defines a set of time series which have similar characteristics for each block. Description of type i contains one or more time series reflecting the typical behavior of the time series (etalons) for the type: $C_i^E \in C$, $i = 1, ..., N_{Etalons}$,

$N_{Etalons}$ - number of etalons.

7. Measurement template is an extended description of a time series type. The measurement template for the type j includes a set of object states description, which cover possible data source states L_j and a description of a set of possible states sequences R_j. So the measurement template can be defined as: $H_j = (L_j, R_j)$.

Table 2. Quantitative and qualitative data presentation model

N	Level	Description	Basic algorithms	Result
Quantitative and qualitative data presentation model				
1	$L^2 0-0$	Statistical description of the data	Statistical processing methods	Vector of the statistical characteristics
2	$L^2 0-1$	Description in reduced feature space	Dimensionality reduction algorithms	Values vectors
3	$L^2 0-2$	Description as data groups	Cluster analysis methods	Data groups
3	$L^2 0-3$	Description as data classes	Methods of classifiers construction	Data classes
Different data types dependencies description models				
4	$L^2 1-0$	Dependencies in the values of quantitative and qualitative data	Association methods	Association rules

8. An image O_k of type k consists of two components. The first component, namely O_k^+, contains a template of a time series type together with a set of time series defined for type k: $O_k^+ = (H_k, C_k^E)$. The second component O_k^- contains time series representing the examples of deviational behavior regarding the template.

9. To describe the data source state transition graph in the case of multidimensional measurements the set of possible states of the data source and the set of admissible state transitions observed in the various time series are defined.

10. Dependencies in the measurements behavior are presented with associative rules: $X \Rightarrow Y$, $X \cap Y = \varnothing$, where X and Y are data source states. Association rules like $X \overset{Relation}{\Rightarrow} Y$ can be used to describe the transition graph.

The model for quantitative and qualitative data presentation is shown in Table 2, model for combined representation of different types of data is shown in Table 3.

Table 3. Model for combined representation of different types of data

N	Level	Description	Basic algorithms	Result
Model for combined representation of different types of data				
1	$L^3 0-0$	Calculated characteristics	Methods for solving computational problems	Values vector
2	$L^3 0-1$	Description in reduced feature space	Dimensionality reduction algorithms	Values vector
3	$L^3 0-2$	Description as data groups	Cluster analysis methods	Data groups
3	$L^3 0-3$	Description as data classes	Classification methods	Data classes
Different data types dependencies description models				
4	$L^3 1-0$	Different data types dependencies description	Algorithms for identifying dependencies	Association rules
5	$L^3 1-1$	Data groups dependencies description	Algorithms for identifying dependencies	Association rules
Combined spatial data description models				
6	$L^3 2-1$	Spatial-temporal data sets	Algorithms for identifying dependencies	Association rules
7	$L^3 2-2$	Regular data grids	Algorithms for data grids	Data grids

To determine the algorithms of processing and analysis of multi-dimensional measurements on each of the levels a classification of measures is developed. The main types of time series are as follows: calculable / measured, slowly changing / fast changing / specialized, without breaking / with a break of given order of derivative, typed / without a type, stationary / non-stationary / piecewise stationary. As classification criteria for slowly changing measurements following criteria's have been chosen: complexity (global / local / weighted), behavior (convexity / concavity), variability, proximity to the curve of a given order (polynomial), characteristic points (number of maximums, minimums, and predetermined level crossings), curvature (maximum, minimum, median). The classification of adaptive and non-adaptive representation of time series of the data has been developed. For presentation of adaptive time series following presentations can be used: sorted coefficients, piecewise polynomial representation (piece-wise linear approximation (interpolation / regression), adaptive piecewise constant approximation), singular value decomposition, symbolic representation (natural language / string), trees. For presentation of non-adaptable time series: wavelets (orthonormal: Haar, Daubechies; biorthogonal: symlets, coiflets), spectral representation (discrete Fourier transform, discrete cosine transform), piecewise aggregate approximation. As an exploratory procedure it is proposed to use a set of procedures designed to ensure the definition of the type of time series and the way for its representation based on developed classification criteria. Also exploratory procedures include methods and algorithms that are applicable to the time series in accordance with classification of algorithms, build on the base of specified time series types.

3   Geospatial Data Representation Models

To represent the measurements and the results of their processing in a GIS environment, an integration of processing results presentation models and geospatial data presentation and displaying models is made [8]. Description of the model for geospatial data representation is shown in Table 4. It uses two types of geospatial objects - points, which include data sources, they are binding measurements, quantitative and qualitative data, and the area corresponding to a limited part of the space environment, which is a binding set of measurements processing results.

Table 4. Geospatial data representation models

№	Information type	Description
1	Type information catalog	Contains basic data about catalog type, data supplier, base data type (SXF, S-57)
2	Metadata catalog	Contains catalog of meta information
3	Spatial object type information	Contains information about spatial object type
4	Spatial object metadata	Contains spatial object metadata
5	Spatial object attribute type information	Contains information about spatial object attribute type
6	Spatial object attribute information	Contains information about spatial object attribute
7	Spatial object geometry information	Contains information about spatial object geometry

4   Case Study

The proposed approach of data and processing results presentation is used in the implementation of an adaptive measurements processing and analysis of oceanographic data in geographic information system of conditions lighting. The conditions lighting system is a situation assessment system, that provides end-users with integrated information about the situation and the state of the environment in the interests of making justified decisions in a particular situation. The system provides access to geospatial and meteorological data, technical facilities and objects, to the environment description data, data obtained from external sources. It provides the solution of mathematical problems (problems of the theory of search, sonar, radar, etc.) and the tasks of modeling. In the course of the pilot study was carried out data processing on the Arctic region. Baseline data included measurements obtained using different facilities for the period from 1870 to 2010. The total number of processed measurements is about 2,000,000. Example of sample input data is shown in Figure 1. Components that implement the processing and analysis of oceanographic data were integrated into the system. Figure 2 shows the distribution of the measurements of temperature and salinity over the years.

Latdeg	Longdeg	Year	Mnth	Dy	Hr	Sound	d/p	TEMP	PSAL
68.4	41.83333	1870	7	22	0	-9	0	9	-9
68.4	41.83333	1870	7	22	0	-9	36	6.4	-9
68.4	41.83333	1870	7	22	0	-9	45	5.7	-9
68.4	41.83333	1870	7	22	0	-9	54	5.6	-9
68.65	43.41667	1870	7	22	0	-9	0	7	-9
68.65	43.41667	1870	7	22	0	-9	45	4.9	-9
68.65	43.41667	1870	7	22	0	-9	55	4.7	-9
69.1	37.8	1870	7	31	0	-9	0	9.5	-9
69.1	37.8	1870	7	31	0	-9	36	6.9	-9
69.11667	38.18333	1870	7	31	0	-9	0	10.6	-9
69.11667	38.18333	1870	7	31	0	-9	36	7.4	-9
69.16667	38.91667	1870	7	31	0	-9	0	11.5	-9
69.16667	38.91667	1870	7	31	0	-9	36	6.4	-9
69.25	39.33333	1870	7	31	0	-9	0	11.1	-9

Fig. 1. Sample of input data

To meet the challenges of processing and analysis of oceanographic data, the system of lighting of conditions includes the following components:

- the service to receive data from centers providing measurements obtained by Argo buoys and other means of measurements of the ocean parameters is a part of the server interfaces to external systems;
- the server of modeling and mathematical models includes services for processing management as well as services with the implementation of the algorithms used in the processing and analysis;
- the server of hydrometeorological information includes services providing the results of calculations of oceanographic parameters in a given area upon request.

Ontology of the system of lighting of conditions is expanded to include the following members:

- objects involved in gathering, processing and analysis of oceanographic data, such as objects that describe means of measurement, centers providing measurement data, etc.
- measurements obtained from different sources of data, and processing results of single measurements and results of the joint analysis of the measurements;
- formal descriptions of algorithms and groups of algorithms used in the processing and analysis of measurements;
- formal description of completed processes.

The system's knowledge base includes logic patterns derived from the analysis of accumulated oceanographic data showing the relationship between the changes in various parameters of the aquatic environment, as well as changes of parameters depending on the geographical location and the time of measurements.

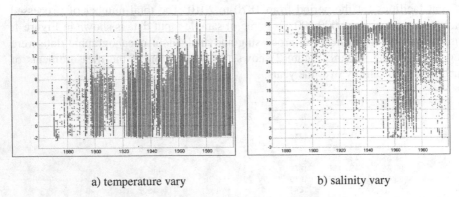

a) temperature vary   b) salinity vary

Fig. 2. Measurements distribution over the years

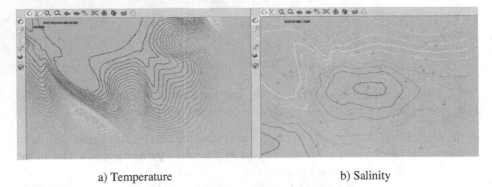

a) Temperature   b) Salinity

Fig. 3. Measuring facilities and environment parameters example images in the aquatic environment lighting system

Execution of processing and analysis chain for oceanographic data in order to build regular grids involves two main phases: data verification, under which runs the harmonization and integration of data, and the stage of regularization data, solving the problem of data fusion. The main purpose of the data verification step is systematic storage, analysis and processing of data in order to prepare them to deal with the problem of constructing regular grids. The main objective of the regularization stage is to build a regular grid of accumulated measurements and evaluation of the accuracy of data in the grid. Results of data processing produced by the proposed approach are shown on Figure 3.

Evaluation of the results was carried out on the basis of a comparison of the calculated values of temperature and salinity in grid using the proposed approach and the results obtained by the expert. Results of the comparison showed that the accuracy of the results has not worsened, while experts estimate the processing time was reduced by 20%.

5   Conclusion

Application of the proposed models of formalized data description within the system of adaptive processing and analysis of multi-dimensional yielded the following positive effects:

- The integration of formal multidimensional measurements presentation models and JDL model, as well as geospatial objects data models, allowed to create a single representation model of various data used in GIS;
- The ability to build dynamic processing of multi-dimensional measurements with the accumulated knowledge in the automatic mode, as it allowed to use all the knowledge about the measurements, which were obtained from the solution of other problems;
- The hierarchical model allows gradually obtain and add the knowledge about the measurements to the knowledgebase, with each successive level reducing the amount of stored data and increasing it's informational content;
- The formalized representation of measurements and results of their processing allows the accumulation of knowledge about measurements and ability to reuse it both within the developed system and in other systems of processing and analysis of oceanographic data.

The further direction of development of this approach involves the expansion of the considered model in the description of data and knowledge on the functioning of the processing and analysis system for multi-dimensional measurements.

References

1. Zhukova, N.A., Pankin, A.V.: Principles of managing the processing and analysis of multi-dimensional measurements in IGIS. In: Proceedings of the Information Technologies in Management, Saint-Petersburg, October 9-11 (2012)
2. Smith, H., Fingar, P.: Business Process Management (BPM): The Third Wave. Meghan-Kiffer Press (2003)
3. Mathias, W.: Business Process Management: Concepts, Languages, Architectures. Springer (2007)

4. http://www.opengeospatial.org/standards
5. http://www.w3.org/2005/Incubator/geo/XGR-geo-ont-20071023/
6. Steinberg, A.N., Bowman, C.L., White, F.E.: Revisions to the JDL data fusion model. In: The Joint NATO/IRIS Conference, Quebec (1998)
7. Llinas, J., Bowman, C., Rogova, G., Steinberg, A., Waltz, E., White, F.: Revisiting the JDL data fusion model II. In: Proceedings of the Seventh International Conference on Information Fusion, Stockholm, Sweden (2004)
8. Koh, S.V.: Software tools for data integration in GIS tasks solution. In: Proceedings of the Information Technologies in Management, Saint-Petersburg, October 9-11 (2012)

Transformation of SBVR Business Rules
to UML Class Model

Stuti Awasthi and Ashalatha Nayak

Dept. of Computer Science and Engineering,
Manipal Institute of Technology, Manipal University
Manipal, Karnataka, India, 576104
stutiawasthi27@gmail.com, asha.nayak@manipal.edu

Abstract. Multiple attempts have been made these days to automate the creation of class diagrams by providing structured English statements as input. The resulting diagrams are of close proximity to what the user wants. This paper is one such attempt to transform business designs written in OMG's (Object Management Group) standard SBVR (Semantics of Business Vocabulary and Rules) framework, into a set of classes in UML (Unified Modeling Language) Class Model using a theoretical approach. SBVR provides a set of specific rules which are processed in order to get class diagram of close proximity. It involves the transformation of "Structured English" into a set of UML Class Model with SBVR as a mediator. Further, the results of the approach are validated using VeTIS tool.

Keywords: SBVR(Semantics of Business Vocabulary and Rules), Class Diagram, Business Rules, UML (Unified Modeling Language), OOA (Object-Oriented Analysis).

1   Introduction

Object-oriented analysis (OOA) looks at the problem domain, with the aim of producing a conceptual model of the information that exists in the area being analysed. Analysis models do not consider any implementation constraints that might exist, such as concurrency, distribution, persistence, or how the system is to be built [5]. The development of a business model is the stage where a Business Analyst (BA) designs and puts the constraints on the system. The OMG's SBVR [2] is an approach which allows creating the business design in terms of business vocabulary and rules in natural language format. SBVR was originally developed to share the business semantics between different communities practicing UML and its variants [9].

SBVR provides a way to capture specifications in natural language and represent them in formal logic so they can be machine-processed. Methodologies used in software development are typically applied only when a problem is already formulated and well described. The actual difficulty lies in describing the problems and getting expected functionalities. This is due to the anomalies and ambiguities in natural language specifications. Example is, *We now have dress shirts on sale for men with 16 necks*. The anomaly here is whether shirts are available in 16 different collar

H.D. Pfeiffer et al. (Eds.): ICCS 2013, LNAI 7735, pp. 277–288, 2013.

designs or 'a' shirt has 16 collars in it. The statement will make legitimate sense to a human but same is not the case with a computer.

Stakeholders involved in software development can express their ideas using a language very close to them, but they usually are not able to formalize these concepts in a clear and unambiguous way. SBVR can be used in order to overcome this problem enabling natural language to well represent and formally define problems and requirements [3]. The contribution of this paper is to analyse all those requirements and presenting a methodology which allows business people to convert their business designs into UML Class Diagram in a theoretical approach.

Main challenge in transforming the Business Model is the detection of automatable business rules and their automation. "Automating" a rule means to enforce the rule through automation. In general, an enforcement policy needs to be specified for each rule and putting an obligation on the system or process as to exactly how, when, and where the system or process will enforce the SBVR rules. That is, there should be a rule or a set of rules for each automatable rule about how each of them will be enforced by the system and this paper, tries to develop those set of rules. This is non-trivial, as there are often several options available, and it is a system design choice which one to use. Additionally, the enforcement may be complex, involving many steps and coordinated activity to enforce the rule. This information is generally not in the automatable rule itself, but involves other considerations too such as complexity of the English language, construction of the SBVR rules etc.

Here in this paper, we take SBVR rules as our input and transform it into a class model using grammar and parse tree approach. We then validate the proposed design using MagicDrawsoftware [8] with VeTIS plugin. VeTIS plugin was proposed by the Department of Information Systems, Kaunas University of Technology, Lithuania. Additionally, the proposed approach is validated using the unified approach proposed by Bahrami [1]. In this way, the proposed design algorithm can be checked for its theoretical and practical correctness.

The paper is organized as follows: Section 2 presents the *background* information onOOA and SBVR. In Section 3, previous *related work* is described. The *methodology* describing the parsing of SBVR rules to facilitate the transformation into Class Diagram is explained in Section 4. The *experimental results* are presented in Section 5 to provide an example with VeTIStool [7]. In this section, various other approaches by which we automatically transform the business vocabulary and rules written in SBVR into UML Class Diagram are presented. Finally, Section 6 concludes the paper.

2   Background

OOA

Object-oriented analysis (OOA) is the process of extracting the needs of a system to satisfy the user's requirements. The goal of OOA is to understand the domain of the problem and the system's responsibilities by understanding how the users will use the system. This is accomplished by constructing several models of the system. Models concentrate on describing what the system does rather than how it does. Separating the behavior of a system from the way it is implemented requires viewing the system from the user's perspective rather than that of the machine [1].

SBVR

The Semantics of Business Vocabulary and Business Rules (SBVR) [2] is an adopted standard of the Object Management Group (OMG)[9] intended to be the basis for formal and detailed natural language declarative description of a complex entity such as a business. SBVR is intended to formalize complex compliance rules, such as operational rules for an enterprise, security policy, standard compliance, or regulatory compliance rules. Such formal vocabularies and rules can be interpreted and used by computer systems [2].

Figure 1 shows the model driven architecture in OMG. SBVR comes under the business model section of the MDA (Model Driven Architecture) which can undergo transformation in order to generate elements required to generate Class Models.

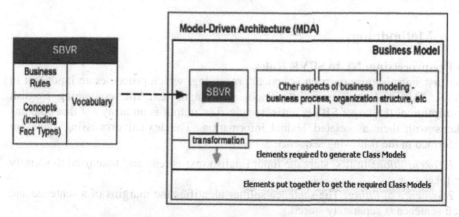

Fig. 1. Position of SBVR in Model-Driven Architecture in OMG [2]

3   Related Work

In the last decade, a number of software tools have been designed and implemented to facilitate in automatic generation of UML class model from natural languages, such as SBeaVeR [10] and VeTIS [7]. In the following, we provide a brief survey of these approaches.

Paper [4] attempts dynamic generation of the OCL (Object-Constraint Language) constraints from the NL (Natural Language) specification provided by the user. The input text in natural language is parsed to form a set of business rules which are finally translated to OCL expressions as the output.

Paper [6] presents a methodology to generate UML AD (Activity Diagram), SD (Sequence Diagram) and CD (Class Diagram) from the SBVR model driven business design. The paper attempts to bridge the gap between business people and IT people, by converting business designs into platform independent UML Activity Diagram, Sequence Diagram and Class Diagram.

SBeaVeReditor [10] is an Eclipse based plugin that allows business modelers and analysts to create fact-oriented business models and rules based on OMG's SBVR standard. SBeaVeR provides a tool for formalizing the semantics of business knowledge using the Structured English notation.

VeTIStool [7] is SBVR-compliant plug-in for the CASE tool MagicDraw, which can also be used as a standalone tool. The user interface of VeTIS was developed on the basis of SBeaVeR an Eclipse plug-in, which was a part of the Digital Business Ecosystem project. VeTIS tool is used for the definition of Business Vocabularies and Business Rules using controlled natural language (a subset of English language) and for transformation of SBVR specifications into UML class diagrams with OCL constraints.

However, our survey reveals that these tools are limited to verify the syntax and type checking of the already written SBVR rule. In this paper a methodology has been proposed to use SBVR rules taken from the natural language by using a technique presented in paper [6]. Then, we generate a class model based on the SBVR rules and further verify our investigations in the experimental results.

4   Methodology

A) Pre-processing: NL to SBVR Rules

The first stage in our approach is lexical processing, which processes an input text in SBVR. In lexical processing the input is a plain text file containing English description of the target SBVR business rule. The output is an array list that contains tokens with their associated lexical information. The lexical processing steps are performed in the following sequence:

1.Tokenization: In first step, the input English text is read and tokenized to identify the tokens.

2.Sentence Splitting: The sentence splitter identifies the margins of a sentence and each sentence is separately stored.

3.Identification: After performing the previous steps all nouns and verbs are identified. These identified nouns form the class names whereas verbs form the methods of that class. The Fact Types and Quantifiers are also identified during this phase.

The following mapping rules assist in identifying different SBVR elements for the subsequent stages. Basic SBVR elements such as noun concept, individual concept, object type, verb concepts, etc. are identified from the English input [4].

- All proper nouns are mapped to the individual concepts
- All common nouns appearing in subject part are mapped to the noun concepts or general concept
- All common nouns appearing in object part are mapped to object type
- All action verbs are mapped to verb concepts
- All auxiliary verbs and noun concepts are mapped to the fact types
- The adjectives and possessive nouns (i.e. ending in's or coming after 'of') are mapped to the attributes
- All articles and cardinal numbers are mapped to quantification

The above rules are applied to the English text and the output is stored in an array list. The example shown in Figure 2 highlights the proposition of basic SBVR elements in a typical SBVR rule.

Ex: A person's age must be 18 years.

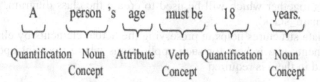

Quantification Noun Attribute Verb Quantification Noun
Concept   Concept   Concept

Fig. 2. Semantic analysis of English text

The second stage in our approach is parsing the SBVR rules corresponding to natural text into various elements. These elements are stored in the form of a table. Figure 2 gives the semantic analysis of English text based on which the SBVR rules are generated to give the statement a less ambiguous meaning. This table will depict various classes, their attributes, relationships between other classes etc., which can assist a Business Analyst in drawing a class diagram manually.

B) Transformation of SBVR Rules into Class Diagram

In this paper, two algorithms are proposed to convert business rules into class diagram. In the first algorithm Algorithm-A1, the business rules are parsed one by one for nouns, verbs, fact types and quantifiers after defining the type of the business rule.

Figure 3 present the composition of business rules into various components. Here BR is Business Rule, TBR is Type of Business Rule, SIS is for Search in Statement, Q is Quantifier, N is Noun, V is Verb, and FT is Fact Type.

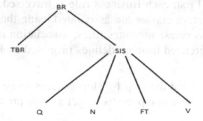

Fig. 3. Division of business rules into various categories

The various phases to parse the specified business rules are explained as follows:

Phase 1:

The given SBVR business rules as input are parsed one at a time. During parsing, the type of the business rule is specified as one of the following types.

1. Structural Business Rules
2. Definitional Business Rules
3. Operative Business Rules
4. Behavioral Business Rules

Phase 2:

Now, for each Business Rule (BR), every word is parsed to identify its category as Quantifiers, Nouns, Fact Types or Verbs. The category to which the word belongs to is pushed into its respective stack. The word which doesn't belong to either of the specified categories is pushed into a temporary stack. Each time a word is pushed into a stack its respective counter is incremented. At the end of this activity, we get a stack of all elements together which will be used to create the class diagram. Let us name this stack as 'main_stack'.

The stack data structures helps in preserving the actual elements by eliminating the repetitive elements and is best for such applications where frequent updating of the elements stored in them is required.

Phase 3:
First, for the noun stack each element is popped out of the main_stack one by one and is matched with the class rules. If the popped element qualifies the rules for being a class then the element is pushed into class stack else if the element is a noun but not qualified to be a class, it is pushed into the attribute stack. If both the conditions for the noun fail, the element is pushed back into the main_stack.

Similarly, the remaining elements in the main_stack are matched with respective rules one by one for verb, fact type and quantifier and the qualifying elements are pushed into their respective stacks. If the rules are satisfied, for verb, the element is pushed into the method stack.

The fact types and quantifiers are used for defining the association between two classes. After Phase 3, all the stacks such as noun, verb, fact type and quantifier stacks are cleared. These steps are repeated for all the business rules. Note that before pushing an element into any of the stacks, it is checked if the element is already present in that stack. If present, it is not pushed to the stack.

In the second algorithm, each of the stacks is used to create the class diagram, first using the noun stack based on the rules. Algorithm A2 gives a flow of how all the steps are to be followed. First, each Business rule is traversed for the fact type. If the fact type exists, its respective classes are associated using the predefined association rules. The qualifying class rules, attributes rules, association rules and quantifiers for the business rules are referenced from guidelines proposed by Bahrami [1].

Implementation
The following algorithms give the step by step process to be followed to extract the components and then use these components to get a close proximity class diagram for the given rules.

Algorithm A1
```
start

Step 1: for (each BR k of n)
        {
        Current_rule ← rules[k]
        Type = TBR (Current_rule)

        // search in statement BR

Step 2: while (! End_of_line)
        {
                for each word w
                {
```

```
      if (w=noun)
      Stack_noun← push (w)
      N_count = N_count+1

      else if (w=verb)
      Stack_verb← push (w)
      V_count = V_count+1

      else if (w=fact_type)
      Stack_ft← push (w)
      Ft_count = Ft_count+1

      else if (w=Quantifier)
      Stack_q← push (w)
      Q_count=Q_count+1
      }
}
```

Step 3: The elements are put into their respective stacks

```
for ( i=0; i<N_count ; i++ )
    {
    Element ← pop (Stack_noun)
    N_count = N_count - 1;

    if ( Class_rules ( Element ))
       Classes ← push (Element)
         C = C+1
    else
       Attributes ←push(Element)
       A = A+1
    }

    for (i=0; i<V_count;i++)
    {
    Element ← pop (stack_verb)
    V_count = V_count - 1
    Methods ← push(Element)
    M=M+1
    }

clear_stack(stack_noun)
clear_stack(stack_verb)
} // end the top most for loop

end
```

Algorithm-A1 shows how the SBVR rules are parsed to get various elements for the class diagram. Here BR is a Business rule, n is the total number of business rules, TBR is type of business rule Structural, Definitive, Operative or Behavioral. The Current_rule is the rule being parsed presently. The variable 'w' holds the word for the Current_rule from the beginning to end to check for the qualifiers. N_count, V_count, Ft_count and Q_count keep the count of nouns, verbs, fact type and quantifiers respectively.

Algorithm A2:

```
start
Step 1: create classes
    while (classes stack not empty)
    {
    Current_class←pop(classes)
    Create_class (Current_class)
    C = C-1
    }

Step 2: insert attributes into empty classes created
while (attributes stack not empty)
{
Current_attribute←pop(attributes)
for (each BR k of n BRs)
{
Current_rule← rules[k]
search Current_attribute in Current_rule

if (Current_attribute in Current_rule)
{
Current_class← pop (classes)
if (Current_class in Current_rule)
    {
       insert_attribute(Current_class,
    Current_attribute)
          C = C - 1
          A = A - 1
    }
}
}

Step 3: insert methods into their respective classes
while (methods stack not empty)
{
Current_method←pop(methods)
    for (each BR k of n BRs)
    {
    Current_rule← rules[k]
```

```
      search Current_method in Current_rule
      if (Current_method in Current_rule)
      {
      Current_class← pop (Classes)

      if (Current_class in Current_rule)
      {
         insert_method (Current_class, Current_method)
            C = C - 1
            M = M - 1
      }
  }
  }


Step 4:find relations between the classes

search each Fact_type in BRs
if (Fact_type exists in Current_BR)
{
for each class in Class stack
if (Current_class exists in Current_BR)
{
  if the same fact_type exists in the other BR
  create_association(Classes)
}
}

Step 5: use the stack_quantifiers to set the association
        rules between the classes associated

  end
```

Algorithm A2 comprises of 5 steps. In the first step, the nouns are popped of the noun stack and classes are created. In step 2 the attributes are inserted into the classes created in the previous step using the Business Rules. In step 3 the methods are inserted into their respective classes. In step 4 and 5 the fact type and quantifiers are used to create associations between the classes.

The summary of the algorithms are as follows. Algorithm-A1 is for transformation of SBVR rules into various elements. This algorithm shows how to process a SBVR rule to get class diagram components. Algorithm-A2 uses the components extracted above to create the final class diagram. In other words, at the end of algorithm A1 we get various components required to draw the relevant class diagram. Algorithm A2 is used to use these components to get associations and relationship between the classes.

5   Experimental Results

VeTIS tool [7] is used for the definition of Business Vocabularies and Business Rules using controlled natural language (a subset of English language) and for transformation of SBVR specifications into UML class diagrams with OCL constraints. The SBVR rules developed from the natural language become the input for deriving the various components of the class diagram. MagicDraw [8] software is used to create different types of UML diagrams. The VeTIS tool is a plugin which is installed on MagicDraw. The resultant diagram generated by VeTIS is created in MagicDraw project.

Figure 4 illustrates our methodology by taking an example of a business rule stated as *Example Business rule 1*. When the proposed algorithms are applied to the given business rule(s), the class diagram in Figure 4 is obtained. Based on the given BR, the reader can verify the results by creating a class diagram using the proposed algorithms or by treating the BRs as problem statement and manually creating the class diagram.

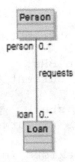

Fig. 4. Example 1 Class diagram

Example Business rule 1: It is necessary that a person has exactly one loan.

Here, Person and Loan qualify as classes. The association is formed by the terms 'necessary' and 'has'. Figure 4 shows the class diagram obtained for the above business rule. The class diagram has been validated using VeTIS tool. Figure 5 shows the snapshot of VeTIS editor with our example business rule as input. Figure 6 depicts the corresponding class diagram generated. The above example clarifies the procedure for obtaining the class diagram and validation process [7].

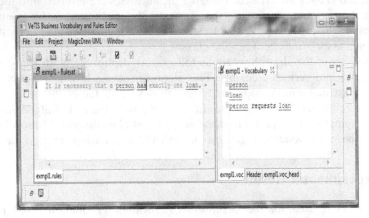

Fig. 5. Business rule in the VeTIS tool

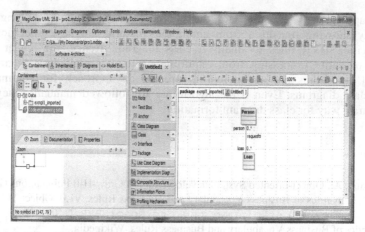

Fig. 6. Class diagram generated by the business rule specified in the VeTIS tool

Another example of a business rule is stated below as *Example Business rule 2* with its class diagram in Fig.7.

Example Business rule 2:

It is necessary that a __person__ owns at least 1 __account__.
It is necessary that a __person__ owns at most 5 __account__.
It is necessary that the __account__ is owned by exactly one __person__.

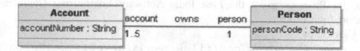

Fig. 7. Example 2 Class diagram

A similar diagram will be generated if the proposed approach is applied for the given BRs. However, the *accountNumber* and *personCode* attributes are generated by VeTIS tool only. When the business rules have been defined in the VeTIS editor it has to be exported to MagicDraw UML tool so that it can be pictorially represented in the form of a class diagram. At present, our theoretical approach is limited only for generation of class diagram from the business rules and not directly from the natural language specifications.

6   Summary

This paper first briefs about how natural language can be converted into more specific, business rules using existing approaches. Then, a theoretical approach is applied to extract various components from SBVR rules. Algorithm A1 deals with extracting the various elements of the class diagram from the SBVR rules. Algorithm A2 deals with how each of these extracted elements can be used to create a class diagram. Further, the results obtained via these algorithms have been verified using a plugin, VeTIS used with the MagicDraw software.

This paper gives a direction for the people working in software engineering community showing transition from ambiguous natural language specification to more specific SBVR rules to finally a class diagram. Hence, the reader gets an overview of entire processing from requirements to the final class diagram. The theoretical algorithms are an attempt to present an approach for automation of creating class diagrams. Even if the algorithms are followed manually, they give a more accurate way for class diagram formation.

References

1. Baharami, A.: Object-oriented System development. McGraw-Hill Publications (1999)
2. SBVR: Semantics of Business Vocabulary and Business Rules, v1.0. Object Management Group (January 2008), http://www.omg.org/spec/SBVR/1.0/PDF
3. Semantics of Business Vocabulary and Business Rules, Wikipedia, http://en.wikipedia.org/wiki/Semantics_of_Business_Vocabulary_and_Business_Rules
4. Bajwa, I.S., Lee, M.G., Bordbar, B.: SBVR Business Rules Generation from Natural Language Specification. In: Proceedings of Artificial Intelligence for Business Agility, AAAI 2011 Spring Symposium, USA (2011)
5. Object-orientedanalysis and design, Wikipedia, http://en.wikipedia.org/wiki/Object-oriented_analysis_and_design
6. Raj, A., Prabhakar, T.V., Hendryx, S.: Transformation of SBVR Business Design to UML Models. In: Proceedings of the First India Software Engineering Conference, ISEC, Hyderabad, India (2008)
7. VeTIS User Guide, http://www.magicdraw.com/files/manuals/VeTISUserGuide.pdf
8. MagicDraw homepage, https://www.magicdraw.com/
9. UML 2.3 homepage, http://www.omg.org/spec/UML/2.3/
10. SBeaVeR host webpage, http://SBeaVeR.sourceforge.net/

Representation of the Event Bush Approach in Terms of Directed Hypergraphs

Cyril A. Pshenichny and Dmity I. Mouromtsev

Intellectual Systems Laboratory, National Research University of Information Technologies,
Mechanics and Optics, Kronverksky Prospect, 49, St. Petersburg 197101, Russia
cpshenichny@yandex.ru, d.muromtsev@gmail.com

Abstract. The paper discusses the relation between a novel approach of knowledge engineering, the event bush, and the formalism of directed hypergraph. Despite the seemingly obvious similarity, the relation appears to be far from transparent. However, if formulated accurately, it may give a handy demonstration tool for the event bush approach and open a new avenue of research in directed hypergraph theory.

Keywords: event bush, directed hypergraph, vulgar bush, knowledge engineering.

1   Introduction

Traditional knowledge representation tools (OWL, KIF, RDF, ER, N3 and many others) operate with objects considered either as classes (i.e., types: concept types or relation types) or as individuals [3]. This has proven to be efficient for a wide range of tasks like building data models and search algorithms or description of those domains in which things, their properties and relations are considered fixed – e.g., production units of a factory or biological species. Nevertheless, there are tasks in natural-scientific and technical domains that urge us to analyze not relations between the objects but relations between their combinations in a form of statements. The latter relations are based on relations between the objects but cannot be reduced to these. For instance, considering the notion, "when the bus comes, people at the bus stop stand up", we deal not with relations between the objects "bus", "to come", "people", "stop", "to stand up", but between the events "bus comes" and "people stand up". Conceptual graphs add action to statements and treat verbs, which denote it, as names of specific classes (e.g., classes "to come" or "to stand"). However, this can be only a part of solution, because in some cases one needs to consider operations between statements per se. These operations could be not only logical, as was shown by Rieger [7] who attempted to formalize some common-sense relations between events. Methods that organize knowledge in a form of statements united by operations make up a separate group and unlikely can be substituted by traditional tools of knowledge representation, mainly because they allow an object to change its properties and relations and even transform to another object. Such a change can be considered as

H.D. Pfeiffer et al. (Eds.): ICCS 2013, LNAI 7735, pp. 289–300, 2013.

event. This gave ground to Pshenichny and Kanzheleva [5] to recognize object-based and event-based methods, which should address static and dynamic knowledge, correspondingly (herewith, not only classical ontologies but also conceptual graphs, dynamic and process ontologies must be regarded as object-based, as they view dynamic knowledge in a "static" way). Event-based methods include influence diagrams, sequence diagrams, Petri nets, event/probability trees, Bayesian belief networks, causal loops and other approaches. They allow to show scenarios of evolution of some domains and relations between the events.

However, while the semantics of object-based methods is well defined, that of event-based ones remains quite loose. Even the statements there are not always explicitly put. To cope with this shortcoming and suggest an event-based method that would be as detailed semantically as the object-based ones, the event bush was proposed [6]. It will be described below based on the most recent publication [5].

This method, suggested first mainly for the geosciences, is gaining importance in other fields, e.g., decision support in marketing [4]. Nevertheless, its formal basement is still being elaborated. Therefore, it looks pertinent to compare this method with existing mathematical and logical formalisms used in the knowledge engineering to find out whether and how they relate to each other. Graphs have been commonly used to visualize both static and dynamic knowledge representation tools. Among the graphs, based on structural similarity, directed hypergraph [2] is the first candidate to be examined for representation of the event bush. The present paper aims to do this, proceeding from syntactic fundamentals of the method to its connectives and inference and, finally, discussing pros and cons of hypergraph formalization of the event bush.

2   Basic Syntax of the Event Bush

The event bush was designed to address a particular kind of modeling environment, often observed in the Earth science domain but also occurring in many others – that of directed alternative changes [5]. This type of environment can be metaphorized as an arena for various, intertangling processes, which are thought to originate from a limited number of causes or sources (but not less than two), tend to leave a definable result and can be followed as more or less distinct scenarios.

Formalisms describing such environments are expressed graphically as constructions having many "roots" or net-shaped, allowing various paths at least between some of the nodes, and permitting local cycles, i.e., notations strongly resembling graphs and likely interpretable as such. Many of the abovementioned event-based notations apply to these environments, excluding tree-shaped (event/probability trees, Bayesian trees) and completely cyclic ones (e.g., causal loops).

The environment of directed alternative changes is described by the event bush as a set of events of four types arranged in certain order (defined as multiflow structure – see Fig. 1) and connected by special connectives.

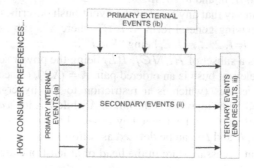

Fig. 1. Syntax of basic blocks of event bush – a multiflow structure

The events of the event bush fall into following classes, or types.

(ia) Primary internal events. These are primary, non-unique inputs representing the "passive conditions". These would determine any further course of events ("happenings").

(ib) Primary external events meaning the "invading agents". They may affect basic inputs or influence their further, indirect manifestations, thus "shaping up" different "happenings".

(ii) Secondary events that result from primary internal inputs with or without the contribution of primary external ones – the "happenings" proper formulated in a strict concise way indicating their core features determined by the causes, following the principle "one more cause – one change of event".

(iii) Tertiary events that denote end results, or products, generated either by primary internal or by secondary events, with or without primary external ones. Tertiary events document the completed "happenings".

Any event in event bush is uniquely characterized by the set of subjects, set of predicates and type *(ia, ib, ii* or *iii)* and, in some cases, generality. Two primary events may not have similar subject.

This structure can be projected on the formalism of directed hypergraph.

In mathematics, hypergraph is a generalization of graph in which an arc can connect any number of vertices. Formally, a hypergraph is a pair $G = (V,E)$, where $V = \{v_1, v_2,..., v_n\}$ is the set of vertices (or nodes) and $E = \{E_1, E_2,..., E_m\}$, with $E_i \subseteq V$ for $i = 1,..., m$, is the set of non-empty subsets of V called hyperedges, or arcs [2]. However, as "E" is reserved in the event bush semantics for "event", i.e., node, henceforth we will denote hypergraph as $G = (E,A)$, E standing for the set of events, or nodes, and A, for the set of arcs, or hyperedges. Also, not to confuse "hypergraph" and "hyperarc head" (see below), the symbol "H" will be used for the latter, and the hypergraph proper, though commonly denoted as "H" in the literature, is henceforth designated as "G".

In the event bush, E is a set consisting of non-empty disjoint subsets *Ia, Ib, II* and *III*, where *Ia* are nodes describing primary internal events, *Ib*, nodes describing primary external events, *II*, nodes describing secondary, and *III*, tertiary events. Every node in the event bush belongs simultaneously, first, either to *Ia*, or to *Ib*, or to *II* or to

III, and then, at least to one arc. Nodes belonging to *Ia* are denoted as ia, those belonging to *Ib*, ib, to *II*, ii, and to *III*, iii.

In general, one can say that any node of the event bush describes an event, which is a verbal expression having general form "*S is P*" where *S* is subject, and *P*, predicate or predicates: $E=\{E_i, i=1, 2, \ldots, n\}$, E_i being "*S is P*".

A, by definition, is a subset of $P(E)\setminus\{\varnothing\}$, $P(E)$ being the power set of *E*.

Directed arc in the event bush is an ordered pair, $A = (X,Y)$, of necessarily non-empty disjoint subsets of vertices (which is a restriction to the hypergraph theory which normally allows *X* and *Y* be empty); *X* is the *tail* of *A* while *Y* is its *head*. In the following, the tail and the head of arc *A* will be denoted by $T(A)$ and $H(A)$, respectively.

The subsets *Ia*, *Ib*, *II* and *III* can be defined as follows,

Ia includes nodes that necessarily are in the head of at least one arc and may not be in the tail of any arc;

Ib includes nodes that necessarily are in the head of at least one arc if this head includes one more node of different type and may not be in the tail of any arc;

II includes nodes that necessarily are in the head of at least one arc and in the tail of at least one arc;

III includes nodes that necessarily are in the tail of at least one arc and may not be in the head of any arc.

The arcs of event bush meet the following requirements:

- not a single node can be in the tail and head of similar arc;
- not a single ia node can be in the head of any arc;
- not a single ib node can be in the head of any arc;
- every ia node must be in the tail of at least one arc;
- every ib node must be in the tail of at least one arc, and in this case the tail consists of two nodes and includes, along with the ib, one ia or ii node;
- not a single iii node can be in the tail of any arc;
- every iii node is in the head of only one arc and no other nodes are in the head of this arc;
- every ii node is in the tail of at least one arc and in the head of at least one arc;
- if an ai node is alone in the tail of an arc, only one ii or iii node can be in the head of the same arc;
- if an ai node is in the tail of an arc not alone, it may be accompanied there only by one ib or one ii nodes;
- two ai nodes cannot occur in the tail of one arc;
- no arc with empty tail or head is allowed;
- no arc that simultaneously has a head having more than one node and a tail having more than one node is allowed.

In terms of backward and forward arcs (b-arcs and f-arcs), the above requirements can be reformulated as follows (here we have to add simple one-to-one arcs not considered by Gallo et al. [2]).

Allowed one-to-one-arcs include

- $T(ia)$, $H(ii)$,
- $T(ia)$, $H(iii)$,

- $T(ii)$, $H(ii)$,
- $T(iii)$, $H(ii)$.

Allowed b-arcs include

- $T(ia, ib)$, $H(ii)$,
- $T(ii, ib)$, $H(ii)$,
- $T(ii, ii)$, $H(ii)$,
- $T(ii, ii, ..., ii)$, $H(ii)$.

Allowed f-arcs include $T(ii)$, $H(ii, ii, ..., ii)$.

3   Event Bush Connectives and Inference

Inference in the event bush still needs to be elaborated but in general it is governed by the connectives defined as follows, taking e as an individual event, or, mathematically speaking, an element of E.

Flux connective (Fig. 2, a) describes one event (e_i) producing another (e_j):

$$e_i \ Flux \ e_j.$$

Influx connective (Fig. 2, b) describes two events (e_i, e_j) producing another (e_k), but playing different roles (this will be described below):

$$e_i, \ e_j \ Influx \ e_k.$$

Furcation connective (Fig. 2, c) describes production of multiple events (e_{i+1}, e_{i+2}, ..., e_n) by one (e_i):

$$e_i \ Furcation \ e_{i+1}, \ e_{i+2}, \ ..., \ e_n.$$

Conflux connective (Fig. 2, d) describes production of one event (e_n) by multiple events (e_i, e_{i+1}, e_{i+2}, ..., e_{n-1}):

$$e_i, \ e_{i+1}, \ e_{i+2}, \ ..., \ e_{n-1} \ Conflux \ e_n.$$

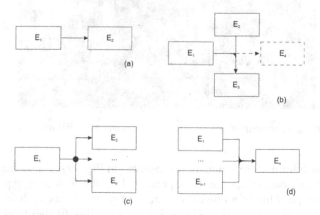

Fig. 2. Graphic notation for the connectives of the event bush: (a), flux, (b), influx, (c), furcation, (d), conflux,; e_1, e_2, ..., e_n are connected events

Depending on the of event in the bush (*ia*, *ib*, *ii* or *iii*) and type of change (there can change subject, predicate or both; also, some particular modes of change are specified) there exist different modi of the connectives.

Flux connective:

$$ii\ S_iP_k\ Flux\ Modus\ 1\ ii\ S_jP_k,$$
$$ii\ S_iP_k\ Flux\ Modus\ 2\ ii\ S_jP_l,$$
$$ii\ S_iP_j\ Flux\ Modus\ 3\ ii\ S_{Pj}P_{Si},$$
$$ii\ S_i{\sim}P_k\ Flux\ Modus\ 4\ ii\ S_iP_k,\ ii\ S_iP_k\ Flux\ Modus\ 4\ ii\ S_i{\sim}P_k,$$
$$ia\ S_iP_j\ Flux\ Modus\ 5\ ii\ S_iP_j,\ ii\ S_iP_j\ Flux\ Modus\ 5\ iii\ S_iP_j,\ ia\ S_iP_j\ Flux\ Modus\ 5\ iii\ S_iP_j,$$
$$ia\ any\ S_iP_j\ Flux\ Modus\ 6\ ii\ some\ S_iP_j,$$
$$ii\ any\ S_iP_j\ Flux\ Modus\ 6\ ii\ some\ S_iP_j.$$

Furcation connective:

$$ii\ S_iP_jP_kP_l\ Furcation\ Modus\ 1\ ii\ S_iP_j{\sim}P_kP_l,\ ii\ S_iP_jP_k{\sim}P_l,$$
$$ii\ S_iP_l\ Furcation\ Modus\ 2\ ii\ S_jP_l,\ ii\ S_kP_l.$$

Conflux connective:

$$ii\ S_iP_k,\ ii\ S_jP_k\ Conflux\ Modus\ 1\ ii\ S_{Pk}P_{Si}P_{Sj},$$
$$ii\ S_iP_i,\ ii\ S_jP_i\ Conflux\ Modus\ 2\ ii\ S_{SiSj}P_i,$$
$$ii\ some\ S_iP_j,\ ii\ some\ S_iP_j\ Conflux\ Modus\ 3\ ii\ any\ S_iP_j.$$

For detail, the reader is referred to [5].

An example of event bush used in volcanological studies is given below. It describes an environment presented in Fig. 3.

Fig. 3. Described environment: slope of Mt. Etna, Sicily, near Rifugio Sapienza; photo by C. Pshenichny

The event bush for this environment is presented in Fig. 4 (see also Appendix).

The event bush connectives can be represented by the following types of arcs of directed hypergraph. Flux is a one-to-one arc. Subtypes of flux include $T(ia)$, $H(ii)$; $T(ia)$, $H(iii)$; $T(ii)$, $H(ii)$ and $T(iii)$, $H(iii)$. Six modi of flux fit these subtypes: $T(ia)$, $H(ii)$ – one case of modus 5, one case of modus 6, $T(ia)$, $H(iii)$ – one case of modus 5, $T(ii)$, $H(ii)$ – modi 1, 2, 3, 4, one case of modus 6, $T(iii)$, $H(ii)$ – one case of modus 5. Similarly other connectives can be treated.

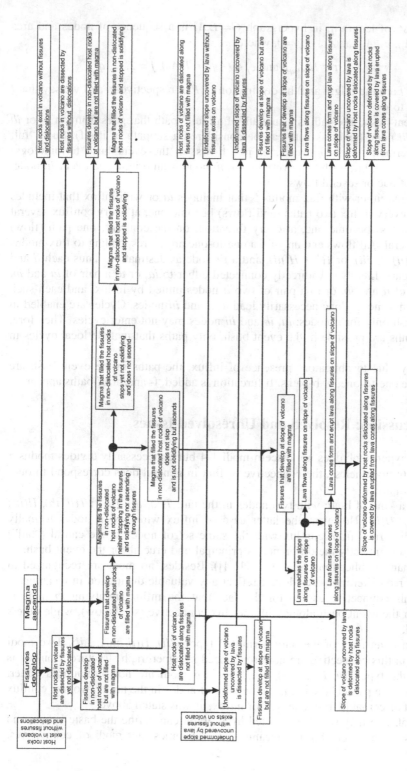

Fig. 4. The event bush explaining how the observed bodies were formed and what other bodies had to cogenetically form below the surface or could alternatively form on and below the surface in the described environment. See textual description in Appendix.

A path *Pst*, of length *q*, in hypergraph $G=(E,A)$ is a sequence of nodes and arcs $Pst=(e_1=s, A_{i1}, e_2, A_{i2}, ..., A_{iq}, e_{q+1}=t)$, where:

$$s \in (A_{i1}), t \in H(A_{iq}), \text{ and } e_j \in (A_{ij-1}) \cap T(A_{ij}), j = 2, ..., q.$$

Nodes *s* and *t* are the origin and the destination of *Pst*, respectively, and we say that *t* is connected to *s*.

In the event bush, any path is a subpath of a "full" path that goes from an *ai* or *ib* node to an *iii* node (and this path can be considered a subpath of itself). Such "full" paths" are termed flows of events, or just flows, in the event bush terminology. Therefore, one can put forth that every *ai* and *ib* node is an origin, and every *iii* node, destination of one or several flows.

Flows begin either with flux modus 5 that includes *ia* or with influx that includes *ib*. Then, at every influx two paths (and flows) become one, at every conflux several paths (and flows) become one, at every furcation, on the contrary, one path (flow) becomes several. All flows end up with a one-to-one arc corresponding to flux modus 5 – either *T(ii), H(iii)*, or *T(ia), H(iii)*, and a *iii* node as destination. Thus each *ii* and *iii* node appears directly or indirectly connected either to *ia*, or to a pair of *ia* and *ib*, or to a pair of *ii* and *ib*, or to a pair of two *ii* nodes united by influx, and each such pair and each *ia* node alone necessarily lead to *ii* and *iii* nodes. Cycles are enabled in the event bush only for ii nodes; *ia, ib* and *iii* nodes may not enter cycles. Therefore, no cyclic paths are possible in the event bush, only paths that include local cycles in the *ii* zone.

Obviously, due to obligatory presence of influx, the paths in the event bush are either simple one-to-one, or b-paths. If furcation is added, f- and/or bf-paths appear.

4   Discussion: Resolved and Unresolved Issues

Obviously, hypergraph does not discern modi 1-4 but unnecessarily divides modus 5 and modus 6; nonetheless, the connective of flux in general does correspond to one-to-one type of arc.

Influx is a kind of b-arc with two nodes in the tail: *T(ia, ib), H(ii)*; *T(ii, ib), H(ii)*; and *T(ii, ii), H(ii)*. However, the latter case of influx with two ii nodes formally cannot be discerned from conflux with the same set of nodes in the causal ("tail") part, while semantically the difference is principal and crucial for the event bush, as shown in many publications (see, e.g., [1; 4]). Besides, no modi are recognized in influx, so three different b-arcs do not reflect any valuable distinction in meaning or usage of this connective except for the fact that the influx involving *ib* node is essential for the event bush along with the flux connective (see below), while others are optional.

Furcation is an f-arc of the only permitted type, *T(ii), H(ii, ii, ..., ii)*. Two modi recognized in this connective are not pronounced in hypergraph structure. Conflux is a b-arc of the type *T(ii, ii, ..., ii), H(ii)*. Its two modi are not pronounced either. Moreover, in one particular case, if the number of nodes in the tail is two, the conflux per se is not discernable from one type of influx, as was stated above.

In general, one may say that directed hypergraph can define the basic syntax and connectives of the event bush but cannot, its semantics and modi of connectives.

Therefore, generally speaking, directed hypergraph in its present form does not support inference in the event bush and cannot be used for formal definition of this method. However, given the extension of directed hypergraph accounting for semantics of verbal expressions in vertices, the modi of all event bush connectives may be defined and discerned. If such extension is elaborated (with all the corollaries constraining the paths in the graph), the event bush could be considered a particular case of semantically-sensitive directed hypergraph. At present, the event bush in terms of directed hypergraph can be termed as "vulgar bush" having the set of connectives the same as the event bush except the conflux (which is optional), but no semantic rules, which make up the virtue of this approach [6]. However, yet in this perverted form, it can be useful at least for demonstration purposes, and this alone urges us to examine further how it fits the directed hypergraph formalism.

By the definition of event bush, it must include flux and influx and may, though not necessarily does, have furcation and conflux. In the vulgar bush, conflux should be omitted, as shown above. However, this definition may be constrained further: flux modus 5 involving ia and iii nodes and influx involving ib node are required for the event/vulgar bush. In terms of hypergraph, this can be put down as follows: subsets $T(ia)$, $H(iii)$ and $T(ia, ib)$, $H(ii)$ are necessarily non-empty. This, in turn, implies that at least one more flux, $T(ii)$, $H(iii)$, is not empty too. Other arcs may be empty (sub)sets. Here the hypergraph notation definitely helps define things more precisely. Hence, the event bush in general case can be viewed as a kind of BF-hypergraph, but in absence of conflux and furcation it becomes a B-hypergraph.

Multiple arcs with the same set of nodes are not allowed, i.e., the event bush cannot be multigraph. Symmetric image for an event bush is a hypergraph but is not an event bush.

As is seen from the above, connectives are arbitrary sets of nodes and can therefore contain an arbitrary number of nodes. Even in the simplest case, if the bush has only flux and influx, yet these two have different cardinality and therefore cannot form a K-uniform hypergraph.

The same time, the event bush/vulgar bush can be regarded as clutter, where no arc appears as a subset of another arc, though several arcs may have similar tail or similar head, i.e., in terms of sets theory, arcs of the event bush do intersect.

Further implications can be made from the directed hypergraph formalism to the event bush approach, but this seems pertinent with development of ad hoc extension of the said formalism accounting for semantic rules of the event bush.

5   Conclusions

1. The event bush basic syntax can be well defined in terms of directed hypergraph, and this gives ground to introduce a concept of vulgar bush, which may be useful for demonstration purposes and some straightforward applications..

2. Nevertheless, semantics of the event bush cannot be adequately represented in terms of directed hypergraph, because the latter is ignorant to the event formation rules and, hence, fails to define the modi of the event bush connectives.

3. If an extension of directed hypergraph theory is developed to account for semantics of verbal expressions that may be associated with its vertices, this will open an

opportunity to entirely formalize the event bush as a semantically restricted (or semantically sensitive) directed hypergraph.

Acknowledgements. The authors are deeply obliged to Dmitry Ignatov whose enthusiasm largely fueled up the creation of this paper; also, Sergey Nikolenko, Anthony Yakovlev, Oksana Kanzheleva and Roberto Carniel should be thanked for their vigorous discussion of the issue in 2008-2010.

References

1. Carniel, R., Pshenichny, C., Khrabrykh, Z., Shterkhun, V., Pascolo, P.: Modeling Models: Understanding of Structure of Geophysical Knowledge by Means of the Event Bush Method. In: Marschallinger, R., Zobl, F. (eds.) IAMG Proceedings, Salzburg. Mathematical Geosciences at the Crossroads of Theory and Practice, Salzburg, pp. 1336–1350 (September 2011)
2. Gallo, G., Longo, G., Nguyen, S., Pallottino, S.: Directed hypergraphs and applications. Discrete Applied Mathematics 42, 177–201 (1993)
3. Martin, P.: Knowledge Representation in CGLF, CGIF, KIF, Frame-CG and Formalized-English. In: Priss, U., Corbett, D.R., Angelova, G. (eds.) ICCS 2002. LNCS (LNAI), vol. 2393, pp. 77–91. Springer, Heidelberg (2002)
4. Mouromtsev, D.I., Pshenichny, C.A., Yakovlev, A.V.: Semantic and structural delineation of market scenarios by the event bush method, Submitted to Decision Support Systems and Electronic Commerce
5. Pshenichny, C.A., Kanzheleva, O.M.: Theoretical foundations of the event bush method. In: Sinha, K., Gundersen, L., Jackson, J., Arctur, D. (eds.) Societal Challenges and Geoinformatics. GSA Special Paper 482, pp. 139–165 (2011)
6. Pshenichny, C.A., Nikolenko, S.I., Carniel, R., Vaganov, P.A., Khrabrykh, Z.V., Moukhachov, V.P., Akimova-Shterkhun, V.L., Rezyapkin, A.A.: The Event Bush as a Semantic-based Numerical Approach to Natural Hazard Assessment (Exemplified by Volcanology). Comp. Geosc. 35, 1017–1034 (2009)
7. Rieger, C.: The common-sense algorithm as a basis for computer models of human memory, inference, belief and contextual language comprehension. Theoretical Issues in Natural Language Processing, Cambridge (1973)

Appendix

The event bush plotted in Fig. 4 can be represented in text form as a succession of the following expressions.

(ia Host rocks exist in volcano without fissures and dislocations) Flux Modus 5 (iii Host rocks exist in volcano without fissures and dislocations)

(ia Host rocks exist in volcano without fissures and dislocations, ib Fissures develop) Influx (ii Host rocks in volcano are dissected by fissures yet not dislocated)

(ii Host rocks in volcano are dissected by fissures yet not dislocated) Flux Modus 5 (iii Host rocks in volcano are dissected by fissures yet not dislocated) (ii Host rocks in volcano are dissected by fissures yet not dislocated) Flux Modus 2 (ii Fissures develop in non-dislocated host rocks of volcano but are not filled with magma)

(ii Fissures develop in non-dislocated host rocks of volcano but are not filled with magma) Flux Modus 5 (iii Fissures develop in non-dislocated host rocks of volcano but are not filled with magma)

(ii Fissures develop in non-dislocated host rocks of volcano but are not filled with magma, ib Magma ascends) Influx (ii Fissures that develop in non-dislocated host rocks of volcano are filled with magma)

(ii Fissures that develop in non-dislocated host rocks of volcano are filled with magma) Flux Modus 3 (ii Magma fills the fissures in non-dislocated host rocks of volcano)

(ii Magma fills the fissures in non-dislocated host rocks of volcano neither stopping in the fissures in non-dislocated host rocks of volcano and solidifying nor ascending through fissures in non-dislocated host rocks of volcano) Furcation Modus 1 (ii Magma that filled the fissures in non-dislocated host rocks of volcano stops in the fissures in non-dislocated host rocks of volcano yet does not solidify and does not ascend through fissures in non-dislocated host rocks of volcano, ii Magma that filled the fissures in non-dislocated host rocks of volcano does not stop in the fissures and yet does not solidify in non-dislocated host rocks of volcano but ascends through fissures in non-dislocated host rocks of volcano)

(ii Magma stops in the fissures in non-dislocated host rocks of volcano and yet does not solidify) Flux Modus 4 (ii Magma that stopped in the fissures in non-dislocated host rocks of volcano solidifies in the fissures in non-dislocated host rocks of volcano)

(ii Magma that stopped in the fissures in non-dislocated host rocks of volcano solidifies in the fissures in non-dislocated host rocks of volcano) Flux Modus 5 (iii Magma that stopped in the fissures in non-dislocated host rocks of volcano solidifies in the fissures in non-dislocated host rocks of volcano)

(ii Fissures develop in non-dislocated host rocks of volcano but are not filled with magma) Flux Modus 3 (ii Host rocks of volcano are dislocated along fissures not filled with magma)

(ii Host rocks of volcano are dislocated along fissures not filled with magma) Flux Modus 5 (iii Host rocks of volcano are dislocated along fissures not filled with magma)

(ia Undeformed slope without fissures, uncovered by lava exists on volcano) Flux Modus 5 (iii Undeformed slope without fissures, uncovered by lava exists on volcano)

(ia Undeformed slope without fissures, uncovered by lava exists on volcano, ii Fissures develop in non-dislocated host rocks of volcano but are not filled with magma) Influx (ii Undeformed slope of volcano uncovered by lava is dissected by fissures)

(ia Undeformed slope without fissures, uncovered by lava exists on volcano, ii Host rocks of volcano are dislocated along fissures not filled with magma) Influx (ii Slope of volcano uncovered by lava is deformed by host rocks of volcano dislocated along fissures not filled with magma) (ii Slope of volcano uncovered by lava is deformed by host rocks of volcano dislocated along fissures not filled with magma) Flux Modus 5 (iii Slope of volcano uncovered by lava is deformed by host rocks of volcano dislocated along fissures not filled with magma)

(ii Fissures develop in non-dislocated host rocks of volcano but are not filled with magma, ii Magma ascends through fissures in non-dislocated host rocks of volcano) Influx (ii Fissures that develop at slope of volcano are filled with magma)

(ii Fissures that develop at slope of volcano are filled with magma) Flux Modus 5 (iii Fissures at slope of volcano are filled with magma)

(ii Fissures that develop at slope of volcano are filled with magma) Flux Modus 2 (ii Lava reaches the slope of volcano)

(ii Lava reaches the slope along fissures on slope of volcano yet neither flowing form fissures nor forming lava cones along fissures) Furcation Modus 1 (ii Lava that reached the slope along fissures on slope of volcano flows along fissures on slope of volcano from fissures and does not form lava cones on slope of volcano, ii Lava that reached the slope along fissures on slope of volcano does not flow along fissures on slope of volcano but forms lava cones along fissures on slope of volcano)

(ii Lava forms lava cones along fissures on slope of volcano) Flux Modus 3 (ii Lava cones form and erupt lava along fissures on slope of volcano)

(ii Lava cones form and erupt lava along fissures on slope of volcano) Flux Modus 5 (iii Lava cones form and erupt lava along fissures on slope of volcano)

(ii Lava cones form and erupt lava along fissures on slope of volcano) Flux Modus 2 (ii Lava flows along fissures on slope of volcano from fissures)

(ii Slope of volcano uncovered by lava is deformed by host rocks of volcano dislocated along fissures not filled with magma, ii Lava forms lava cones along fissures on slope of volcano) Influx (ii Slope of volcano deformed by host rocks of volcano dislocated along fissures not filled with magma is covered by lava erupted from lava cones along fissures)

(ii Slope of volcano deformed by host rocks of volcano dislocated along fissures not filled with magma is covered by lava erupted from lava cones along fissures) Flux Mode 5 (iii Slope of volcano deformed by host rocks of volcano dislocated along fissures not filled with magma is covered by lava erupted from lava cones along fissures)

Concept Lattices of a Relational Structure

Jens Kötters

Abstract. Conceptual patterns can be described by graphs, entailment by graph homomorphism. The mapping of a pattern to its set of instantiations, represented as a table, constitutes one half of a Galois connection. The join operation is the infimum in a complete lattice of tables, and a most descriptive pattern can be assigned to each table by means of a categorial product construction. This construction constitutes the other half of the Galois connection. In this approach, relational structures assume the role of formal contexts in standard Formal Concept Analysis (FCA). Concepts arise as connected components of powers of these relational structures. The ordered set of these concepts may be conceived as a navigation space.

Keywords: Formal Concept Analysis, Relational Structures, Category Theory, Databases.

1 Introduction

The idea of using concept lattices to browse data can be traced back to [7]. In [7], a set of attributes is considered a query, and the set of objects having all the attributes (which is a concept extent) is the corresponding result set. The downward (upward) edges in a lattice's line diagram indicate the ways in which a query can be refined (weakened) to effect a minimal change in the result set. The capability to successively modify queries in this fashion is thought to make data more accessible to the information seeker.

More advanced applications of lattice-based browsing make use of conceptual scales to incorporate and distinguish between different types of values in the data. Relational scales can be used to account for inter-object relations in the data. The reader is referred to [1,3] for recent applications that treat relational data. In this paper, we describe mathematically a navigation space akin to those underlying the mentioned systems, but it is obtained directly from a relational structure and not by means of relational scaling.

For an example, consider the family tree in Fig. 1. The nodes represent the family members A(nne), B(ob), C(hris), D(ora) and E(mily). The arcs say that Anne is the m(other) of Bob and Chris, and Bob is the f(ather) of Dora and Emily. This graph defines (and visually represents) a relational structure $\mathcal{F}$ with underlying set $\{A, B, C, D, E\}$, unary relations $\male$ and $\female$ and binary relations m and f. We assume that $\mathcal{F}$ also has a p(arent) relation p, which is not drawn here. In Fig. 2, the nine nodes arranged as a cube form a particular concept lattice.

The concept intents, drawn to the right of each concept, are relational structures representing conjunctive queries [2]. The black nodes designate the subject(s)

H.D. Pfeiffer et al. (Eds.): ICCS 2013, LNAI 7735, pp. 301–310, 2013.

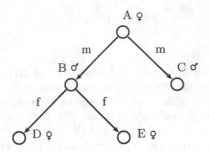

Fig. 1. Example: family tree $\mathcal{F}$

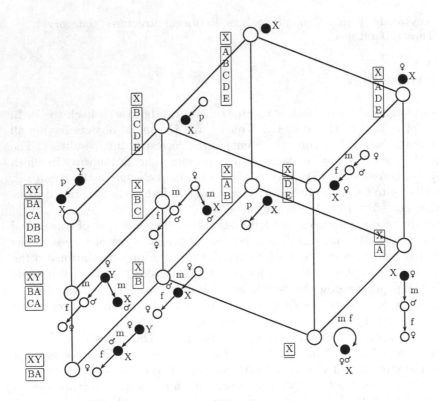

Fig. 2. Example: concept lattice $\mathfrak{C}_{\mathcal{F}}[\{x\}]$ (plus three concepts from $\mathfrak{C}_{\mathcal{F}}[\{x, y\}]$)

of a query, variable names are assigned to them (technically by a function ν). The white nodes correspond to existentially quantified variables. The top concept and its lower neighbors (from left to right) are 'person', 'child', 'parent' and 'female'. The concept extents, drawn to the left of each concept, are result sets of the intent queries. Each result arises from a homomorphism from the intent to $\mathcal{F}$. Consider now the three lower neighbors of 'child'. The right one can be identified with 'granddaughter', but note that the most precise description (the intent) tells us more. The middle one can *not* be identified with 'uncle' because we do not (and can not) express that the males are different persons. We could make up a name "parentship" for the left one, the intent has two free variables x, y and the extent consists of all instances of 'parentship'. This concept does not belong to the "cube" $\mathfrak{C}_{\mathcal{F}}[\{x\}]$ (the lattice of all concepts definable with one free variable x), it belongs to the concept lattice $\mathfrak{C}_{\mathcal{F}}[\{x, y\}]$. Only three concepts of $\mathfrak{C}_{\mathcal{F}}[\{x, y\}]$ are shown in Fig. 2.

Graphs are a natural candidate for the formal representation of queries over RDF. Chandra and Merlin's result on query optimization by graph folding [2] may exemplify on a more general level the benefits of such representation. Pattern Structures [5] formalize the idea of representing concept intents by some kind of pattern rather than by an attribute set. In [5], the authors mention Conceptual Graphs and formalize chemical graphs (see also [4]) as examples for patterns. The notions of homomorphism which accompany these graphs make their approach seem very similar to the one presented here. A difference is that in [5], extents are still sets of objects, while here we use tables for their representation (although one-column tables are naturally identified with object sets). One could argue that such a construction is no longer a concept lattice, but in fact we have identified concepts in the foregoing example.

We will stipulate that concepts are described by connected graphs. However, the Galois connection in Sect. 5 extends to all windowed structures (i.e. conjunctive queries, Sect. 3), it does not harm to allow even infinite ones. The concept lattices lie embedded in the complete lattice that arises from this Galois connection (Sect. 6). Before the Galois connection is defined, the preordered class of windowed structures (Sect. 3) and the complete lattice of tables (Sect.4) are introduced independent of each other.

2   Preliminaries

A relational signature is a set $\mathcal{S}$ of relation symbols. The arity of a symbol $R \in \mathcal{S}$ is a natural number $|R| \geq 1$. A relational structure over the signature $\mathcal{S}$, also called an $\mathcal{S}$-structure, is a pair $\mathcal{A} = (A, (\mathcal{A}(R))_{R \in \mathcal{S}})$, where $\mathcal{A}(R) \subseteq A^{|R|}$ for all $R \in \mathcal{S}$. The set A is called the underlying set of $\mathcal{A}$ and can also be denoted by $|\mathcal{A}|$. A homomorphism from an $\mathcal{S}$-structure $\mathcal{G}_1$ to an $\mathcal{S}$-structure $\mathcal{G}_2$ is a map $\varphi : |\mathcal{G}_1| \rightarrow |\mathcal{G}_2|$ such that for all $R \in \mathcal{S}$ and $(x_1, \ldots, x_{|R|}) \in \mathcal{G}_1(R)$ we have $(\varphi(x_1), \ldots, \varphi(x_{|R|})) \in \mathcal{G}_2(R)$.

The product $\prod_{i \in I} \mathcal{G}_i$ of $\mathcal{S}$-structures is defined by

$$\left|\prod_{i \in I} \mathcal{G}_i\right| = \underset{i \in I}{\times} |\mathcal{G}_i|$$

and, for all symbols R of $\mathcal{S}$,

$$(x_1, \ldots, x_{|R|}) \in (\prod_{i \in I} \mathcal{G}_i)(R) :\Leftrightarrow \forall i \in I : (x_1(i), \ldots, x_{|R|}(i)) \in \mathcal{G}_i(R) .$$

The I-th power of a structure $\mathcal{G}$ is the product $\times_{i \in I} \mathcal{G}$ and is denoted by $\mathcal{G}^I$. The product of relational structures is a product in the sense of category theory:

Proposition 1. *Let $(\mathcal{G}_i)_{i \in I}$ be a family of $\mathcal{S}$-structures. Then each projection $\pi_i : \times_{i \in I} |\mathcal{G}_i| \to |\mathcal{G}_i|$ is a homomorphism from $\prod_{i \in I} \mathcal{G}_i$ to $\mathcal{G}_i$. Furthermore, for each $\mathcal{S}$-structure $\mathcal{H}$ and family $(\varphi_i : \mathcal{H} \to \mathcal{G}_i)_{i \in I}$, there exists a unique $\varphi : \mathcal{H} \to \prod_{i \in I} \mathcal{G}_i$ such that $\varphi_i = \pi_i \circ \varphi$ for all $i \in I$.*

When we talk about the nodes of $\mathcal{A}$, what we mean are the elements of $|\mathcal{A}|$. A sequence $(a_1, \ldots, a_n)$ of nodes of $\mathcal{A}$ is called a path from a_1 to a_n in $\mathcal{A}$, if for all $1 \leq i < n$ there exists an $R \in \mathcal{S}$ such that $\{a_i, a_{i+1}\} \subseteq \{x_1, \ldots, x_{|R|}\}$ for some $(x_1, \ldots, x_{|R|}) \in \mathcal{A}(R)$. We call a structure $\mathcal{A}$ connected if there exists a path from a to b for all $a, b \in |\mathcal{A}|$. We define an equivalence relation

$$a \sim b \Leftrightarrow \text{there exists a path from } a \text{ to } b$$

over $|\mathcal{A}|$. A connected component of $\mathcal{A}$, or simply a component of $\mathcal{A}$, is an $\mathcal{S}$-structure $\mathcal{C}$ for which $|\mathcal{C}|$ is a class of $\sim$ and $\mathcal{C}(R) = \mathcal{A}(R) \cap |\mathcal{C}|^{|R|}$ for all $R \in \mathcal{S}$.

Throughout the paper, we will use Var to denote a countably infinite set of variables. By ι (or any variety such as $\tilde{\iota}, \iota_1, \ldots$) we shall always denote an inclusion map from some set X_1 to some set X_2, i.e. a map with $\iota(x) = x$ for all $x \in X_1$, where $X_1 \subseteq X_2$ is implied. The sets X_1 and X_2 will be clear from the context.

3   Windowed Structures

Definition 1. *Let $\mathcal{S}$ be a relational signature. A **windowed $\mathcal{S}$-structure** is a triple $(X, \nu, \mathcal{G})$ consisting of a set $X \subseteq$ Var, an $\mathcal{S}$-structure $\mathcal{G}$ and a map $\nu : X \to |\mathcal{G}|$.*

Definition 2. *Let $W_1 = (X_1, \nu_1, \mathcal{G}_1)$ and $W_2 = (X_2, \nu_2, \mathcal{G}_2)$ be windowed $\mathcal{S}$-structures. A **homomorphism** $\varphi : W_1 \to W_2$ **of windowed structures** is a structure homomorphism $\varphi : \mathcal{G}_1 \to \mathcal{G}_2$ with $\varphi \circ \nu_1 = \nu_2 \circ \iota$, where $X_1 \subseteq X_2$ is assumed.*

Homomorphisms of windowed $\mathcal{S}$-structures are closed under composition. Also, the identity id $: |\mathcal{G}| \to |\mathcal{G}|$ is a homomorphism of any windowed $\mathcal{S}$-structure $(X, \nu, \mathcal{G})$ onto itself. These two facts imply that windowed $\mathcal{S}$-structures with homomorphisms form a category. Furthermore, they imply that the following relation on the class of windowed $\mathcal{S}$-structures is a preorder:

Definition 3 (Homomorphism Preorder). *For windowed S-structures W_1 and W_2, we set*

$$W_1 \lesssim W_2 :\Leftrightarrow \exists \varphi : W_1 \to W_2 .$$

The homomorphism preorder induces an equivalence relation on the class of windowed S-structures:

Definition 4 (Homomorphic Equivalence). *For windowed S-structures W_1 and W_2, we set*

$$W_1 \simeq W_2 :\Leftrightarrow W_1 \lesssim W_2 \wedge W_2 \lesssim W_1 .$$

Definition 5. *The **product** of a family $((X_i, \nu_i, \mathcal{G}_i))_{i \in I}$ of windowed S-structures is the windowed S-structure*

$$\prod_{i \in I}(X_i, \nu_i, \mathcal{G}_i) := (\bigcap_{i \in I} X_i, \nu_I, \prod_{i \in I} \mathcal{G}_i) ,$$

where $\nu_I(x) := (\nu_i(x))_{i \in I}$.

The product of the empty family is $(\mathrm{Var}, \nu_\emptyset, (\emptyset, (\{\emptyset\}^{|R|})_{R \in S}))$, where $\nu_\emptyset(x) = \emptyset$ for all $x \in \mathrm{Var}$.

The product is indeed a product in the category theoretical sense, as the following proposition shows:

Proposition 2. *Let $((X_i, \nu_i, \mathcal{G}_i))_{i \in I}$ be a family of windowed S-structures. Each projection $\pi_i : \bigtimes_{i \in I} |\mathcal{G}_i| \to |\mathcal{G}_i|$ is a homomorphism from $\prod_{i \in I}(X_i, \nu_i, \mathcal{G}_i)$ to $(X_i, \nu_i, \mathcal{G}_i)$. Furthermore, for each windowed S-structure $(Y, \mu, \mathcal{H})$ and family $(\varphi_i : (Y, \mu, \mathcal{H}) \to (X_i, \nu_i, \mathcal{G}_i))_{i \in I}$, a unique $\varphi : (Y, \mu, \mathcal{H}) \to \prod_{i \in I}(X_i, \nu_i, \mathcal{G}_i)$ exists such that $\varphi_i = \pi_i \circ \varphi$ for all $i \in I$.*

Proof. From Prop. 1 we obtain $\pi_i : \prod_{i \in I} \mathcal{G}_i \to \mathcal{G}_i$ for $i \in I$. The definition of ν_I provides $\pi_i \circ \nu_I = \nu_i \circ \iota_i$ for all $i \in I$ (see the right circuit in Fig. 4). This proves the first claim. Now let $(\varphi_i : (Y, \mu, \mathcal{H}) \to (X_i, \nu_i, \mathcal{G}_i))_{i \in I}$ be a family of homomorphisms on some $(Y, \mu, \mathcal{H})$. In particular, we have $\varphi_i \circ \mu = \nu_i \circ \tilde{\iota}_i$ for all $i \in I$ (outer circuit). Also, Y must be a subset of each X_i, and so we have an inclusion map $\iota : Y \to \bigcap_{i \in I} X_i$. The equations $\tilde{\iota}_i = \iota_i \circ \iota$ (upper circuit) hold trivially. Again from Prop. 1 we obtain $\varphi : \mathcal{H} \to \prod_{i \in I} \mathcal{G}_i$ with $\varphi_i = \pi_i \circ \varphi$ (lower circuit). Altogether, we obtain

$$\pi_i \circ \varphi \circ \mu = \varphi_i \circ \mu = \nu_i \circ \tilde{\iota}_i = \nu_i \circ \iota_i \circ \iota = \pi_i \circ \nu_I \circ \iota$$

for all $i \in I$, and thus $\varphi \circ \mu = \nu_I \circ \iota$ (left circuit). Note that this last equation can not be inferred from the commutativity of the diagram!

We have shown that φ is a homomorphism from $(Y, \mu, \mathcal{H})$ to $\prod_{i \in I}(X_i, \nu_i, \mathcal{G}_i)$. From $\varphi_i = \pi_i \circ \varphi$ it follows that $\varphi(x) := (\varphi_i(x))_{i \in I}$, so φ is unique. □

The coproduct $\coprod_{i \in I}(X_i, \nu_i, \mathcal{G}_i)$ of windowed graphs is identical to a pushout of S-structures, if all X_i are the same. In the general case, it is constructed from the disjoint union of the $\mathcal{G}_i$, $i \in I$, by identifying all nodes $\nu_i(x)$ and $\nu_j(x)$ where $x \in X_i \cap X_j$.

The product and coproduct can be understood as infimum and supremum in the homomorphism preorder. This is made precise in the corollary:

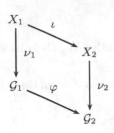

Fig. 3. Windowed graph morphism

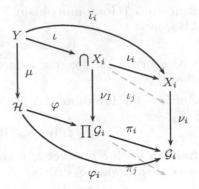

Fig. 4. Product of windowed graphs

Corollary 1. *Let* $(W_i)_{i \in I}$ *be a family of windowed S-structures. The following equivalences hold for all windowed S-structures W:*

$$W \lesssim \prod_{i \in I} W_i \Leftrightarrow \forall i \in I : W \lesssim W_i , \tag{1}$$

$$\coprod_{i \in I} W_i \lesssim W \Leftrightarrow \forall i \in I : W_i \lesssim W . \tag{2}$$

4   Tables

Definition 6. *Let G be a set. A **table over** G is a pair (X, Λ), where $X \subseteq \mathrm{Var}$ and $\Lambda \subseteq G^X$. The set of all tables over G is denoted by $\mathrm{Tab}(G)$.*

Definition 7. *For tables (X_1, Λ_1) and (X_2, Λ_2) over G, we define*

$$(X_1, \Lambda_1) \leq (X_2, \Lambda_2) :\Leftrightarrow X_2 \subseteq X_1 \wedge \Lambda_1 \circ \iota \subseteq \Lambda_2 .$$

Proposition 3. *The pair $(\mathrm{Tab}(G), \leq)$ is a complete lattice. The infimum of a family $((X_i, \Lambda_i))_{i \in I}$ of tables is given by the join operation*

$$\bigtimes_{i \in I} (X_i, \Lambda_i) := (\bigcup_{i \in I} X_i, \Lambda_I) , \tag{3}$$

where $\Lambda_I := \{\lambda : \bigcup_{i \in I} X_i \to G \mid \forall i \in I : \lambda \circ \iota_i \in \Lambda_i\}$. The supremum is

$$\bigtimes_{i \in I} (X_i, \Lambda_i) := (\bigcap_{i \in I} X_i, \bigcup_{i \in I} (\Lambda_i \circ \iota_i)) . \tag{4}$$

5   Galois Connection

Definition 8. *Let $\mathcal{D}$ be an S-structure. The **solution (in $\mathcal{D}$)** of a windowed S-structure $(X, \nu, \mathcal{G})$ is a table over $|\mathcal{D}|$, given by*

$$(X, \nu, \mathcal{G})' := (X, \mathrm{Hom}(\mathcal{G}, \mathcal{D}) \circ \nu) . \tag{5}$$

The **description (over $\mathcal{D}$)** for a table (X, Λ) over $|\mathcal{D}|$ is a windowed $\mathcal{S}$-structure, given by

$$(X, \Lambda)' := (X, \nu_\Lambda, \mathcal{D}^\Lambda) , \tag{6}$$

where $\nu_\Lambda(x) := (\lambda(x))_{\lambda \in \Lambda}$ for $x \in X$. The two operations thus defined are both denoted by the same sign and are called the **derivation operations with respect to $\mathcal{D}$**.

Proposition 4. Let $\mathcal{D}$ be an $\mathcal{S}$-structure. The derivation operators w.r.t. $\mathcal{D}$ form a Galois connection. That is, the following equivalence holds for all windowed $\mathcal{S}$-structures $(X, \nu, \mathcal{G})$ and for all tables (Y, Λ) over $|\mathcal{D}|$:

$$(X, \nu, \mathcal{G}) \lesssim (Y, \Lambda)' \Leftrightarrow (Y, \Lambda) \leq (X, \nu, \mathcal{G})' . \tag{7}$$

Proof. The left side of the statement can be transformed into the right side by a series of equivalences (explained below):

$$(X, \nu, \mathcal{G}) \lesssim (Y, \nu_\Lambda, \mathcal{D}^\Lambda) \Leftrightarrow \forall \lambda \in \Lambda : (X, \nu, \mathcal{G}) \lesssim (Y, \lambda, \mathcal{D})$$
$$\Leftrightarrow \Lambda \circ \iota \subseteq \mathrm{Hom}(\mathcal{G}, \mathcal{D}) \circ \nu$$
$$\Leftrightarrow (Y, \Lambda) \leq (X, \mathrm{Hom}(\mathcal{G}, \mathcal{D}) \circ \nu) .$$

To see the first equivalence, use that $(Y, \Lambda)'$ is the product of all $(Y, \lambda, \mathcal{D})$, $\lambda \in \Lambda$, and then apply Cor. 1. For the second equivalence, note that the statements on either side assert that for each $\lambda \in \Lambda$ there exists $\varphi : \mathcal{G} \to \mathcal{D}$ with $\lambda \circ \iota = \varphi \circ \nu$. The last equivalence follows from Def. 7. $\qquad \square$

In Props. 5 and 6 we state some consequences of (7) which are well-known in their general form. Proofs can e.g. be found in the introductory chapter of [6]. These carry over to our case of a Galois connection involving a preordered class (note Prop.5(iii), however).

Proposition 5. Let $\mathcal{D}$ be an $\mathcal{S}$-structure. The following holds for all tables T, T_1 and T_2 over $|D|$, and for all windowed $\mathcal{S}$-structures W, W_1 and W_2:

$(i)\ T \leq T''$ $\qquad\qquad\qquad$ $(i')\ W \lesssim W''$

$(ii)\ T_1 \leq T_2 \Rightarrow T_2' \lesssim T_1'$ $\qquad$ $(ii')\ W_1 \lesssim W_2 \Rightarrow W_2' \leq W_1'$

$(iii)\ T' \simeq T''' = T'''''$ $\qquad\quad$ $(iii')\ W' = W'''$

Proposition 6. Let $\mathcal{D}$ be an $\mathcal{S}$-structure. The following holds for all families $(W_i)_{i \in I}$ of windowed $\mathcal{S}$-structures and for all families $(T_i)_{i \in I}$ of tables over $|\mathcal{D}|$, respectively:

$$\left(\coprod_{i \in I} W_i \right)' = \bowtie_{i \in I} W_i' ,$$

$$\left(\bigtimes_{i \in I} T_i \right)' \simeq \prod_{i \in I} T_i' .$$

6   Concepts and Lattices

As in Formal Concept Analysis, we proceed to define a set of pairs which are stable under the Galois connection,

$$\mathfrak{L}_{\mathcal{D}} := \{(T, W) \mid T \in \mathrm{Tab}(\mathcal{D}) \wedge T' = W \wedge W' = T\} , \tag{8}$$

and define an order on that set,

$$(T_1, W_1) \leq (T_2, W_2) :\Leftrightarrow T_1 \leq T_2 \Leftrightarrow W_2 \lesssim W_1 . \tag{9}$$

The second equivalence in (9) follows from Prop. 5(ii)(ii'). Note that the elements of $\mathfrak{L}_{\mathcal{D}}$ are precisely the pairs (W', W''), or equivalently the pairs (T'', T'''), generated by the windowed $\mathcal{S}$-structures W and tables $T \in \mathrm{Tab}(\mathcal{D})$, respectively (see Prop. 5(iii)(iii')). For $X \subseteq \mathrm{Var}$ we define

$$\mathfrak{L}_{\mathcal{D}}[X] := \{((X, \Lambda)'', (X, \Lambda)''') \mid (X, \Lambda) \in \mathrm{Tab}(\mathcal{D})\} \tag{10}$$
$$= \{((X, \nu, \mathcal{G})', (X, \nu, \mathcal{G})'') \mid (X, \nu, \mathcal{G}) \text{ windowed } \mathcal{S}\text{-structure}\} .$$

The following definition of concept is suggested:

Definition 9. *A* **concept** *is a pair* $(T, (X, \nu, \mathcal{G})) \in \mathfrak{L}_{\mathcal{D}}$ *for which all nodes* $\nu(x)$, $x \in X$, *belong to the same component of* $\mathcal{G}$. *The set of all concepts of the relational structure* $\mathcal{D}$ *is denoted by* $\mathfrak{C}_{\mathcal{D}}$.

In analogy to (10), we define

$$\mathfrak{C}_{\mathcal{D}}[X] := \mathfrak{L}_{\mathcal{D}}[X] \cap \mathfrak{C}_{\mathcal{D}} . \tag{11}$$

We may identify a concept intent with the component containing $\nu(X)$.

Theorem 1. *The ordered set* $(\mathfrak{L}_{\mathcal{D}}, \leq)$ *is a complete lattice. Infimum and supremum are given by*

$$\bigwedge_{i \in I}(T_i, W_i) = \left(\bigbowtie_{i \in I} T_i, (\coprod_{i \in I} W_i)'' \right) , \tag{12}$$

$$\bigvee_{i \in I}(T_i, W_i) = \left((\bigboxtimes_{i \in I} T_i)'', (\prod_{i \in I} W_i)'' \right) . \tag{13}$$

For all $X \subseteq \mathrm{Var}$, *the suborders* $(\mathfrak{L}_{\mathcal{D}}[X] \cup \{\top\}, \leq)$ *and* $(\mathfrak{C}_{\mathcal{D}}[X] \cup \{\top\}, \leq)$, *where* $\top$ *denotes the maximum of* $(\mathfrak{L}_{\mathcal{D}}, \leq)$, *are* $\bigwedge$-*sublattices of* $(\mathfrak{L}_{\mathcal{D}}, \leq)$.

Proof. The formulas for the infimum and supremum follow from Prop. 6. Now let $(C_i)_{i \in I}$ be a family in $\mathfrak{L}_{\mathcal{D}}[X] \cup \{\top\}$ and $C := \bigwedge_{i \in I} C_i$, and let us further define $C_i =: ((X, \Lambda_i), (X, \nu_i, \mathcal{G}_i))$. If $C_i = \top$ for all $i \in I$, the infimum is $\top$. Else (3) simplifies to

$$\bigbowtie_{i \in I}(X, \Lambda_i) = (X, \bigcap_{i \in I} \Lambda_i) , \tag{14}$$

which means in particular that $C \in \mathfrak{L}_{\mathcal{D}}[X]$. We write $C =: ((X, \Lambda), (X, \nu, \mathcal{G}))$.

If $(C_i)_{i \in I}$ is a family in $\mathfrak{C}_{\mathcal{D}}[X] \cup \{\top\}$, we have to show in addition that $C \in \mathfrak{C}_{\mathcal{D}}$ if $C_i \in \mathfrak{C}_{\mathcal{D}}$ for some $i \in I$. In this case, there exists $\varphi : \mathcal{G}_i \to \mathcal{G}$. Homomorphisms preserve paths, so C must also be a concept. $\square$

7   Construction

In this section, a brute force construction algorithm for $\mathfrak{C}_{\mathcal{D}}[X]$ is given (X and $\mathcal{D}$ finite), where intents are (and need be) computed up to homomorphical equivalence only. A key observation is that all concept intents are components of powers of $\mathcal{D}$, complemented by some assignment from X to the nodes (cf. (6)). It can be shown that conversely, each windowed structure $(X, \nu, \mathcal{C})$, where $\mathcal{C}$ is a component of a power of $\mathcal{D}$ and ν is chosen arbitrarily, is homomorphically equivalent to some concept intent. If we pick one of these windowed structures from each $\simeq$-class, we have determined all concept intents up to homomorphical equivalence. Note that the components $\mathcal{C}$ can be taken from the power structures $\mathcal{D}^1, \ldots, \mathcal{D}^n$, $n := ||D|^X|$, because a power $\mathcal{D}^{\Lambda}$, $\Lambda \subseteq |\mathcal{D}|^X$, is isomorphic to $\mathcal{D}^{|\Lambda|}$. The following terminating condition can be proven: If we compute the powers $\mathcal{D}^1, \ldots, \mathcal{D}^n$ in sequence and reach some $\mathcal{D}^i$, $1 < i \le n$, such that every windowed structure obtained from a component of $\mathcal{D}^i$ is homomorphically equivalent to one computed earlier, then the set of concept intents (up to isomorphism) is complete. To build the line diagram (or check for homomorphical equivalence), it may be more convenient to compare extents. If data is stored in a database, extents could be computed by the query engine (this would involve translating windowed structures into some other form of query).

The nine concepts of $\mathfrak{C}_{\mathcal{D}}[\{x\}]$ from our initial example are obtained from the ten components of $\mathcal{F}$ and $\mathcal{F}^2$ (see Figs. 1 and 5). Thirty windowed structures are obtained from these components (as many as there are nodes), each can be folded onto some equivalent graph in Fig. 2. Higher powers of $\mathcal{F}$ do not yield any further concepts.

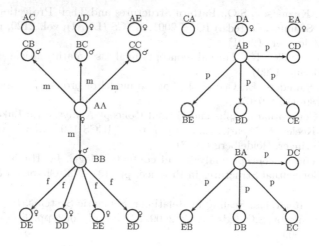

Fig. 5. Example: family tree (squared)

8 Conclusion

We have generated concept lattices directly from a relational structure. The representation of concept intents and extents by graphs and tables establishes connections to graph theory and database theory with their proven formalisms. This gives hope that notions and results from these areas may produce new insights into questions related to lattice-based navigation, and thus guide the development of applications. The similarity of the model to well-known Pattern Structures requires further, detailed comparison. The model will also have to be compared with other formal approaches dealing with relational data, including Concept Graphs [9,8] and Relational Semantic Systems [10].

References

1. Azmeh, Z., Huchard, M., Napoli, A., Hacene, M.R., Valtchev, P.: Querying relational concept lattices. In: Proc. of the 8th Intl. Conf. on Concept Lattices and their Applications (CLA 2011), pp. 377–392 (2011)
2. Chandra, A.K., Merlin, P.M.: Optimal implementation of conjunctive queries in relational databases. In: Proceedings of the Ninth Annual ACM Symposium on Theory of Computing, STOC 1977, pp. 77–90. ACM, New York (1977)
3. Ferré, S.: Conceptual Navigation in RDF Graphs with SPARQL-Like Queries. In: Kwuida, L., Sertkaya, B. (eds.) ICFCA 2010. LNCS, vol. 5986, pp. 193–208. Springer, Heidelberg (2010)
4. Ganter, B., Grigoriev, P.A., Kuznetsov, S.O., Samokhin, M.V.: Concept-Based Data Mining with Scaled Labeled Graphs. In: Wolff, K.E., Pfeiffer, H.D., Delugach, H.S. (eds.) ICCS 2004. LNCS (LNAI), vol. 3127, pp. 94–108. Springer, Heidelberg (2004)
5. Ganter, B., Kuznetsov, S.O.: Pattern Structures and Their Projections. In: Delugach, H.S., Stumme, G. (eds.) ICCS 2001. LNCS (LNAI), vol. 2120, pp. 129–142. Springer, Heidelberg (2001)
6. Ganter, B., Wille, R.: Formal concept analysis: mathematical foundations. Springer, Berlin (1999)
7. Godin, R., Saunders, E., Gecsei, J.: Lattice model of browsable data spaces. Inf. Sci. 40(2), 89–116 (1986)
8. Wille, R.: Conceptual Graphs and Formal Concept Analysis. In: Lukose, D., Delugach, H., Keeler, M., Searle, L., Sowa, J. (eds.) ICCS 1997. LNCS, vol. 1257, pp. 290–303. Springer, Heidelberg (1997)
9. Wille, R.: Formal concept analysis and contextual logic. In: Hitzler, P., Schärfe, H. (eds.) Conceptual Structures in Practice, pp. 137–173. Chapman & Hall/CRC (2009)
10. Wolff, K.E.: Relational Scaling in Relational Semantic Systems. In: Rudolph, S., Dau, F., Kuznetsov, S.O. (eds.) ICCS 2009. LNCS, vol. 5662, pp. 307–320. Springer, Heidelberg (2009)

Representing Median Networks with Concept Lattices

Uta Priss

Ostfalia University of Applied Sciences
Wolfenbüttel, Germany
www.upriss.org.uk

Abstract. Median networks have been proposed as an improvement over trees in phylogenetic analysis. This paper argues that concept lattices represent essentially the same information as median networks but with the advantage that there is a larger FCA research community and a variety of available software tools. Therefore evolutionary analysis is an interesting new application domain for FCA.

1 Introduction

The field of phylogenetics tries to establish evolutionary relations among groups of organisms - usually in form of evolutionary trees. For example, by sampling DNA from organisms and looking at differences evolutional changes can be reconstructed. For obvious reasons most of the DNA is extracted from currently living organisms, thus any reconstruction of phylogenetic trees is somewhat hypothetical. There are established means for inferring such trees (for example, involving "genetic distances", statistical maximum parsimony and maximum likelihood) but in cases where parallel mutations or reversals occur, it is difficult to decide on the exact sequences of the mutations. For example, the left-hand side in Figure 1 shows two possible trees for the changes between 1, 2, 3 and 4. As Sykes (2001, p. 178) explains, in such cases it is often not necessary to ultimately decide which change occurred first, i.e., whether 4 derived from 1 via 2 or via 3. Instead of deciding which of the trees is correct, one can use a graph as shown in the right half of Figure 1 which summarises both possible trees. Not only simplifies this the analytic process, it can also lead to more readable diagrams. Bandelt et al. (1995) have developed the construction of such graphs into a method using median networks as explained in the next section.

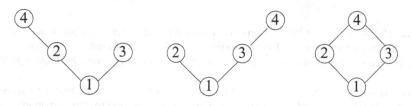

Fig. 1. Two possible trees (on the left) are summarised in one graph (on the right)

H.D. Pfeiffer et al. (Eds.): ICCS 2013, LNAI 7735, pp. 311–321, 2013.

Since the graph on the right-hand side of Figure 1 is a lattice and since trees can be embedded into lattices, the question arises as to whether Formal Concept Analysis[1] (FCA) can be used instead of or in addition to median networks. One advantage of using FCA is that FCA has a larger research community than median networks/graphs[2]. Furthermore, there exist more well-tested software tools for FCA[3] compared to median networks and, for example, Bandelt et al. (2000) still discuss "manual construction" of median networks alongside some algorithms.

From the viewpoint of FCA, it is interesting to establish a further application of FCA in the field of genetics or bioinformatics (for median networks this was first suggested by Priss (2012)) and a further connection with a similar or related graphical representation method. This extends previous research showing similarities between FCA and other fields, for example, Priss and Old (2008) show that concept lattices are similar to lattice-based methods developed in information retrieval and computational linguistics. The following section provides further details about median networks. Section 3 discusses how the phylogenetic data can be modelled with FCA and what is different or similar to how the data is modelled with median networks. The paper finishes with a concluding section.

2   Median Networks and Phylogenetics

This section provides a brief introduction to the application area of this paper[4]. Median graphs are undirected graphs where any three vertices have a unique median. More precisely, an *interval* between two vertices x, y in a graph is defined as $I(x, y) = \{v \mid d(x, y) = d(x, v) + d(v, y)\}$ where $d()$ is the usual distance function in a graph. In other words, an interval consists of the vertices on the shortest paths between two vertices. A graph is called a *median graph* if the following property holds: $\forall_{x,y,z} : |I(x, y) \cap I(x, z) \cap I(y, z)| = 1$. That means that there is a unique vertex (called *median*) that belongs to shortest paths between any two of three vertices.

Below is a brief summary of the close relationship between median graphs, distributive lattices and median semilattices (mostly following Bandelt (1984)). In this paper we are using the dual of the usual definition of a median semilattice (which we call a "reverse" median semilattice) because it fits better with the constructions in the next section. A *reverse median semilattice* is a join-semilattice such that every principal filter $\{x \mid x \geq a\}$ is a distributive lattice and any three elements have a lower bound whenever each pair of them does.

---

[1] Because FCA has been a topic of this conference for many years, this paper does not provide an introduction to FCA. Information about FCA can be found, for example, on-line (http://www.fcahome.org.uk) and in the main FCA textbook by Ganter & Wille (1999).

[2] As confirmed by retrieving about 10000 hits for a search for "formal concept analysis" on Google Scholar, as opposed to 1200 for "median network" and 900 for "median graph".

[3] See http://www.upriss.org.uk/fca/fcasoftware.html

[4] Based on Bandelt et al. (1995 and 2000), Sykes (2001) and Wikipedia pages.

- The covering graph of any finite distributive lattice is a median graph.
- A finite graph G is the covering graph of a finite distributive lattice $\Longleftrightarrow G$ is a median graph with two vertices 0 and 1 such that every other vertex lies on a shortest path between them.
- In a distributive lattice, Birkhoff's *median operation* can be observed: $m(a, b, c) = (a \wedge b) \vee (a \wedge c) \vee (b \wedge c) = (a \vee b) \wedge (a \vee c) \wedge (b \vee c)$ which also fulfills the axioms of a *median algebra*.
- Every median graph is a covering graph of a reverse median semilattice with largest element a where a is any fixed vertex.
- The covering graph of a reverse median semilattice S is a median graph provided that S is discrete, i.e., all intervals are finite.
- A discrete lattice L with 0 is distributive $\Longleftrightarrow$ the covering graph of L is median.
- Tree graphs are median graphs.

Although the relationship between median graphs and lattices is mathematically well-understood, there are still open questions left with respect to how FCA can be used to generate meaningful concept lattices from the data.

As mentioned in the introduction, in the field of phylogenetics, it is attempted to infer evolutionary trees from observed characteristics of species. Trees are considered best if they are *most parsimonious* which means that the number of presumed evolutionary changes is minimal. For example, in the right-hand side of Figure 1 it would be more parsimonious to assume that 2 evolved directly from 1 instead of evolving from 1 via 4 and 3. A goal of phylogenetic analysis is to compute all "most parsimonious trees" for a given data set, thus out of all possible trees the ones with minimal number of changes. Unfortunately, this is a computationally complex task. *Median networks* (or *Buneman graphs*) are median graphs where each vertex represents a species and each edge represents a genetic change. Bandelt et al. (1995) argue that since a median network is guaranteed to contain all most parsimonious trees, it is a preferred representation of evolutionary change and a significant improvement over other methods which artificially construct a tree from the data using statistical methods (see also Sykes (2001) and Bandelt et al. (1995 and 2000)).

Figure 2 shows an example of a median network on the left-hand side. The example is hypothetical and not based on real data. On the left side are white mice versus brown mice on the right. The top two vertices represent large mice, the other ones small mice. The bottom two vertices represent tailless as opposed to tailed mice. The vertex on the left in the middle is empty (*latent*) because no species in the data displays these characteristics. This vertex is generated from the data because without it, it would not be a median graph and not contain all most parsimonious trees. Without assuming that small tailed white mice are latent, the difference between small and large white mice would have coincided with loss of tail whereas in brown mice first the size changed, then the tail was lost. Not all possible combinations are latent. For example, the existence of large tailless mice is not implied by the data. The right-hand side of Figure 2 is explained in the next section.

The median network in Figure 2 summarises all possible evolutionary trees. If one assumes that the root of the trees is the top left vertex, four trees are possible. For example, large white mice could have first become small and then brown or first become

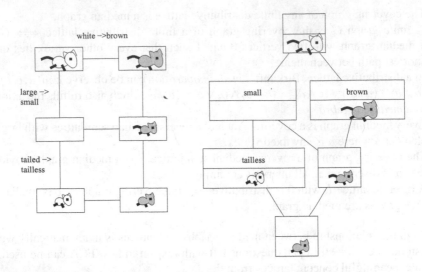

Fig. 2. Median network with latent vertex (left) and concept lattice (right)

brown and then small. While the sequence between the changes in colour and size is not known, the change in size definitely preceded the loss of tail. If one assumes that the change in size for white mice occurred before the change in colour, then the change in colour is an instance of *parallel mutation* because large white mice became brown independently of small white mice becoming brown. If the change in colour occurred first, then the change in size would be parallel mutation. If no parallel mutations or reversals were to occur in some data, then its median network would automatically be a tree. But considering the examples by Bandelt et al. (1995 and 2000), most data sets tend to contain at least some parallel mutations.

If the sample size is large, an unmodified median network may be too complex to be graphically represented. Bandelt et al. (1995) suggest a method for reducing median networks based on weight and frequency. In order to construct a median network, one summarises all changes that occur simultaneously with respect to the sample species as "weight". For example, if colour changes in mice always correspond to changes in ear size (hypothetically), then one would not draw separate edges for colour and ear changes. Instead one would record one change but with a higher weight. Graphically this can be represented by drawing a longer edge.

In the same manner, if several species have the exact same characteristics, one creates only one vertex for this group of species but records a higher frequency for this vertex. This can be graphically represented by a larger node for the vertex. Using frequencies and weights one can reduce the network by eliminating some of the edges which are less likely to have occurred. Bandelt et al. (1995) state that in all examples they considered so far even reduced networks still contained all most parsimonious trees, but there is no guarantee that that is always the case.

Characteristics in phylogenetic analysis are often binary, i.e., having two possible values. In the example in Figure 2, the characteristics are naturally binary (such as large or not large). Other characteristics can be made binary. For example, although DNA sequences can be of four values (A, G, C or T), Bandelt et al. (1995) argue that it is unusual for more than one change to occur at the same site in a set of closely related species. Thus it is sufficient to record for each site whether a change occurred or not, ignoring the value of the change.

A median network contains all most parsimonious trees independently of where the root of the tree is. There are methods for determining the root or evolutionary ancestor of a set of species although it might not be easy and the root might be latent. One method is to compare a set of species with an *outgroup* or *reference group* which is more distantly related to all the other species than they are too each other.

3   Modelling with FCA

It is straightforward to represent the example on the left-hand side in Figure 2 as a concept lattice as presented on the right-hand side. One advantage of using FCA is the availability of established mathematical vocabulary for describing the phylogenetic phenomena. Important phylogenetic notions can be directly translated into FCA terminology. Series of evolutionary changes that are unambiguous correspond to attribute implications in the lattice. For example, the implication from "tailless" to "small" in the lattice in Figure 2 corresponds to the evolutionary loss of tail occurring after the change in size. Latent species correspond to concepts that do not contain objects in their contingent. Each meet-reducible concept in the lattice corresponds to a choice point between different possible trees.

Table 1 shows a more complex example using mitochondrial data from Ward et al. (1991) which was also used by Bandelt et al. (1995). In FCA terms it represents a many-valued context. The second row from the top shows the default values for each column. A dot in the matrix means that the default value occurs. A letter indicates a change. As can be seen in the table, only one type of change occurs in each column. For example, in the first data column the default value is "T" which is changed to "C" in three rows. No changes to "A" or "G" occur in the first data column. As discussed by Bandelt et al. (1995) this is usually the case. Therefore such tables can be interpreted as binary matrices or single-valued contexts by only considering whether the default value or a change occur and ignoring the type of change.

In FCA terminology, the formal objects in Table 1 are 28 mitochondrial lineages. The right-hand column indicates the frequency of the lineages. For example, lineage 1 occurred in 3 individuals. A total number of 63 individuals was involved in the study. The formal attributes encode the positions where the DNA sequences occur in the human reference sequence. If one encodes the attributes so that each cross represents the positions where an object differs from the reference group then the top of the lattice will correspond to the root of the possible evolutionary trees. This is because, as discussed in the previous section, comparison with a reference group can be used to determine the root. Using FCA the preprocessing of summarising objects with identical row values and attributes with identical column values is not really necessary because such objects (or equivalently attributes) would be grouped into the contingent of a single concept automatically in the concept lattice.

Table 1. Nuu-Chah-Nulth mitochondrial lineages (Ward et al., 1991) as a formal context

	69	88	91	106	124	149	162	166	190	194	200	219	233	247	251	255	267	271	275	296	301	302	304	319	339	344	
	T	C	C	G	C	T	C	T	G	T	C	C	C	C	G	C	C	C	T	G	T	T	C	T	T	A	
1	.	.	.	.	.	.	C	A	.	T	.	.	.	.	.	T	.	.	.	.	.	.	.	.	.	.	3
2	.	.	.	.	.	.	.	A	.	T	.	.	.	.	.	T	.	.	.	.	.	.	.	.	.	.	2
3	.	.	.	.	.	.	.	.	.	T	.	.	.	.	.	T	.	.	.	.	.	.	.	.	.	.	1
4	.	.	.	.	.	.	.	.	.	T	.	.	.	.	.	T	.	.	.	.	.	.	.	C	.	.	1
5	.	T	.	.	A	.	T	.	.	T	.	.	.	.	.	.	T	.	.	A	.	.	.	C	.	.	2
6	.	T	.	.	A	.	.	.	.	T	.	.	.	.	.	.	.	.	.	A	.	.	.	C	.	.	2
7	C	T	.	.	A	.	.	.	.	T	.	.	T	.	.	.	.	.	.	A	.	.	.	C	.	.	1
8	.	T	.	.	A	.	.	.	.	T	.	.	.	.	.	.	T	.	.	A	.	.	.	C	.	.	2
9	C	T	.	.	.	.	.	.	.	T	.	.	.	.	.	.	T	.	.	A	.	.	.	C	.	.	2
10	.	T	.	.	.	.	.	.	.	T	.	.	.	.	.	.	T	.	.	A	.	.	.	C	G	.	1
11	.	T	.	.	.	.	.	.	.	T	.	.	.	.	.	.	T	.	.	A	.	.	.	C	.	.	5
12	.	T	.	.	.	.	.	.	.	.	.	.	.	.	.	.	T	.	.	A	.	.	.	C	.	.	9
13	.	T	.	.	.	.	.	A	.	.	.	.	.	.	.	.	T	.	.	A	.	.	.	C	.	.	1
14	.	T	.	.	.	.	.	.	.	T	.	.	.	.	T	T	.	.	.	A	.	.	.	C	.	.	1
15	.	T	.	.	.	.	.	.	.	T	.	.	.	.	T	T	.	.	.	A	C	.	.	C	.	.	2
16	.	.	.	.	.	.	.	.	.	T	T	.	.	.	.	.	.	.	.	C	.	.	T	C	.	.	1
17	.	.	T	.	.	.	.	.	.	T	.	.	.	.	.	.	.	.	C	.	.	.	.	C	.	.	1
18	.	.	T	.	.	.	.	.	.	T	.	.	.	.	.	.	.	.	.	C	.	.	.	C	.	.	2
19	.	.	.	T	.	.	.	.	.	T	.	.	.	.	.	.	.	T	.	C	.	.	.	C	.	.	1
20	.	.	.	.	.	C	.	.	.	T	.	.	.	A	.	.	.	.	.	C	.	.	.	C	.	.	3
21	.	.	.	.	.	.	.	.	.	T	.	.	.	.	.	.	.	.	.	C	.	.	.	C	.	.	3
22	C	.	.	.	.	.	.	.	.	T	.	.	.	.	.	.	.	.	.	C	.	.	.	.	.	.	3
23	.	.	.	.	.	.	.	.	.	T	T	.	.	.	.	.	.	.	.	.	C	.	C	T	.	.	1
24	.	.	.	.	.	.	.	.	.	T	.	.	.	.	.	.	.	.	.	.	C	.	C	T	.	.	7
25	.	.	.	.	.	.	.	.	.	T	T	.	.	.	.	.	.	.	.	.	C	.	C	T	.	.	3
26	.	.	.	.	.	.	.	.	.	.	T	.	.	.	.	.	.	.	.	.	C	.	C	T	.	.	1
27	.	.	.	.	.	.	C	.	C	.	.	.	.	.	.	.	.	.	.	.	.	.	.	.	.	.	1
28	.	.	.	.	.	.	C	.	C	.	.	.	T	.	.	.	.	.	.	.	.	.	.	.	.	.	1

Because the median network and concept lattice for Table 1 are fairly complex, we will first discuss a network and lattice derived for a simpler context of the same type before discussing the one in Table 1. Figure 3 shows a concept lattice for a data table discussed by Bandelt et al. 2000 (using HVS I data by Vigilant et al.). Two attributes are called *compatible* in Bandelt's terminology if they are lattice-theoretically comparable or their meet is the bottom node. Bandelt calls a set of attributes a *clique* if the attributes are pairwise compatible and the set is maximal with respect to inclusion. In other words, cliques represent maximal trees. In Figure 3 one clique contains all attributes except 16243 and another clique contains all attributes except 16294 and 16239. These are the only two cliques in Figure 3. Bandelt et al. describe a fairly complicated algorithm for deriving the median network using cliques, peripheral elements and torsos (where the *torso* data matrix consists of the non-compatible attributes).

Figure 4 shows a median network for the data in Figure 3. In contrast to Bandelt et al. (2000), the attributes, frequencies and weights are omitted in the figure. This means that all nodes are of the same size and the length of the edges does not carry meaning. The lattice in Figure 3 is not distributive and thus not a covering graph of a median graph. Nevertheless if one omits the bottom node from the lattice then its covering graph and the median network in Figure 4 differ only by one vertex: the vertex next to the one labelled with "8" in Figure 4. Using the statements about reverse median semilattices from the last section, an algorithm for converting a concept lattice as in Figure 3 into a

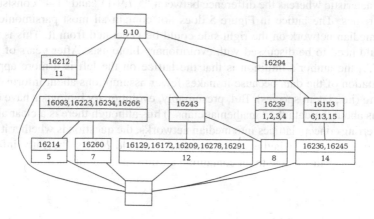

Fig. 3. Concept lattice for HVS I data of Vigilant used by Bandelt et al. (2000)

median network as in Figure 4 consists of omitting the bottom node and then checking every principal filter for distributivity and turning it into a distributive lattice if it is not already one.

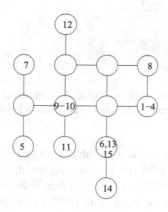

Fig. 4. The median network for Figure 3

The principal filter in Figure 3 that is not distributive is shown on the left in Figure 5 alongside the median network of the torso of Figure 4. In contrast to Figure 2 where both the lattice and the graph produce a latent vertex, in this case the lattice does not have one. The reason is because in Figure 2 the attribute "small" is shared by small tailless white mice and small brown tailed mice whereas in Figure 5 object "12" does not share any attributes with the objects "1, 2, 3, 4". The median network in Figure 5 generates a latent species because the difference between objects "12" and "8" consists only of one characteristic whereas the difference between "5,7,9-11" and "1-4" consists of two characteristics. The lattice in Figure 5 does not contain all most parsimonious trees but the median network on the right side could be generated from it. This is an issue that would need to be discussed with evolutionary biologists. After years of working with FCA, the author's intuition is that the lattice on the left is a more appropriate representation of the data because it makes fewer assumptions about information that is missing (i.e., latent species). But, presumably, evolutionary biologists have different intuitions about the data than mathematicians. Thus although there is a clear algorithm for converting concept lattices into median networks, the question is whether it is really necessary to do so or whether a concept lattice would be a sufficiently informative representation of the data without containing all most parsimonious trees.

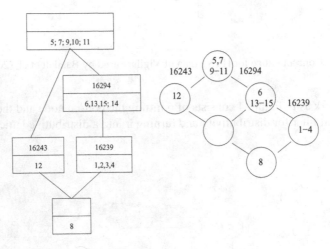

Fig. 5. Concept lattice (left) of the only non-distributive principal filter in Figure 3 and median network (right) which is the "torso" of the median network in Figure 4

Coming back to the data presented in Table 1, Figure 6 shows the reduced median network from Bandelt et al. (1995) for the data. Again, the frequencies and weights are not represented in the diagram. The root of the tree is the node labelled "X". The network contains 10 latent vertices. Ward et al. (1991) identified four clusters among the lineages ($\{1,2\}$, $\{5,6,7,8,...,15\}$, $\{23,24,25,26\}$, and $\{27,28\}$) by deriving a phylogenetic tree using statistical methods. Bandelt et al. (1995) criticise the tree presented by Ward et al. because they believe that one cluster is missing and several other clusters could be modified. The large boxes in Figure 6 are meant to indicate the clusters according to Bandelt et al. who observe that the cluster consisting of 18, 19, 20 and 21 (and possibly also 16, 17 and 22) is missing from Ward's tree and that maybe 3 and 4 should also belong to the cluster of 1 and 2. They argue that the information about the clusters is very clear in the median network but might not be visible in a tree. They further state that these problems are not restricted to Ward's paper but can be observed in other papers as well.

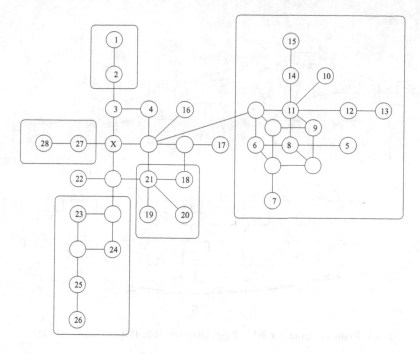

Fig. 6. Reduced median network for Table 1 (following Bandelt et al. (1995))

The median network in Figure 6 is reduced. The reduction algorithm is described by Bandelt et al. in great detail. Effectively the reduction algorithm splits some attributes into versions *a* and *b* so that objects in one cluster have version *a* and the objects in other clusters have *b*. For example, attribute 166 applies to lineages in two different clusters. If the attribute is split into 166a for lineage 1 and 166b for lineages 27 and 28, then the structure of the network is simplified. The reasoning behind this is that if the same change occurs for lineages that are in very different clusters, it is quite likely that the change does not represent a single event but instead happened several times independently. The basis for these decisions are frequencies and weights. We do not have an exact list of which attributes were split in Figure 6. Therefore the attributes that were split in Figure 7 are not necessarily the same as in Figure 6. We chose to split attributes 69, 166 and 190. Furthermore we completely omitted attribute 200 because it applies to almost all objects. The resulting lattice is shown in Figure 7. Structurally, the graphs in Figure 6 and Figure 7 are quite similar although in Figure 7 attribute 16 is closer to the cluster involving 23 to 26 and there is a connection between 4 and 14. We do not know whether either representation is more plausible from a phylogenetic viewpoint.

In order to decide which attributes to split, one needs to first determine which objects form clusters. Figure 8 shows the object ordering (implications) of Table 1. Apart from the already mentioned connection between 4 and 14, the clusters emerging from the object ordering are the same as the ones discovered by Ward et al. and Bandelt et al. Thus we propose an algorithm for reducing concept lattices as follows: determine

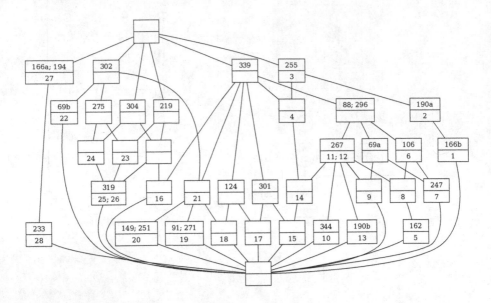

Fig. 7. Reduced lattice for Table 1 (splitting 69, 166, 190 and omitting 200)

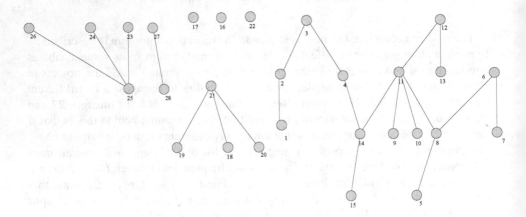

Fig. 8. Object ordering for Table 1

clusters of objects by considering the object ordering. Then investigate attributes that apply to objects belonging to different clusters. If these attributes are high up in the lattice, consider splitting the attribute. The resulting lattice will have fewer line crossings and be more "tree like". We are not necessarily proposing that the attributes are completely automatically selected, but that instead expert advice is considered in the selection process.

4   Conclusion

This paper discusses the representation of phylogenetic data as concept lattices instead of or in addition to median networks. Both concept lattices and median networks contain essentially the same information but FCA has a larger research community. The paper sketches an algorithm for converting a concept lattice into a median network and for reducing a lattice based on clustering of objects. Further discussion with phylogenetics researchers will need to establish in how far they would be willing to accept concept lattices that do not contain all most parsimonious trees as a representation of their data. More experiments with larger data sets are needed to determine the practical feasibility of the suggested algorithms and to compare more examples of median networks and concept lattices with respect to the readability of the diagrams. One aim of the paper is to alert the wider FCA community to this application area. Because median graphs have many interesting properties and applications themselves, establishing a connection between them and FCA could lead to further interesting research (for example, social network analysis or other graph and networking applications).

References

1. Bandelt, H.J.: Discrete Ordered sets whose covering graphs are median. Proceedings of the American Mathematical Society 91, 1 (1984)
2. Bandelt, H.J., Forster, P., Sykes, B.C., Richards, M.B.: Mitochondrial portraits of human populations using median networks. Genetics 141(2), 743–753 (1995)
3. Bandelt, H.-J., Macaulay, V., Richards, M.: Median networks: Speedy construction and greedy reduction, one simulation, and two case studies from human mtDNA. Molecular Phylogenetics and Evolution 16(1), 8–28 (2000)
4. Ganter, B., Wille, R.: Formal Concept Analysis. Mathematical Foundations. Springer, Heidelberg (1999)
5. Priss, U., Old, L.J.: Lattice-based Modelling of Thesauri. In: Lattice-Based Modeling Workshop, Olomouc, Czech Republic (2008),
   http://researchrepository.napier.ac.uk/3477/
6. Priss, U.: Concept Lattices and Median Networks. In: Szathmary, Priss (eds.) Proceedings of the Ninth International Conference on Concept Lattices and Their Applications, Universidad de Malaga, pp. 351–354 (2012)
7. Sykes, B.: The seven daughters of Eve. Bantam Press (2001)
8. Ward, R.H., Frazier, B.L., Dew-Jacer, K., Pääbo, S.: Extensive mitochondrial diversity within a single Amerindian tribe. Proc. Natl. Acad. Sci., USA 88, 8720–8724 (1991)

Txt2vz: A New Tool for Generating Graph Clouds

Laurie Hirsch and David Tian

Sheffield Hallam University, Sheffield, UK
l.hirsch@shu.ac.uk

Abstract. We present txt2vz (txt2vz.appspot.com), a new tool for automatically generating a visual summary of unstructured text data found in documents or web sites. The main purpose of the tool is to give the user information about the text so that they can quickly get a good idea about the topics covered. Txt2vz is able to identify important concepts from unstructured text data and to reveal relationships between those concepts. We discuss other approaches to generating diagrams from text and highlight the differences between tag clouds, word clouds, tree clouds and graph clouds.

Keywords: visualization, concept map, tag cloud, tree cloud.

1 Introduction

Tag clouds are simple visualizations that display word frequency information via font size and colour, that have been in use on the web since 1997. Users have found the visualizations useful in providing an overview of the context of text documents and web sites. Whereas many systems are formed using user provided tags, there has been significant interest in 'word tags' or 'text tags' which are automatically generated using the text found in documents or web sites. For example, the popular tool Wordle has seen a steady increase in usage [1]. Word clouds are based on the frequency of individual words found in the available text after stop word removal. The most frequent words are selected and then presented using various techniques to adjust font, colour, size and position, in a way that is pleasing and useful to the user. The words are commonly sorted alphabetically, although various systems of sorting and arrangement have been proposed and attempts have been made to place similar words together [2] [3]. Word clouds are simple and are commonly presented on web sites with little or no explanation of how they should be used or interpreted. Three distinct tasks have been identified which may be accomplished namely, searching, browsing and "impression formation" whereby "The cloud can be scanned to get a general idea about a subject" [4]. Successful realisation of this last task is the main objective of the Txt2vz tool. Trees have been presented as an easy to read and meaningful format and the term 'tree cloud' has been proposed. A freely available system which generates trees based on the semantic distance between words derived from the original text is also available [5].

Co-occurrence information has long been understood to be an important aid to understanding the meaning of words, and using this information has proved essential to

H.D. Pfeiffer et al. (Eds.): ICCS 2013, LNAI 7735, pp. 322–331, 2013.

many natural language processing and information retrieval tasks [6] [7]. We extract
and use co-occurrence information here as a way of giving context to words presented
to the user and as a way of identifying and highlighting the most important words. We
propose a new method of generating diagrams, based on co-occurrence information
derived from the original text. We suggest the term 'word graphs' for the Txt2vz
generated diagrams since they are not necessarily in tree format and indeed can sometimes be in the form of two or more disjoint graphs. An important feature of the
Txt2vz graphs is that link information is the critical element of graph construction:
co-occurrence links are directly displayed and nodes (words) with the most links are
placed toward the centre of the graph.

2   Description of Txt2vz

2.1   The Overall Methodology

To reduce dimensionality of the document(s) all words are placed in lower case, stop
words are removed and stemming applied, such that only the most frequent form of a
word is preserved. Depending on the size of the document or the collection, this can
still leave a large number of words, and further reduction is achieved by ordering
words according to their frequency or tf-idf (term frequency-inverse document frequency) weighting in the case where the document is part of a collection, and then
selecting the top N words from the sorted list.

   After dimension reduction, every possible pair of the remaining words is analysed
for co-occurrence information. Many techniques have been described for identifying
co-occurrence [6] [7] but we take a relatively simple approach here. A graph is generated by selecting the top K pairs of words from a list of word pairs in descending
order of their significance value defined as follows.

2.2   Significance Measure

We define a measure of significance for a pair (P, Q) of words, based on the number
of occurrences of (P, Q), or more specifically the co-occurrences and the distance
between P and Q where the distance between P and Q is defined to be the number of
words between P and Q:

$$significance(P,Q) = \sum_{i=1}^{M} B^{distance(PQ_i)} \tag{1}$$

where M is the number of co-occurrences of P and Q; distance(PQ_i) is the distance
between P and Q in the ith co-occurrence; $0<B<1$ B is typically set to 0.9. We do not
consider the significance if the distance is beyond a pre-set maximum distance which
has a default of 20 words.

2.3   Graph Generation Algorithm

The significance of each pair of words is computed and all the word pairs are sorted
in descending order by their significance values. An undirected graph is then built by
selecting the top K word pairs in the rank and creating an edge between the two words
of each pair. The degree of each node (word) i.e. the number of edges attached to each
node, can be used as an indication of the importance of that word. The most significant word is the word with the largest degree. Different colours are used to group
words of similar importance and node and font sizes are used to highlight the importance of the words within the graph.

The type of graph produced can be partly determined by the user. In particular we
have provided an adjustment facility whereby the user can change the number of
words (N) to analyse; the number of links to display (K) and the maximum distance
allowed between words when calculating co-occurrence (shown in figure 1).

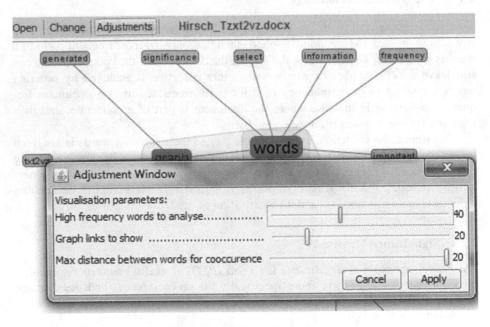

Fig. 1. Txtvz adjustment window

Algorithm
1. Tokenize the text and apply dimension reduction using lower case, stop words
   and word stemming.
2. Order the words according to frequency or tf-idf where the document is part of
   a collection.
3. Create a set of words W by selecting the top N words from the ordered list
   where N has a default value of 40 but can be adjusted by the user.

4. Analyse every possible pair of words from W and assign a co-occurrence value to each pair. For each case where both words occur within a maximum distance of 20 words (this value can also be adjusted by the user) we add a value to the co-occurrence metric for the pair determined by:

$$0.9^{wordDistance}$$

where *wordDistance* is simply the number of intervening words. Note: a decaying function is used such that words occurring closer to each other add more to the co-occurrence value.

5. Create an ordered list of the word pairs based on the co-occurrence value for each pair.

6. Generate a graph by selecting the top K word pairs from the sorted list of pairs where K is a value that can be set by the user, but with a default value of 20.

7. The number of links attached to each node is used as a further indication of the importance of a particular word.

As an initial example if we use the text taken from the ICCS'13 call for papers (http://iccs2013.hbcse.tifr.res.in/call-for-papers) and show the top 10 pairs.

Table 1. ICCS'13 CFP co-occurrence values

Word Pair		Co-occurrence value
data	stem	8.277417941925659
conceptual	structure	7.837253642946149
papers	conference	6.667569657552907
papers	accepted	6.645902314469966
papers	called	6.5704478047496115
papers	phd	5.7306734434201365
conceptual	knowledge	4.425574377763965
data	concept	4.275419763354885
called	workshops	4.2135456501
dot	chair	4.205350186716726

The word pairs are used directly to create the graph shown in figure 4. Each unique word generates a node and each pair generates an edge.

3   Examples

We begin by presenting diagrams generated from the ICCS'13 call for papers which contains 493 words. We compare the graph produced by Txt2vz with the ones generated by the Wordle [8] and tree cloud [5] approaches.

Fig. 2. Wordle word cloud of ICCS'13 CfP

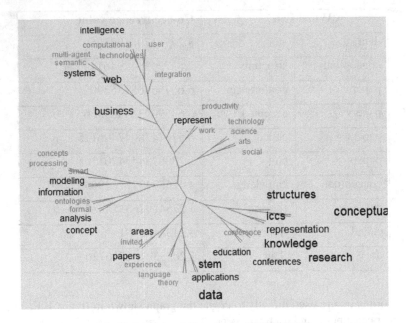

Fig. 3. Tree cloud diagram for ICCS'13 CfP

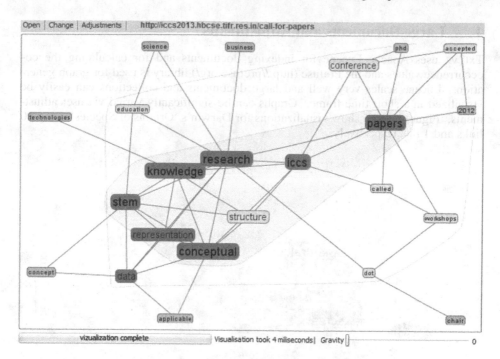

Fig. 4. Txt2vz graph cloud for ICCS'13 CfP

The three diagrams shown in figures 2, 3 and 4 include many common words, but the presentation is different in a number of respects. Which format is the 'best' is not a discussion we plan to resolve in this paper. However, we can identify that certain types of information can be obtained from the different formats. For example the fact that 'conceptual', 'knowledge' and 'structures' are related is not shown on the Wordle diagram whereas it is evident from the positioning of the words in the tree cloud and made clear via the arcs in the Txt2vz graph. Links between words in a Txt2vz graph indicate recognizable connections and nodes with a higher number of links are emphasized using large font sizes and by positioning these nodes at the centre of the graph. In figure 4 the word 'chair' has only one link and appears smaller and to the edge of the window whereas 'iccs' has 7 links, is larger and placed toward the centre of the graph. The point we wish to emphasise here is that the Txt2vz diagram clearly shows how words link to each other and uses that information to highlight important topic words with a high number of links, rather than only using word frequency information as in the tree cloud. For example, the Txt2vz diagram makes it easy to see that 'conceptual' is directly related to a number of other words and this is not obvious from the other two diagramming systems. You can test your own documents at txt2vz.appspot.com

4   Large Documents and Adjustments

Txt2vz uses Apache Lucene for indexing documents and for calculating the co-occurrence values and the Prefuse (http://prefuse.org/) library is used for graph generation. Lucene scales very well and large documents and collections can easily be visualized in a short time frame. Graphs can be significantly varied via user adjustments. Figure 5 and 6 show visualizations for Darwin's 'Origins of Species' using 50 links and 1 link respectively.

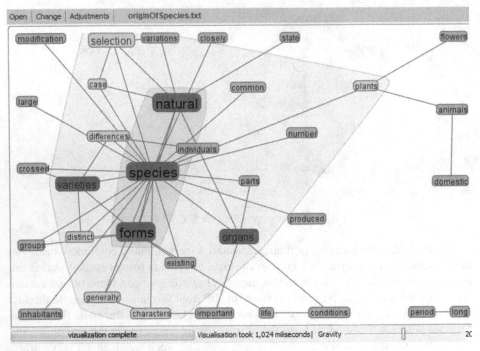

Fig. 5. Origins with 50 links

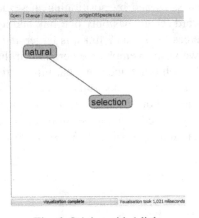

Fig. 6. Origins with 1 link

Txt2vz also offers an alternate radial graph format which uses the Docuburst library[9] and we show the visualization of this paper in radial graph format (figure 7).

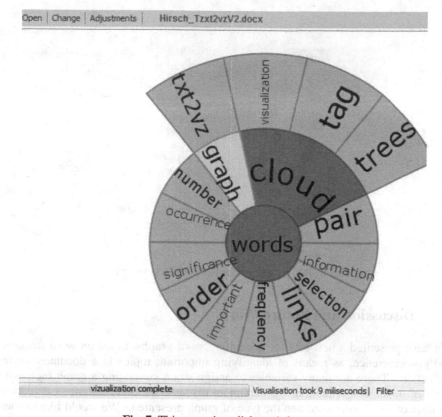

Fig. 7. This paper in radial graph format

5 Document Collections

We believe that the graphs produced by Txt2vz might be especially useful to people who need a visual summary of large collections of documents and as mentioned above, Lucene makes this perfectly possible. The example shown in figure 8 was generated in less than 10 seconds from 389 documents from the training set for the Reuters-21578 [10] 'crude' category containing news stories concerning crude oil.

The key topics words are identified as having the largest number of links ('dlrs', 'mln' and 'oil') and are located near the centre of the graph and surrounded with a shaded area.

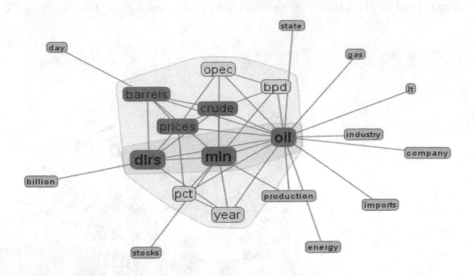

Fig. 8. Txt2vz diagram for Reuters category "crude"

6   Discussion and Future Work

We have presented a new tool for generating word graphs, based on word frequency and co-occurrence, as means of identifying important topics in a document or text collection. There are many variables to assign when generating a graph such as the scale of dimension reduction, the maximum distance for co-occurrence calculations, number of word pairs used and the type of graph presented. We would like to spend more time evaluating the usefulness of the tool as perceived by human subjects. We would also like to investigate the feasibility of using Txt2vz as part of web search engine such that a user could be presented with a quick visual summary of the content of the pages pointed to by the result links.

References

1. Viégas, F.B., Wattenberg, M.: Tag Clouds and the Case for Vernacular Visualization. ACM Interactions XV(4) (July/August 2008)
2. Hassan-Montero, Y., Herrero-Solana, V.: Improving Tag-Clouds as Visual Information Retrieval Interfaces. In: InSciT 2006 (2006)
3. Viégas, F.B., Wattenberg, M., van Ham, F., Kriss, J., McKeon, M.: Many Eyes: A Site for Visualization at Internet Scale. In: Proc. of IEEE InfoVis 2007 (2007)
4. Bateman, S., Gutwin, C., Nacenta, M.: Seeing Things in the Clouds: The Effect of Visual Features on Tag Cloud Selections. In: Proc. of the 19th ACM Conference on Hypertext and Hypermedia, pp. 193–202. ACM Press, New York (2008)

5. Gambette, P., Véronis, J.: Visualising a text with a tree cloud. In: Proceedings of 11th IFCS Biennial Conference, pp. 561–570 (2009)
6. Lin, D.: Using collocation statistics in information extraction. In: Proceedings of the Seventh Message Understanding Conference, MUC-7 (1998)
7. Veling, A., Van der Weerd, P.: Conceptual grouping in word co-occurrence networks. In: Proceedings of the IJCAI 1999, vol. 2, pp. 694–699 (1999)
8. Viégas, F.B., Wattenberg, M., Feinberg, J.: Participatory Visualization with Wordle. IEEE Transactions on Visualization and Computer Graphics 15, 1137–1144 (2009)
9. Collins, C., Carpendale, S., Penn, G.: DocuBurst: Visualizing Document Content using Language Structure. In: Computer Graphics Forum, Proceedings of Eurographics/IEEE-VGTC Symposium on Visualization (EuroVis 2009), vol. 28(3), pp. 1039–1046 (June 2009)
10. Reuters-21578 at http://www.daviddlewis.com/resources/testcollections/reuters21578/

Author Index